**James Tobin**

James Tobin won the National Book Critics Circle Award for his first book, *Ernie Pyle's War*, and the J. Anthony Lukas Work-in-Progress Award from Harvard's Nieman Foundation and Columbia University for *First to Fly*. Twice nominated for the Pulitzer Prize in journalism, Tobin also holds a Ph.D in history. He lives in Ann Arbor, Michigan.

## Also by James Tobin

*Ernie Pyle's War*

*Great Projects*

# FIRST TO FLY

## *The Unlikely Triumph of Wilbur and Orville Wright*

JAMES TOBIN

JOHN MURRAY

First published by Free Press, a division of Simon & Schuster, Inc.
in the United States of America

First published in Great Britain in 2003 by John Murray (Publishers)
A division of Hodder Headline

Paperback edition 2003

1 3 5 7 9 10 8 6 4 2



A CIP catalogue record for this title is available from the British Library

ISBN 0-7195-5738 0

Typeset in Adobe Goudy

Printed and bound in Great Britain by Clays Ltd, St Ives plc

John Murray (Publishers)
338 Euston Road
London
NW1 3BH

*For my parents*
*James and Dorothy Tobin*
*with thanks and love*

# CONTENTS

*Prologue*

# Decoration Day, 1899

"I FELT ITS PATHETIC PREEMINENCE IN A STREET OF MEAGER HOMES."

7 Hawthorn Street, Dayton, Ohio, about 1900

HIS FATHER AND SISTER had gone to Woodland Cemetery to plant flowers at the grave of his mother. His younger brother was busy elsewhere. It was a holiday, and the house was quiet. He could take care of the letter he had been meaning to write.

He sat, took out paper and pen, and wrote:

> The Smithsonian Institution
> Washington.
>
> Dear Sirs:
>
> I have been interested in the problem of mechanical and human flight ever since as a boy I constructed a number of bats of various sizes. . . . My observations since have only convinced me more firmly

> that human flight is possible and practicable. It is only a question of knowledge and skill as in all acrobatic feats.

He was thirty-two years old and unmarried. He was lean and bald. His lips were thin and he usually held them tightly shut; he may have been self-conscious about his teeth, which had been smashed in a hockey game years ago. His ears flared. Only his eyes saved him from outright homeliness. Ten years later, when he was one of the most famous men in the world, reporters had trouble getting him to say much about himself. So they remarked on his eyes as indicative of the character within. "This man is strange and cold," one said, "but of a coldness that is smiling and sympathetic. . . . The countenance is remarkable, curious—the head that of a bird, long and bony, and with a long nose . . . the eye is a superb blue-gray, with tints of gold that bespeak an ardent flame." He was a shrewd observer of people and of nature. Yet he once told his sister that "my imagination pictures things more vividly than my eyes."

The subject of his letter was not the sort of thing he would mention casually to a neighbor. On this block of Hawthorn Street, on the west side of Dayton, Ohio, the houses stood so close together you could name the song somebody was playing on the piano three doors down, and make a fair guess who was playing. Any news traveled fast, and the news that somebody hoped to fly like a bird would travel faster than usual. Aspirations here tended toward the sensible. The neighbors included four carpenters, two day laborers, a machinist, a printer, a motorman, a market vendor, an insurance salesman, three widows, and two clergymen. One of the clergymen was the young man's father—quite a prominent man, a bishop of the Church of the United Brethren in Christ, a stout widower who kept house with his schoolteacher daughter and two of his four grown sons. Their home was number 7, second from the corner of Hawthorn and West Fourth.

"When I saw this house," a visitor said later, "I felt its pathetic preeminence in a street of meager homes." It was narrow but it extended way back on the lot, with white clapboards and green shutters. At the front was the parlor, with a slant-top writing desk; a cherrywood rocking chair with horsehair upholstery; a chaise longue and a settee. In back of the parlor was the sitting room, dominated by a tall cherry bookcase, its contents suggesting a family of enthusiastic readers with broad interests, from Charles Dickens and James Fenimore Cooper to a six-volume *History of France* and Charles Darwin's *Origin of Species*. Next came a spacious dining room, with a long sideboard, a drop-leaf table, and six walnut chairs. At the rear of the house was the kitchen. Upstairs there were four small bedrooms. Out in back was a "summer kitchen"—a detached shed—and an outhouse. Along the front and left side of the house there was a long porch. The young man and his brother had built it in 1892. They kept it simple, leaving off the usual gingerbread trim.

The young man continued his letter:

> Birds are the most perfectly trained gymnasts in the world and are specially well fitted for their work, and it may be that man will never equal them, but no one who has watched a bird chasing an insect or another bird can doubt that feats are performed which require three or four times the effort required in ordinary flight. I believe that simple flight at least is possible to man and that the experiments and investigations of a large number of independent workers will result in the accumulation of information and knowledge and skill which will finally lead to accomplished flight.

He once remarked that for a person endowed with greater gifts than others, but lacking in the push needed for conventional success, there was "always the danger" that he would "retire into the first corner he falls into and remain there all his life."

This is what had happened to him, and the letter was part of his effort to get out of his corner. He possessed extraordinary gifts. Yet he had lived more than half the average span of an American man of his time without doing or making anything he could call his own. He lived in his father's house. The woman in his life was his sister. The children he loved belonged to his older brothers. He gave most of his time to a storefront business without caring for business or money. His only advanced education had come from his father's books. He had put his powerful mind to work in only one cause—an obscure church controversy, also his father's. But he did not even have his father's faith.

He was "an enthusiast," he wrote:

> . . . but not a crank in the sense that I have some pet theories as to the proper construction of a flying machine . . . I am about to begin a systematic study of the subject in preparation for practical work to which I expect to devote what time I can spare from my regular business. I wish to obtain such papers as the Smithsonian Institution has published on this subject, and if possible a list of other works in print in the English language . . . I wish to avail myself of all that is already known and then if possible add my mite to help on the future worker

who will attain final success. I do not know the terms on which you send out your publications but if you will inform me of the cost I will remit the price.

Yours truly,
Wilbur Wright

BY JUNE 1, traveling by rail, the letter arrived in Washington, where it was carried to the turreted, red-sandstone headquarters of the Smithsonian Institution, on the southern edge of the long green known then as Smithsonian Park. There the letter was opened, sorted with others of its type, and taken through echoing halls to the office of Richard Rathbun, second in command at the Smithsonian, whose duties included the oversight of the Institution's correspondence with scholars, scientists, and the merely curious in every part of the world.

Rathbun, an expert on marine invertebrates, handled many such inquiries each week. Like Wilbur Wright, people everywhere regarded the Smithsonian as a fountainhead of scientific and cultural information, much of it published in the Institution's own periodicals. A man once wrote to ask for all Smithsonian publications on geology, biology, botany, the National Museum, the Bureau of American Ethnology, Indians, International Exchanges, the National Zoological Park, the Astrophysical Observatory, "and any other interesting subjects." He was told that compliance with his request would require the shipment of several thousand volumes. With the financial support of the U.S. Congress and a host of private benefactors, the Smithsonian was the best-endowed, most prestigious institution of science, culture, and learning in the entire nation. Its exhibits, repositories, storehouses, laboratories, and libraries were known throughout the world, and all of these existed to fulfill the terms of the 1846 will of the Institution's founding benefactor, the Englishman James Smithson, who had called for an American institution to foster "the increase & diffusion of knowledge." Every legitimate question was to receive a careful answer. So people wrote by the hundreds every year.

The letter from Dayton would have occasioned no special notice in Rathbun's office but for one salient characteristic. It raised the question of mechanical flight. This was a topic of consuming interest to the fourth secretary of the Smithsonian himself, Samuel Pierpont Langley.

In public stature and prestige, Langley was the most prominent scientist in the United States. His best friend was Alexander Graham Bell, the inventor of the telephone. Langley was a frequent guest at the White House. He dined regularly with the historian Henry Adams, grandson and great-grandson of presidents; and John Hay, who had been personal secretary to Abraham Lincoln and was now secretary of state. Langley corresponded with the likes of Rudyard Kipling and the philosopher Charles Sanders Peirce; as a young man he had listened for hours to the philosophical discourses of the great British historian Thomas Carlyle.

In the entire world only a handful of men with any standing in science had suggested that human flight was possible. Langley not only had said so, but had done more than anyone else to bring the possibility within reach. He was now the leading flight experimenter in the world, and the pursuit of human flight had become the passion of his life.

Wilbur Wright would have to contend with the doubts of his neighbors in Dayton. Secretary Langley, far more grandly and self-consciously, was assailing the arguments of Sir Isaac Newton and several of the leading mathematicians and physicists of the day, who said basic laws of logic and physics rendered human flight highly unlikely if not utterly impossible. Wright hoped to "add my mite" in the search for a solution. Langley aspired to join the pantheon of history's greatest scientists.

To build his case, Langley had undertaken a long series of experiments in aerodynamics that culminated in 1896 with the flights of two substantial flying machines—unmanned—over the Potomac River. He called them "aerodromes," his own coinage from the Greek, meaning "air runner."

Among the handful who saw the first unmanned flight of a Langley aerodrome was Alexander Graham Bell, who captured the only photographs of the event. That evening he jotted a note to Langley—"I shall count this day as one of the most memorable of my life"—and to the editor of *Science* he sent a resounding endorsement of the secretary's achievement: "No one could have witnessed these experiments without being convinced that the practicability of mechanical flight had been demonstrated." Certainly Bell was convinced, and he went back to his own flight experiments with new zeal.

"A 'flying-machine,' so long a type for ridicule, has really flown," Langley declared. "It has demonstrated its practicability in the only satisfactory way—by actually flying, and by doing this again and again, under conditions which leave no doubt." As Langley's engineer later put it, the model aerodromes known as No. 5 and No. 6 were "the only things of human construction that had ever really flown for any considerable distance."

Now, three years later, with all the resources of the Smithsonian at his command, and the largest appropriation for research ever granted by the U.S. War Department, Langley was directing an entire staff in the design and construction of a much larger machine—a machine that would carry a man. President William McKinley had taken an interest in the project, and Theodore Roosevelt, now governor of New York, had played a key role in assuring the funding. Langley expected to conduct the first trials by the end of the year.

If the secretary had been in his office that day, Richard Rathbun might have mentioned the carefully written and well-informed request for information on mechanical flight. But Langley was in Europe. He went abroad nearly every summer, to meet with his international peers in the world of science; to raise the Smithsonian's profile as an institution of world rank; and to indulge his own love for the sites of European high culture.

So Rathbun sent back a brief reply to Dayton with "a list of works relating to aerial navigation, which will probably best meet your needs," plus several Smithsonian pamphlets.

Wilbur Wright's letter was filed and forgotten.

*Chapter One*

# "The Edge of Wonder"

"LIKE A LIVING THING"

The first flight of Langley's unmanned Aerodrome No. 5, May 6, 1896

RALPH WALDO EMERSON once remarked that in all nature, birds were the "reality most like to dreams." They were not only graceful and free. They also seemed to pierce the veil between this difficult world and whatever ethereal regions might lie beyond it. So every culture has cherished a collective dream—a myth—about humans finding a path into the sky.

One event more than any other promised to bring the magic within man's grasp. It had happened only three years before Wilbur Wright wrote his letter, on the afternoon of May 6, 1896, when Secretary Langley's unmanned model Aerodrome No. 5 first flew over a remote reach of the Potomac River, thirty-five miles south of Washington, D.C.

Langley had watched from the bank. Alexander Graham Bell sat in a rowboat out on the river, near the houseboat with No. 5 mounted on its catapult on the roof. The unmanned machine looked like a giant white dragonfly, sixteen feet from nose to tip, with one pair of wings forward and another behind. Earlier versions had failed to fly. But this time, when Langley's men started the steam engine and launched the craft over the water, it remained aloft and sailed into a graceful circle. "Like a living thing," Langley remembered, it "swept continuously through the air . . . and as I heard the cheering of the few spectators, I felt that something had been accomplished at last, for never in any part of the world, or in any period, had any machine of man's construction sustained itself in the air before for even half of this brief period."

"We may live to see airships a common sight," he wrote later, "but habit has not dulled the edge of wonder, and I wish that the reader could have witnessed the actual spectacle."

No. 5 had vindicated the labor and hope of ten years. Langley believed it was the herald of a new age, proclaiming that the myth was about to come to life.

HE NEVER INTENDED to be an inventor. He was a scientist, a student of nature, and deeply proud of it. Indeed, he had made himself a scientist against rather long odds, after a false start.

He had been a boy in Boston in the 1840s, the descendant of New England Puritans and the son of a prosperous wholesale merchant who belonged to the city's "aristocracy of trade." At the end of his life, when Langley tried to record everything he could recall from his earliest years, he remembered a day when he was punished for refusing to recite, "an early breaking down of my will." He remembered his first doubt in the existence of God—"a doubt which has never entirely left me"—when a mosquito stung him just as he prayed it would not. He recalled a night trip on the horse ferry to Martha's Vineyard, and the view through his father's telescope as workers put the capstone on the monument to the veterans of Bunker Hill. These were fragments without a pattern. In better times, when he spoke of his childhood, the events he recalled had to do with his life's work—with becoming acquainted with nature, especially the birds and the stars. Langley developed a multifaceted curiosity that would take him deep into literature, history, and art. But his calling was to look skyward.

"I cannot remember when I was not interested in astronomy," he later told a friend. "I remember reading books upon the subject as early as at nine, and when I was a boy I learned how to make little telescopes, and studied the stars through them. Later I made some larger ones, and . . . I think myself they were very good for a boy."

> One of the most wonderful things to me was the sun, and . . . how it heated the earth. I used to hold my hands up to it and wonder how the rays made them warm, and where the heat came from and how. I asked many questions, but I could get no satisfactory replies . . . I remember, for instance, one of the wonders to me was a common hotbed. I could not see how the glass kept it warm while all around was cold, and when I asked, I was told that "of course" the glass kept in the heat; but though my elders saw no difficulty about it, I could not see why, if the heat went in through the glass, it could not come out again.

After Boston High School, where he excelled in mathematics and mechanics, he apprenticed himself to a civil engineer in Boston, then to an architect. He loved the study of the stars. But he had been born a few years too early. Stargazing was common in the seafaring towns of New England, but scientific astronomy was chiefly a gentleman's hobby when Langley was a boy. Observatories were few, and their potential for useful knowledge was little appreciated. When he was nine years old, people stared in awe and fear at the Great Comet of 1843, with its brilliant tail spreading fifty degrees across the sky. The comet spurred support for astronomical studies, and an observatory was established at Harvard. But when Langley was finishing school, astronomy could support only a tiny handful of professional practitioners.

Langley's younger brother John went to Harvard and became a professor of chemistry. But Samuel became stuck at a series of drafting tables—first in Boston, then in St. Louis and Chicago. In 1861, his brother joined the Union Navy as a surgeon. But Samuel stayed put in a civilian job ill-suited to his intellectual gifts and passionate interests. No explanation survives of why one Langley brother from Boston, the home of abolitionism, served the Union cause but not the other.

As a job, architectural drafting was not a bad match for Langley's skills in drawing and mechanics. Yet the work apparently left him empty. He did not marry, so his time off was completely his own, and it is clear that he devoted much of it to reading deeply in astronomy. He observed the skies whenever he could, with whatever instruments he could borrow. Finally, in 1864, at the age of thirty-one, he quit his job and went home to Boston.

If his family considered his decision rash, they must have seen his fixity of purpose in what he did that fall. As the Army of the Potomac beseiged Confederate forces at Petersburg, Langley built a device for observing distant heavenly bodies. He had a Smithsonian monograph on how to make a telescope, and advice from an expert in fine optical equipment, and he enlisted his brother, home from the Navy, as his assistant. Using tools they scrounged from an old barn, the brothers worked for three months on the construction. Samuel ground or reground some twenty mirrors until he had one that he considered acceptable. "My brother's . . . perseverance would not allow us to be satisfied with anything short of a practical degree of perfection," John recalled.

Still "without fixed duties," the brothers then left Boston for a year-long tour of European museums, astronomical observatories, and galleries of art. It may have been on this trip, the first of many that Samuel Langley made to Europe, that he became friendly with the British historian and essayist Thomas Carlyle, who attributed the movements of history to the heroes, the "great men," of each generation, and who advised the young not to "know thyself," as Socrates had taught, but "to know thy work and do it." Langley revisited the philosopher's home time after time over the years, and listened for many hours.

BY THE FALL OF 1865, when Langley returned from Europe, there were fifty observatories in the United States and good positions even for self-taught astronomers. He quickly found a post as assistant to Joseph Winlock, a leading figure in the field, at the Harvard Observatory. After a year he moved to the U.S. Naval Academy at Annapolis, where he was named assistant professor of mathematics. In 1867, he became director of the new Allegheny Observatory and chairman of astronomy and physics at the Western University of Pennsylvania, later to become the University of Pittsburgh. "He had not as yet published anything of note; had not made himself known in the universities; had made no popular addresses; had not pushed himself into notice in any way," said his friend, the historian Andrew Dickson White, founding president of Cornell University. "Yet there was in him something which attracted strong leaders in science, inspired respect, won confidence, and secured him speedy advancement."

Still, the Allegheny post was no prize. Pittsburgh in 1867 was a rough, raw town, just beginning its rise to industrial power, and a long way from Boston, Philadelphia, New York, and Baltimore, the seats of American science and culture. The university boasted a faculty of just twelve, three of whom, like Langley, had no college degree. The observatory was an orphan. The college had taken it off the hands of a bankrupt association of local amateurs. Its telescope was a good one. Otherwise Langley had no scientific apparatus, no library, and no assistant. Besides the telescope, the observatory's furnishings consisted of a table and three chairs.

And the Western University of Pennsylvania was no Harvard or Yale, where a scientist could devote most of his time to pure study. The typical astronomer in such a school, as one said, found "his strength burdened to the limit of endurance, by the routine of daily class instruction."

Langley did not mean to spend his new career at a blackboard. He lost no time in securing two steady sources of outside funding. First, he cultivated a rich railroad man and amateur stargazer, William Thaw, who agreed to supplement Langley's salary and subsidize his projects. Next, Langley arranged to sell the astronomer's one commodity of practical value—accurate time.

In the late 1860s, every city and town in the United States operated according to its own reading of the correct time. For a nation that increasingly depended on efficient railroads, this was intolerable. Observatories, already tracking the movements of the earth with extreme precision, could also track the exact time, and transmit it by telegraph. Langley informed himself on timekeeping at other observatories. Then, with Thaw pulling strings among railroad friends, he beat out competitors for the right to sell astronomical time to the mighty Pennsylvania Railroad and all its subsidiary lines. The fees were enough to equip Langley's observatory handsomely and pay for an assistant. Other customers, including the city of Pittsburgh, were soon added. The time service boosted Langley's standing with practicality-minded administrators and patrons. He could point to it as a substantial down-to-earth benefit of their investment and support.

His faculty colleagues were not so impressed. They complained that Langley was neglecting faculty meetings as well as the classroom, though students' fees paid most of his salary—which, with William Thaw's supplement, was now the highest on campus. Langley retorted that if he neglected the usual faculty duties, it was only to make the observatory "useful not only to science in the abstract but to the University as a seat of learning." In other words, his bid for eminence would pull the lowly university upward with him. He went on to propose that the faculty of the observatory—that is, himself—should become an entity separate from the regular faculty, with the power to run its own affairs. Administrators turned down that proposal, but they did formally declare that Langley would be expected to teach no classes, grade no papers, and play no role in administration. He would operate as a department unto himself, probing the universe full time.

LANGLEY HAD INTENDED to study stars. But in expeditions to witness the total solar eclipses of 1869 and 1870—the first to Kentucky, the second to Spain—he fell captive to the sun. To Langley, an eclipse allowed a full apprehension of humanity's place in a universe of spheres suspended in nothingness. His description revealed his hunger to experience the sublime in nature. As one watched, he said:

> . . . the sun's disk is seen to be slowly invaded by the advancing moon, and as the solar brightness is gradually reduced to a thin crescent, daylight fades with increasing rapidity, and a quite peculiar and unnatural light, hard to describe but which no one forgets who has once seen it, spreads over the landscape. Then, and suddenly, we come to a new sense of the reality . . . of the heavenly bodies, for the moon, which we have been accustomed to see as a disk of distant light on the far background of the starry skies, takes on the appearance of the enormous solid sphere which it is, and a faint glow within its circumference . . . makes its rotundity so perceptible that we feel, perhaps for the first time, the perpetual miracle which holds this great cannonball-like thing from falling.

He had stumbled into solar studies at a promising moment. The recent discovery of a predictable cycle in the appearance of sunspots had spurred intense interest in the possible relationship between sunspot activity and events on Earth, from gravitation to weather to crop cycles. Studies of the sun promised practical knowledge for solving "some of the mightiest problems in our study of the human race itself." And the sun had practical advantages. Langley could do his observing during the day, no small thing in a profession that doomed most of its practitioners to nocturnal lives. Furthermore, as he liked to tell eastern friends, the sun was "perhaps the only celestial object that I can hope to see and observe in any detail and with any regularity in the Pittsburgh area with its coal-burning steel mills and its soot-and-smoke-filled atmosphere."

Astronomers since ancient times had been concerned with the *location* of heavenly bodies. In Langley's time, they were turning to probes of the composition and behavior of stars—not the *where* but the *what,* and *why.* The key tool in the "new astronomy" was the spectroscope, which, with prisms, broke starlight into its component colors and patterns, each of which might reveal secrets about the stars' chemical composition. Langley was all for the new astronomy. But early in his career, he left his spectroscope mostly in storage. Spectroscopy was a business of analysis, not direct observation. It could not satisfy Langley's desire for a primal experience of the heavens. Instead, he chose a program of study that allowed him to look at the sun with his own eyes. He focused on sunspots, those misleadingly named regions "whose actual vastness," Langley said, "surpasses the vague immensity of a dream."

Theories about sunspots diverged wildly. Were they eruptions? Cyclones? Currents? In the era before astronomical photography, painstaking observations and drawings were essential. This required extreme patience even on a clear day, for the earth's atmosphere made the sun's image blur and waver. "One who has sat at a powerful telescope all day is exceptionally lucky if he has secured enough glimpses of the true structure [of the sun] to aggregate five minutes of clear seeing."

Langley's sunspot drawings reveal his fierce self-discipline. His purpose was not to be an artist, not to render a beautiful image or convey his own impression of the thing he observed. It was to be a camera—to record the strange physical features that appeared in the lens exactly as he saw them, without letting his beliefs about their physical nature influence the drawing. Even though Langley might suspect that a sunspot was in fact a solar cyclone, he could not allow himself to depict a sunspot as though it *were* a cyclone. He must draw what he saw. Only if the drawing then resembled a cyclone would it tend to confirm the theory.

The drawings were not meant to be beautiful, but in fact they were, besides being enormously detailed and accurate. Of his hundreds of sunspot render-

"THE ONLY CELESTIAL OBJECT I CAN HOPE TO SEE IN PITTSBURGH"

Langley's freehand drawing of a sunspot

ings, several became classics of solar studies, remaining in textbooks well into the twentieth century. George Everett Hale, a great astronomer of the next generation, once remarked that the more powerful telescopes became, the more their images of sunspots resembled Langley's freehand drawings.

He took pride in his attention to detail and proper method. He recorded every detail of his work in notebooks and insisted his aides do the same. "The user of this book," he wrote in one, "is expected never to commence the entry of a day's observations without writing under 'object' what the general aim of the observations is to be, nor to end the day, under whatsoever pressure of occupation or fatigue, without writing under the heading 'result' a few words (if only a line) to indicate what seems to him then the general tenor and result of the day's work, without waiting for final reduction and analysis." "I can, I hope, honestly say that I spare no personal pains in observation and experiment," he told a friend.

In time, Langley's studies led him to develop an extremely sensitive and delicate thermometer for measuring radiant heat in spectra. He called the device a bolometer. It showed his skill in technology and led to his most important contributions in astronomy. Using it—and only he and a handful of others ever possessed the patience to do so—Langley discovered heretofore unknown regions of the solar spectrum, estimated the heat of various regions of the sun's surface, and calculated the solar constant (the energy of the sun before it strikes Earth's atmosphere). The bolometer bolstered the dawning belief "that heat and light were not two different things, but different effects of the same thing." And Langley's presentations were lucid and winning. "The thoroughness, ingenuity, and beauty of his methods and the clearness of his style in presenting them attracted attention far and wide," said an admirer.

He also mastered less noble means of advancement. He ingratiated himself with prominent men and women. He beefed up his list of publications—which eventually totaled more than two hundred—by composing different versions of the same set of experiments for several journals. He pushed aides to do work for which he received the credit. A good deal of the research at Allegheny was actually performed by Langley's assistant, Frank Very, but it was always Langley's name alone that went on the articles. "He was a man of strong personality," an aide said, "likely to dominate almost any association into which he came: a hard taskmaster, as the real investigator is likely to be, sparing neither means, his assistants, nor himself in the pursuit of his object in research."

He tended to exaggerate, to push his claims too far. He said the moon was colder than it turned out to be. He said the solar constant was higher than it turned out to be. He exaggerated the temperature to which the surface of the Earth would sink without its atmosphere. No one ever charged Langley with intellectual dishonesty. But the pattern of pressing too hard for findings and claims that might attract plaudits is too clear to ignore. "A certain part of Langley," said a careful student of his career in astronomy, "was attracted to the spectacular."

LANGLEY ADORED CHILDREN. When he visited the homes of friends with youngsters, he would take them on his knee and tell them fairy tales, "many of which he would improvise with wonderful tact to please the children." Yet he never married. He was friendly with women, but apparently never became romantically attached to anyone. When the sun went down each day, the gaslight inside the house on the observatory hill burned only for him, alone in his chair with his books.

He read deeply and remembered, it seemed, practically everything. He read the great English novelists and German classics. He was fascinated by fairy tales, folklore, and ancient mythology. A prized possession was his collection of various editions of the *Arabian Nights*. He studied British and French history and biography. Later, in Washington, when Andrew Dickson White gave a series of lectures on the French Revolution, White was astonished at the learning that Langley's questions revealed. "I particularly remember his minute and accurate knowledge of the comparative value of sundry authorities," White recalled. "It was not merely that he had read works of importance in the history of the period . . . but that he had gone extensively into original sources, and especially into the multitude of memoirs."

Though not a churchgoer, Langley thought deeply on questions of ultimate reality. He was "an ardent seeker of religious truth . . ." White said, "equally hostile to dogmatism against and in favor of received opinions." He "loved to talk with men of positive religious views about their own beliefs, and took a deep interest in a Jesuit, or a Jew, or a Buddhist, or a Mohammedan, or, indeed, in any man who thought he had secured any truth and knew the way of life in this world or in the world to come." An aide and friend at Allegheny said, "There was something reaching almost to the transcendental in his inner life."

As the years passed, his work took him into deeper realms of the universe; his reading drew him into the fellowship of far-flung intellectuals, communing through the written word; and he became ever more a man unto himself, less likely to find companionship in Pittsburgh than when he arrived in 1867. Yet he was burdened by "a strong craving for real society," a friend said, "by which he meant intercourse with people of diverse minds and knowledge, all of whom might give him that intellectual companionship for which he hungered." He found few friends among his faculty colleagues, perhaps because he considered them beneath his intellect, perhaps because they continued to resent his high pay and privileges. He attended meetings of Pittsburgh's medical society just to be able to talk with professional men who knew the language of science. On Sunday evenings he would descend his hill and trudge to the back room of a drugstore because a half-dozen souls gathered there to talk about books. To his friend Charles Sanders Peirce, the eminent Johns Hopkins philosopher, who had visited Langley in Pittsburgh, he wrote: "You have seen for yourself how far it is from the companionship a student of science wants; and you will understand what I mean in saying that I am in that respect out in the cold, and would on that account willingly warm myself once or twice at your Baltimore fire."

As factories edged closer to his hilltop, their smoke obscuring the skies, Langley found Pittsburgh more and more stifling. In an 1884 report to the community, he frankly asserted that the observatory lacked "almost every thing outside of its actual apparatus that the ordinary resources of American civilization would provide for it in any large American city but Pittsburgh." Its work—his work—had gone forward despite "a constant struggle with poverty," an assertion that no doubt caused eyes to roll among his lesser-paid colleagues. "There are not only no museums of art, no libraries of reference, no collections of scientific material, but in general, none of those aids to the investigator which are to be found in so many younger and smaller places."

He fled as often as he could. He went to Boston to see his mother, his aunt, and colleagues in the sciences, especially E. C. Pickering, a Harvard astronomer who became a close friend. During summers in Europe, often with Pickering in tow, he visited observatories, met with foreign colleagues, and acquired a taste for fine food and wine. As his own publications mounted, European astronomers increasingly knew Langley's name and reputation. But he cringed whenever he had to explain where he came from. Once a reporter asked him if European scientists didn't pity Langley for being stuck under Pittsburgh's smoky skies. "In the astronomical circles of the Old World," he replied, "it is often the case that the existence of Pittsburgh itself is hardly known, and the Allegheny Observatory is thought, from its name, to be located somewhere on the summit of the Allegheny Mountains."

Friends on other campuses tried to find him a position that would better fit his growing reputation and satisfy his desire for comradeship. C. S. Peirce told

Daniel Coit Gilman, president of Johns Hopkins, that Langley was the right man to direct the great observatory that Gilman wanted to build. Langley would be "a great addition to the university," Peirce advised, "widely known not only as one of the very first men in the New Astronomy but also as a very charming and cultured person." But the Johns Hopkins observatory remained unbuilt. In Cambridge, friends spoke of Langley as the obvious successor to his old superior, Joseph Winlock. But someone else was chosen. He was elected president of the American Association for the Advancement of Science, a post of immense prestige. But he walked the streets of Pittsburgh unrecognized amid people who had no inkling that such a thing as the solar spectrum existed, let alone why anyone would want to measure it.

As the 1880s passed, Langley could not have escaped the sense that his chance to become one of the historic figures of his own field, rather than merely an accomplished practitioner, was slipping away. The developing science of astrophysics was becoming too sophisticated for any scientist who, like himself, lacked proper training in differential equations and calculus. He said he was at best "a learner" in the higher mathematics, which he found a source of both wonder and frustration. Once, he instructed some aides to calculate the equation for a curve, and when they did the work "in a decidedly roundabout way . . . covering many pages with mysterious-looking formulae and equations . . . it was interesting to note the almost reverence with which he afterward turned these pages, really impressed by what looked so profound, and which he did not understand."

In a moment of humility he confessed to a friend: "I know nothing about chemistry and little about chemical physics," and, "I have never been able to flatter myself that I could reach any eminence in . . . mathematics." When he vented his irritation, saying math was inferior to the direct telescopic observation that he did best, another astrophysicist issued a stinging put-down: "Professor Langley . . . shares the views of not a few, possessing like himself marked aptitude for experimental research, but seeming to become actually irritated when physical matters are dealt with in the only way in which they can be satisfactorily analysed. . . . No physicist, certainly no astronomical physicist, can successfully deal with the problems which come before him, without a mastery of at least the elementary methods of mathematical analysis—as the differential and integral calculus, the treatment of differential equations, the calculus of variations, and the like."

If these were the keys to the cosmos, Langley knew he did not possess them.

• • •

IN AUGUST 1886 Langley caught a train for the annual meeting of the American Association for the Advancement of Science in Buffalo. The association had been founded in the 1850s to promote the highest standards of scholarly investigation and to "repress charlatanism." These were the elites of American science, jealous of their reputation as high-minded truth-seekers, ever watchful for cranks and goofball theorists who might undermine public support for science. Still, promising amateurs occasionally were allowed to speak. Among the amateurs on the program at the 1886 meeting was an Illinois farmer and self-taught ornithologist, Israel Lancaster. He claimed that birdlike devices of his own design could stay aloft for as long as fifteen minutes. On a larger scale, he said, such devices could carry a man.

Lancaster's first talk turned into a circus. Scientists laughed at his claims. Five days later the attendees jammed the room for Lancaster's second talk, advertised as a demonstration of artificial birds. But Lancaster showed up with only a device to show the effects of air pressure on plane surfaces. This caused "great disappointment" among his listeners, *The New York Times* reported, and many "questions from the incredulous audience." When Lancaster insisted a flying model could be constructed in minutes, one skeptic challenged him to do so on the spot. Another rose to offer one hundred dollars for a flying model; a third offered one thousand dollars; and this session, too, dissolved in noisy laughing and jeers.

Langley was in the audience, but he didn't laugh. The farmer's presentation had reminded him of the hawks he had watched in his childhood, and now, "I was brought to think of these things again, and to ask myself whether the problem of artificial flight was as hopeless and as absurd as it was then thought to be."

LANGLEY KNEW "the whole subject of mechanical flight was . . . generally considered to be a field fitted rather for the pursuits of the charlatan than for those of the man of science." If he chose to work on it, he would risk his reputation as a serious investigator, and if he failed, as every other experimenter had, he could count on outright ridicule, perhaps even the dismissal of all his legitimate work in astronomy.

So he began cautiously. He would attempt a single step in the direction of artificial flight—"Not to build a flying-machine at once, but to find the principles upon which one should be built." Isaac Newton had believed that artificial flight would require power beyond the capacity of man. This had been true in Newton's own time, perhaps, before the revolution in mechanical power, but was it true still? Langley decided he would try to determine precisely how much power was required "to sustain a surface of given weight by means of its motion through the air."

In the yard of the Allegheny Observatory he directed the construction of a large machine called a whirling table, or whirling arm, for swinging winglike surfaces and stuffed birds in circles. The purpose was to measure the lift exerted by surfaces of various sizes and shapes. A steam engine drove the arm round and round at speeds of up to seventy miles per hour.

In one key test, Langley suspended a simple, one-pound brass plate by a spring from the end of the whirling arm. The weight of the plate, of course, caused the spring to extend a little. Then he set the arm in motion. By Newton's reckoning of centrifugal force, he noted, "It might naturally be supposed that, as it was drawn faster, the pull would be greater." But in fact, "the contrary was observed, for under these circumstances the spring *contracted*, till it registered less than an ounce. When the speed increased to that of a bird, the brass plate seemed to float on the air; and not only this, but taking into consideration both the strain and the velocity, it was found that absolutely less power was spent to make the plate move fast than slow." This result he found "very extraordinary," not to say paradoxical. After all, in any other form of transportation, from the oxcart to the locomotive, it was easier—that is, it required less power—to go slow than to go fast. But in the air, the experiment now suggested, the reverse was true.

To make sense of this puzzle, Langley remembered scenes from childhood. He thought of a stone skipping across water. Moving fast, the stone bounced off the water's surface, but when the stone's momentum slowed, it penetrated the surface and sank. He thought of a skater on the thin ice of a pond in spring. If the skater slowed down or stopped, he would plunge through. But if he maintained his speed, he could skim safely across. Thin ice might not hold much weight, but like every other substance in nature, it possessed a certain amount of inertia—that is, it resisted anything trying to push it out of its place. The same must be true, Langley said, of "the viewless air. Like thin ice on a pond, the air could hold up a moving object, but only if the object moved fast enough." Langley deduced that speed was the key to human flight.

IN THE FALL OF 1886, just as Langley was beginning his study of aerial locomotion, the regents of the Smithsonian Institution were considering who should succeed the great naturalist Spencer Baird as secretary of the Smithsonian. Baird, at sixty-three, was complaining of exhaustion and a weak heart. Asa Gray, a distinguished Harvard botanist and an influential regent, wanted

Langley to become an assistant secretary under Baird, with the understanding that he would move up upon Baird's retirement or death. Gray spoke to his Harvard colleague, E. C. Pickering, asking him to sound out his friend in Pittsburgh.

Langley was surprisingly cautious. On the one hand, the overture exceeded his fondest hopes for recognition and intellectual companionship. Yet the actual job that Baird was offering Langley was assistant secretary for international exchange. This meant presiding over the Institution's elaborate effort to distribute its publications around the world and to communicate with scientists and scholars. It was essentially a librarian's job, and a demanding one. He wanted assurances that he could continue his scientific work. "I have no wish or ambition," he told Baird, "to tempt me from giving most of my time to physical investigation—at least now, while I enjoy exceptional facilities for this, together with a freedom which I could not expect in any subordinate position."

> My professional life here [in Pittsburgh] is . . . a very pleasant one, in most respects, nor have I had occasion to leave the work of my predilection to increase my income. At the same time both my professional and domestic life here are exceptionally isolated, and I have felt the need of some change which would bring with it, along with society, new occupation, if that could be of a kind not wholly dissociated from my accustomed pursuits.

Baird said to come ahead; he could do his research. Langley made arrangements to continue as director of the Allegheny Observatory on a part-time basis, and left for Washington. Eight months later, Baird died, and the regents asked Langley to succeed him. He was inaugurated as the third secretary of the Institution in 1887.

EVEN AS LANGLEY was conducting the whirling-arm experiments, he was groping for a way to move beyond them—to construct a device that would fly by itself. By instinct and experience, he was an observer, not a theorist or designer. He needed something to observe, to gauge its behavior and make adjustments that might lead to a more promising next step. But the only things that actually flew were birds, and birds, he found, did not yield their secrets easily. The medium in which they operated—the ever-shifting air—was invisible. It was hard enough to track the movements of birds' wings, harder still to understand them when he had no way of measuring the forces acting upon them. Now that he was in Washington, he had new means of investigation at his disposal. He examined birds' anatomy, even the skeletons of pterodactyls in dusty Smithsonian storerooms. But it was no simple problem to reproduce the remarkable shape and flexibility of a bird's wing, or its connection to the body. It was especially daunting to make a mechanism that would reproduce the "instinctive control of the wing," which presumably allowed a bird "to meet the requirements of flight that are varying from second to second, and which no automatic adjustment can adequately meet."

When he began in 1887, he was not yet familiar with the hand-held toy flyers that a young Frenchman, Alphonse Pénaud, had devised some years earlier—marvelous contraptions that had amused thousands of children around the world, including Wilbur Wright. (These were the "bats" Wright mentioned in his letter to the Smithsonian.) Though they stayed up no more than ten seconds or so, they were the first devices that ever kept their balance in the air. Langley needed something like them, if only to get some idea of what *not* to do.

One of his strengths—born of courage or sheer curiosity or some combination of the two—was a capacity to begin working in the face of an inexplicable and uncharted problem. He started from scratch. Still good with his hands, he constructed a series of hand-held devices similar to Pénaud's. The first, vaguely resembling the shape of a bird, was "a light wooden frame with two propellers, each driven by a strand of twisted rubber." He built more than forty hand-launched models in all. He tried the early ones on the grounds of the Allegheny Observatory, the later ones on the green plain of Smithsonian Park. Clerks and scientists alike would peer out the windows at their leader; one said, "It was a very amusing sight to behold the dignified Secretary . . . rushing about with coat-tails flying, chasing and dodging the little toy planes."

Model by model, adjustment by adjustment, the devices improved, but not enough. Pénaud had reported flights of up to fifteen seconds with his marvelous little toys, but Langley never could make a model stay up for more than eight seconds, or fly for more than one hundred feet. "No machine in the whole history of invention, unless it were this toy of Pénaud's, had ever . . . flown for even ten seconds," he said later, "but something that will actually fly must be had to teach the art of 'balancing.' " To understand the principles underlying that art, he needed to observe a machine in sustained flight. But to put a machine in flight, he had to understand the underlying principles.

There was only one route out of the paradox. If he meant to keep a machine in the air long enough to allow observation, he needed an engine much more powerful than a twisted strand of rubber. He knew power and speed were

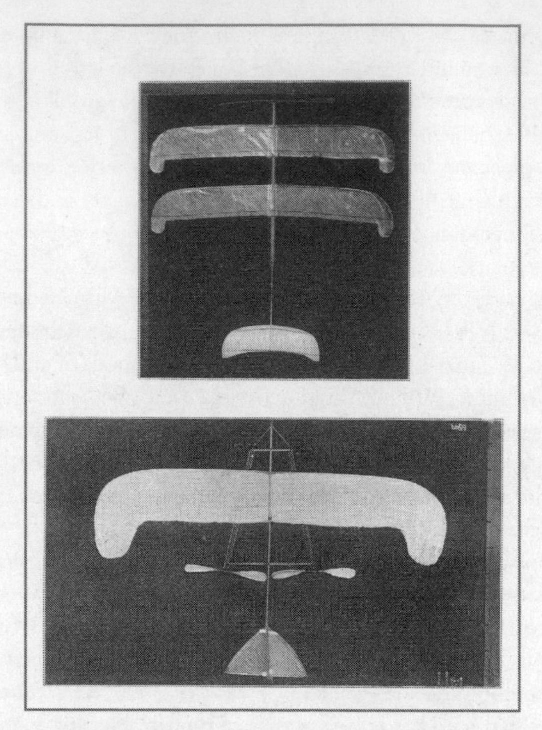

"SOMETHING THAT WILL ACTUALLY FLY MUST BE HAD."

Two of Langley's hand-held model aerodromes

not sufficient, by themselves, to solve the puzzle of flight. "It is enough," he wrote, "to look up at the gulls or buzzards, soaring overhead, and to watch the incessant rocking and balancing which accompanies their gliding motion to apprehend that they find something more than mere strength of wing necessary, and that the machine would have need of something more than mechanical power, though what this something was was not clear." But as a beginning, he would have to concentrate on the task that *was* clear—to build an engine powerful enough to propel a weight through the air, yet light enough not to bring the craft crashing down. How to go up, how to come down, how to turn left and right—those problems, however essential, must wait. On vacations at Alexander Graham Bell's magnificent estate overlooking the Bras d'Or Lakes in Cape Breton, Nova Scotia, Langley and his friend would spend hours on Bell's houseboat, talking and speculating as gulls wheeled overhead. Once, after a long period of observing the birds in silence, the secretary exploded: "Isn't that maddening!"

"What's maddening?" Bell asked.

"The gulls!"

Bell replied, "I was thinking they were very beautiful."

LANGLEY'S METHOD WAS TO try something, test it, fix whatever problems appeared, and try again. Every failure, however frustrating, was welcome, for it identified another problem he might solve if he tried hard enough and often enough. His aide, the astronomer Charles Abbot, once said of Langley: "Whether from natural disposition or from deliberate conviction that time could be saved thereby, or both, his method of attack upon a new experimental problem was to make rough trials at once, to improve the method as experience dictated, and at length reach the final dispositions as the result of correcting this and that detail, rather than by first spending long and careful study over every detail before reducing any part of the work to practice." In engineering, the approach is known simply as "cut and try."

He faced a void. "In designing this first aerodrome," he wrote later, "there was no precedent or example. . . . Everything was unknown." He groped for something to try, searching the catalogues of nature and civilization for some combination of things that move well through fluids. The bird, of course, offered the essential example of the wing and the tail. From the ship he seized the notions of hull and rudder. Wondering what shape the hull of a flying machine might take, he thought of "the lines which Nature has used in the mackerel," and deemed this the best for movement through the air. With these images in mind, he drew a plan.

In a converted shop at the Smithsonian, Langley assembled a team. In one corner, machinists labored on engines and boilers. Other machinists worked on frames, often for more than one model at a time. Carpenters made spars and ribs for wings and tails, and covered them with fine and expensive fabrics, the lightest and strongest to be found—black silk, white silk, even the exceptionally light membrane of cattle intestines called goldbeater's skin. Langley ordered up a smaller version of the whirling table and put it to work testing new shapes.

Slowly, with infinite care and pain, a basic shape emerged, one version supplanting another as Langley and his men struggled to find the right combination of strength and lightness. They would try a new design, test it, then alter "the form of construction so as to strengthen the weakest parts. . . . When the breakdown comes all we can do is to find what is the weakest part and make that part stronger; and in this way work went on, week by week and month by month, constantly altering the form of construction so as to strengthen the weakest parts."

With his keen awareness of history, Langley consulted a philologist about a proper, Greek-derived name for his evolving creation. The term "aerodrome" emerged from these conversations; Langley took it to mean "air runner." He thought of adding the prefix "tachy," to connote the all-important concept of speed. But the philologist, one Basil L. Gildersleeve, suggested that "tachy-aerodrome" would have too many syllables, and Americans would reduce it to "drome" within ten years anyway. So Langley stuck with "aerodrome."

The frames were in the shape of a cross, with a thick hull to hold the engine and arms to either side to hold propellers and wings. Difficult as it was to find a workable frame, it was far harder to design and build a workable engine light enough for Langley's purposes. He believed internal combustion ultimately would provide the best power source for a flying machine, but in 1891, the new technology had not advanced far enough. He tried other sources of power—carbonic acid and compressed air—but soon was forced back to the old standby, steam, heavy but dependable. He set his machinists to work. With each model he would run tests to see if it was powerful enough to sustain the aerodrome in the air. He calculated that he must have an engine with enough power to lift a weight equal to at least 40 percent of the deadweight of the aerodrome in the shop; the wings would provide the additional lifting power in flight. The first design, which Langley labeled No. 0, was never tested. The engine of the second model, called No. 1, could lift scarcely more than 10 percent of the machine's weight. No. 2 made it to 20 percent—encouraging, but the frame and wings were judged too flimsy to withstand a trial.

Langley knew science demanded repeated "backward steps—that is, the errors and mistakes, which count in reality for nearly half, and sometimes for more than half, the whole." But in the aerodrome, the backward steps seemed endless. His memoir of the experiments is larded with countless attempts and tests that ended in disappointment:

"It appeared to be inexpedient to do anything more with it . . ."
" . . . these engines did not give results that were satisfactory . . ."
" . . . a long and tedious series of experiments . . ."
" . . . many unexpected difficulties . . ."
" . . . numerous features of construction . . . which proved useless when rigorously tested . . ."

" . . . difficulties which seem so slight that one who has not experienced them may wonder at the trouble they caused . . ."
"The delays . . . were always greater than anticipated . . ."

Finally tests showed the engine was ready. He knew perfectly well the problems of balance and steering remained unsolved. But he could begin to attack them only if he could watch his craft in actual flight. So "the point was reached where an attempt at actual free flight should be made."

The models had no way to land, so he decided he had to conduct his test flights over water, to minimize the inevitable damage. This led to a long search for a suitable test site. Langley eventually decided on a wide and secluded stretch of the Potomac River near the village of Quantico, Virginia. He decided, too, that the models should be launched from a houseboat, which could be turned to face the wind no matter what the wind's direction.

Attempts began with the coming of warm weather in 1894. Most were smash-ups. The best lasted only a few seconds. The machines, Langley told Bell, would typically "pitch up or down, but ordinarily *up*, showing an excess of power, not a lack of it, like a horse which is always trying to walk on his hind legs and falling backward."

On May 6, 1896, two aerodromes were prepared for testing—No. 5 and No. 6. Shortly after lunchtime, No. 6 met the usual fate. The guy wire between the fore and aft wings caught on a strip that held the wings in place. A wing snapped and the craft flopped into the water, crushing the propellers. The mechanics fished it out, stowed the sodden pieces, then hoisted No. 5 onto the launcher, which they pointed to the northeast, into a "gentle breeze."

Langley stood on shore holding a stopwatch, but "with hardly a hope that the long series of accidents had come to a close." Bell was seated in a small boat near the houseboat, holding a camera. As the mechanics worked, a few locals looked on from shore. The launcher was tripped, and No. 5 slid along the track, then out over the water at a height of about twenty feet. The machine seemed to swoon, sagging three or four feet toward the water. But as it gathered speed it began to rise, and the spectators froze, then cheered. Rising slowly and gracefully at an angle of about ten degrees, the machine tilted slightly to the right and began to describe a great spiral roughly one hundred yards in diameter. It completed one circle and began another. Bell, thunderstruck, remembered to snap the shutter of his camera. Langley glanced at his stopwatch. A minute had passed, "and it still flew on."

By the end of the second circle, when the aerodrome had reached a height of eighty or a hundred feet, the fuel ran out and the propellers slowed. Bell, ex-

"AS IT GATHERED SPEED IT BEGAN TO RISE."

An artist's rendering of the steam-driven Aerodrome No. 5

pecting the craft to capsize and plunge headlong into the water, watched in awe as instead it glided onward in silence, nosing down slightly and finally settling "so softly and gently that it touched the water without the least shock." In ninety seconds No. 5 had flown more than half a mile at a speed of roughly twenty-five miles per hour. That November, No. 6 made a flight over the river of more than five thousand feet, nearly a mile.

THIS SEASON OF TRIUMPH brought loss as well. Shortly after the test of No. 5, William Crawford Winlock, the son of Langley's old patron at Harvard, whom Langley had appointed as a key assistant, died suddenly. A greater blow fell in September 1896, when George Brown Goode, assistant secretary in charge of the National Museum, died after working himself to exhaustion. Goode, a first-rate naturalist much loved inside the Institution and out, was widely regarded as the best museum man in the nation. He and Langley had become assistant secretaries together, and many believed Goode, not Langley, should have succeeded Spencer Baird as secretary. But Goode had become Langley's close friend—"like a brother," Langley said—and Goode's strong hand in the Institution's day-to-day affairs had freed Langley to devote most of his time to his own research. He was "a man who cannot be replaced," the secretary told the regents, "a man who . . . possessed a combination of administrative ability and general scientific knowledge with every element of moral trustworthiness for which I do not know where to look again."

With no wife or children of his own, Langley depended inordinately on a few close friends and their families. So for many months, a friend observed, "the severe strain of his scientific labors and his personal losses tended to a depression of spirits which caused him to shrink from new work." At the request of *McClure's Magazine*, he prepared a detailed account of his experiments. "Nature has made her flying machine in the bird . . . and only those who have tried to rival it know how inimitable her work is, for the 'way of a bird in the air' remains as wonderful to us as it was to Solomon, and the sight of the bird has constantly held this wonder . . . in some men's minds, and kept the flame of hope from utter extinction, in spite of long disappointment." In closing he turned wistful, saying—or *almost* saying—that he was through with his part in the pursuit of flight. "Perhaps if it could have been foreseen at the outset how much labor there was to be, how much of life would be given to it, and how much care, I might have hesitated to enter upon it at all . . . I have brought to a close the portion of the work which seemed to be specially mine—the demonstration of the practicability of mechanical flight—and for the next stage, which is the commercial and practical development of the idea, it is probable that the world may look to others."

In fact he was deeply torn about what to do. His apparent farewell to flight had no sooner appeared in *McClure's* than Langley made it known—privately—that he wanted very badly to carry on the work himself. The difficulty was how to pay for it. To build a man-carrying aerodrome, he believed he would need at least fifty thousand dollars. He asked the regents and received permission to use a reserve research fund for his aeronautical work, but that was only enough to keep the work in low gear. If he pursued manned flight as a moneymaking proposition, he risked a tangle over ethics and propriety. More important, Langley had heard enough about his friend Bell's long struggle to defend his telephone patent to know that he lacked the stomach—and the years—for such a battle.

In the spring of 1897, Bell's aged father-in-law, the Cambridge businessman and philanthropist Gardiner Greene Hubbard, a Smithsonian regent, beseeched Langley to protect his creation. "You have done what no one has ever done before," Hubbard said. "There is in that fact something novel, something new and therefore patentable." Without a patent Langley would lose the credit due him and mankind might lose the benefit, while with one, he would stand a better chance of expanding his experiments. Hubbard even offered to cover the expense of applying for a patent. But the secretary hesitated. "I have passed sixty years, chiefly in quiet pursuits." He could imagine himself making one more large scientific effort. But he could not imagine himself plunging into "years of labor and commercial strife, of a kind for which every habit of my life unfits me . . . I am not of an age to become another man than what I am."

A rich benefactor might fill the need, one who wished only to help the cause of science. But there was none to be found. Twice that year Langley asked the Chicago engineer Octave Chanute, a flight enthusiast and promoter, to be on the lookout for such a patron. "If anyone were to put at my disposal the considerable amount—fifty thousand dollars or more—for . . . an aerodrome carrying a man or men, with a capacity for some hours of flight, I feel that I could build it and should enjoy the task." Chanute offered encouragement but no concrete aid.

Langley's itch to resume the work only grew. His aerodromes weren't little hand-held toys, like Pénaud's, but substantial and weighty. And they had flown for as long as they had fuel! Why not simply increase their dimensions to a size large enough to hold a man? The central problem would not be the form of such an aircraft—that problem he had solved already—but whether an engine could be built to propel it.

Isolated as a medieval scholar, Langley brooded in the nine-towered Smithsonian Building, universally known as the Castle, which appeared to one observer as if "a collection of church steeples had gotten lost, and were consulting together as to the best means of getting home." In the fall of 1897, he drew out his notebook to record "certain private thoughts about a possible extension of my official aerodromic work to a scale which would enable the machine to carry a man."

> I believe that the results already accomplished on May 6/96, and Nov 18/96, make it as nearly certain as any untried thing can be, that with a larger machine of the same model, to carry a man, or men—if the steam engines could be . . . replaced by such gas engines as can probably now be built—flight could be maintained for at least some hours. . . . My idea would be not to experiment further in the quest of an ideal flying-machine, but to take the results already obtained—which it is almost certain will work on a considerably larger scale—and to repeat them in that form with such modifications only as the changed scale and the presence of a man in the machine may demand.

The same machine as No. 5, only bigger, launched from the same sort of houseboat, "but more elaborate," with a more powerful engine, though very light. That would be the key, he was sure—an engine of unprecedented ingenuity, the most powerful lightweight engine in the world. The flat stone skips along the surface of a pond only so long as it moves fast. "Speed, then, is indispensable here." An image from the poetry of Alexander Pope came to his mind:

> Swift Camilla scours the plain,
> Flies o'er the unbending corn, and skims along the main.

"Now, is this really so in the sense that a Camilla, by running fast enough, could run over the tops of the corn? *If* she ran fast enough, yes."

Just before Christmas, Langley wrote again to Octave Chanute, asking for the name of "anyone who is disposed to give the means to such an unselfish end." But the Chicagoan replied: "I know of nobody disposed to give the means for a purely scientific experiment nor do I see what promises of financial profit, or of fame, could be made to a rich man furnishing the funds."

Then events outside the realm of science offered Langley his opportunity.

ON FEBRUARY 15, 1898, in the midst of tense negotiations between the United States and Spain over Spanish misrule in Cuba, an explosion sank the American battleship *Maine* in Havana harbor, killing 260 officers and sailors. Spanish agents were widely blamed, and the ensuing inquiry went forward amid rising cries for a war against Spanish imperialism. Congress approved $50 million to prepare for it. On Saturday, March 19, Senator Redfield Proctor of Vermont, a calm and respected figure just back from an inspection tour of Cuba, confirmed newspaper reports of Spanish brutality. Few doubted that a declaration of war was near.

The following Monday morning, Secretary Langley met with Charles Doolittle Walcott for a long, private talk. Walcott had succeeded the legendary John Wesley Powell, explorer of the Grand Canyon, as director of the U.S. Geological Survey. Langley regarded Walcott as one of the most capable men of his acquaintance. Upon the death of George Brown Goode in 1896, Langley had invited Walcott to run the National Museum during the search for Goode's successor. Walcott himself was everyone's choice for the permanent post. He was not only a paleontologist of the first rank—he later would discover and catalogue the fossils of the Burgess Shale, the greatest fossil trove ever found—but also a shrewd and resourceful politician, well-connected and well-liked throughout Washington. "He was an athletic, breezy type of man," a close associate said, "who would go for a brisk early morning walk in Rock

Creek Park and turn up for breakfast with some influential Representative or Senator, or perhaps with the President. Without apparent guile, and with a cheerful humorous talk, he would put in just the right words to lead his host in the way of promoting some good thing he had at heart." Langley, always anxious about his relations with the capital's power-brokers, needed the help of such a man in the plan he was about to assay.

Some months earlier, Langley had gathered that Walcott would like to become the permanent assistant secretary. But either Langley had misunderstood his friend or Walcott had changed his mind. For in December 1897 he told Langley he must remain at the Survey, even if it meant—as it would—that he could no longer serve the museum part-time. So Langley, disappointed, would have to search for someone else. Their relationship remained warm. But Walcott may have felt a special obligation to do the secretary a favor. And he knew how.

Clearly, the war fever had raised a new possibility in Langley's mind. That morning, after discussion of official items, Langley mentioned his "interest in constructing a man-carrying aerodrome as a possible engine of war for the Government, at least so far as to demonstrate by actual performance that it was capable of carrying a man or men for a flight of an hour or more." Such a project could easily cost fifty thousand dollars, the secretary said, perhaps twice that.

Walcott spoke up. He could talk to McKinley, he said, and to others. Perhaps a committee could be deputized to review the prospects and make a recommendation.

Langley needed no convincing, and Walcott moved with astonishing speed. Before the week was out, he had broached the idea to two important friends—Theodore Roosevelt, assistant secretary of the Navy, and Alexander Meiklejohn, assistant secretary of War—and laid the idea before President McKinley, along with a photograph of No. 5 in flight and a copy of Langley's article in *McClure's*. The President was much pleased with these items, Walcott told Langley. The Army was for it. Roosevelt urged John D. Long, secretary of the Navy, to join in: "The machine has worked. It seems to me worthwhile for this government to try whether it will not work on a large enough scale to be of use in the event of war."

Within a week a secret and unofficial advisory committee had been appointed—two officers from the Army, two from the Navy, and Stimson Brown, a mathematician at the U.S. Naval Observatory. On April 6, the committee spent three hours with Langley at the Smithsonian, examining No. 5 and No. 6 and listening to the secretary review his plan for scaling up to a man-carrying machine. Walcott attended, as did Bell, whose presence, Langley well knew, could not fail to awe the other guests.

Out in the open now, Langley suddenly felt it unseemly to push. He wished to be asked.

He "was by no means eager to assume so much care and responsibility," he assured his visitors. But if his country needed him in time of war, he would undertake this important work. He could make no promise, of course, that a successful flying machine "would be at once an engine of war." Yet he was quite confident it would become one eventually. The cost would be not less than fifty thousand dollars, to be spent entirely at his own discretion, without "the usual restrictions on appropriations." Of course he held no financial interest in the endeavor, but sought only to serve the nation. For three hours he spoke, passed photographs, answered questions.

The committee members could hardly help but find all this persuasive—the esteemed senior scientist, dignified and confident; the otherworldly machines, their fittings gleaming under the laboratory lights; photographs to prove they really had flown; and Bell, a genuine hero of American technology, nodding his approval of this pioneering enterprise. Professor Brown, preparing the committee's report, not only adopted Langley's review of aeronautics but allowed the secretary to edit the document heavily before passing it along to higher authorities.

The committee readily agreed that Langley's project gave all promise of producing a machine capable of wartime reconnaissance; of "communication between stations isolated from each other by the ordinary means of land or water communication," and of serving as "an engine of offense with the capacity of dropping from a great height high explosives into a camp of fortification."

The members cited Octave Chanute's definition of a successful flying machine—that it must have wings capable of supporting it in the air, a motor and propellers to push the craft forward, and the ability to "start up under all conditions," to "alight in safety," and to maintain "equilibrium and dirigibility," that is, to keep its balance and to turn in one direction or another at the operator's command. Langley already had proved he could build sufficient wings and means of propulsion, the committee affirmed. Indeed, he intended to incorporate recent improvements in internal combustion in "a much more efficient form of engine." If a gasoline engine failed, it would be only "a minor difficulty," for steam "has already demonstrated its practicability on even a much larger scale," and would suffice.

As for "alighting safely," the group hinted at only slightly less confidence. They could attest the models had landed in the water without substantial damage. Safe landings on the ground would come when "the attainment of dirigibility and control shall have been completely secured."

And on that question—the operator's ability to steer the craft as he chose—the members repeated what they took to be Langley's view. Perhaps they didn't understand what he had said, or perhaps he had claimed more than he should have. For the committee's report said No. 5 and No. 6 had exhibited control "within the limits possible in such a construction"—a true statement only in the sense that the unmanned models had flown in the direction in which they had been launched, at least for a few seconds. And the committee said control of the manned machine was to be enhanced "by the addition of an intelligence which can intervene, or not, as desired." By that they meant a human pilot. Yet there was no device on the model aerodromes that a pilot could use for steering. Nor did Langley yet know how to make one. So far, the aerodrome resembled a thrown stone, moving through the air only so long as its momentum held out.

The plan was referred to the War Department's Board of Ordnance and Fortification, a committee of Army and Navy officers established in 1888 to beef up national defense, particularly to repair coastal fortifications that had fallen apart in the years after the Civil War. In the 1890s, the board had taken on the responsibility of assessing new kinds of weapons, then all manner of inventions that might be useful in defense and war.

Langley, preparing for a make-or-break presentation to the board, worried things along, phoning Walcott for updates and seeking the lowdown on members of the BOF. After some inquiries, a National Museum staff member assured the secretary that "the gentlemen . . . are honorable, upright and courteous army officers." But in the spring of 1898 they also were on their way to the Caribbean. So, war or no war, Langley went ahead with his usual summer tour of Europe.

LANGLEY KNEW HOW TO BUILD a telescope but not a gasoline engine, the only type he believed would be light enough for his purpose. He needed a talented full-time engineer to serve as his chief assistant. He sent a request for a recommendation to a friend, Robert Thurston of Cornell University, one of the few engineering professors in the U.S. willing to say he thought flying machines feasible. "Have you any young man who is morally trustworthy ('a good fellow') with some gumption and a professional training[?]" Thurston recommended a Cornell senior, Charles Matthews Manly, who was only twenty-two years old but highly promising.

Manly came from an old Virginia family. Several of his forebears had been

Baptist ministers; one of them offered the benediction at the inauguration of Jefferson Davis as president of the Confederacy. Raised in South Carolina, where his father was a minister and college president, Manly as a youngster became fascinated by developments in electricity. Well trained in math and engineering, he possessed an even more valuable trait—a bottomless capacity for hard work. Langley apparently considered no other candidates, and Manly left Ithaca for Washington even before receiving his diploma.

JUST BEFORE NOON on November 9, 1898, Langley entered the ornate stone edifice of the State, War, and Navy Building, just west of the White House. Approval from the Board of Ordnance and Fortification was not the lead-pipe cinch that his meeting with the impromptu committee of the previous spring had been. The fighting with Spain had ended some weeks earlier; the urgency of war had faded. And Langley could not count on a friendly audience.

The chairman of the board was General Nelson A. Miles, commanding general of the Army. Miles had left school at an early age and was not accustomed to the company of astrophysicists. While Langley was preparing engineering drawings at a desk in Chicago, Miles had been killing Confederates at Antietam, Chancellorsville, and Cold Harbor. Since then he had led Army regulars against Cheyennes, Sioux, and Apaches; Pullman strikers in Chicago; and, just lately, Spaniards in Puerto Rico. Now, at the age of fifty-nine, he was believed to aspire to the presidency.

The general strode into the room late, skipped any pleasantries, and invited Langley to make his statement.

The secretary began by emphasizing the scientific foundation he had laid in the late eighties and early nineties, the "long and costly investigations . . . which established the fact that the opinions then held by scientific men as to the impossibility of mechanical flight were unfounded." Next came years of toil to construct engines of "hitherto unheard of lightness," and "when this was done, I found I had . . . but fought my way to the great difficulty—that of balancing and guiding the aerodrome in free air by automatic machinery, when there was no man on board." Yet this, too, had been accomplished, as evidenced by the sheaf of photographs he now presented, showing No. 5 and No. 6 in flight.

Board members interrupted. How fast would a manned craft fly, and for how long?

Twenty-five to thirty miles per hour, Langley replied, with fuel enough for a flight of three hours, though "I think subsequently very much longer periods will be obtained.

"Concerning the use of the aerodrome in war, it is hardly for me, a civilian, to insist upon its utility to a board of military men. But I think I might be justified in saying that anything which, like this, would enable one party to look into the enemy's tactics and movements . . . would tend to modify the present art of war, much as the game of whist might be modified if a player were allowed to look into his opponent's hand."

With an "immediate expenditure" of fifteen thousand to twenty thousand dollars for construction, including a very large houseboat to launch the craft, Langley asserted "with confidence" that "the machine will be completely built and ready for trial within a year"—that is, by the end of 1899—though "I do not desire to convey the impression that it would be able to fly when completely constructed." Time would be needed for "balancing and adjustment, and preliminary trial—a time which cannot be exactly determined, but which . . . is always more likely to take more than can be definitely defined in anticipation." Given this uncertainty, he said, the board should plan on spending no less than fifty thousand dollars for a perfected aerodrome. This was for materials and staff costs only, he stressed; his own time would be offered free of charge.

That afternoon, the board agreed that Langley's project gave "promise of great military value." To hedge their bet, the members allotted only twenty-five thousand dollars, enough to cover Langley's early costs, with the informal understanding that a second allotment would be made after Langley gave evidence of progress.

Someone at the War Department immediately passed the story to reporters. Langley, opening *The Washington Post* the next day, was quietly horrified. Thinking of the hundreds of smash-ups and flubs that had preceded the successes of 1896, he wanted no reporters peering over his shoulder. He had counted on secrecy. Worse, the newspapers said General Adolphus Greely, chief of the Army Signal Corps, would direct the project, with Langley merely giving "the benefit of his devisings and advice." That was out of the question, Langley immediately told the board, as "I could not undertake a position of responsibility without authority, or conduct the operations under any other direction than that of the Board itself." His protest convinced the board, which assured him of "fullest discretion in the work."

The board—indeed, the whole War Department—had its own reason for wishing it had kept its mouth shut. *The Washington Post* greeted the enterprise with mock excitement, as "the flying-machine has always been a favorite dream of ours. . . . We shall do our best to be at the launching. We intend to practice longevity with most industrious enthusiasm in the meanwhile. We should never forgive ourselves were we so careless as to die before the ceremony."

Langley, confident of his purpose, was unfazed. He pressed ahead quickly with his planning. The first task was the engine; he wanted the finest available. Inquiries went to all the leading engineering firms. Most said they could not build an engine light enough yet powerful enough to meet Langley's specifications. But Stephen Balzer, a Hungarian-born engineer who in 1894 had designed and built the first automobile in New York City, said he could create such an engine in a matter of a few months, by the spring of 1899.

*Chapter Two*

# "A Slight Possibility"

"SCRAPS OF WOOD AND METAL AND ABANDONED PIECES OF FLYING TOYS"

The Wright Cycle Company, 1127 W. Third Street, Dayton

IN THEIR SEPARATE WAYS, Samuel Langley and the Wright brothers could claim to be self-made men, but there the resemblance ended. The Langleys were high-toned, well-fixed Boston Brahmins. The Wrights were midwesterners of modest means. Langley grew up among the descendants of Puritans in a world governed by rationalist churches (Congregational and Unitarian) and universities (Harvard and Yale). The Wrights, a generation later, also came from Puritan stock, but of the variety hardened by decades of life on the frontier, surrounded by Scotch-Irish and German farmers. And the family's piety was of a much more active sort.

The father, Bishop Milton Wright, detected a clear pattern in the fabric of his ancestral family, with vivid colors of nonconformism and virtuous combat.

He saw the pattern among his Puritan farmer forebears and in a grandfather who fought the British at Saratoga in 1777. His father, Dan Wright, was a devout Christian who spurned the churches around his homestead on the Indiana frontier "because those he otherwise harmonized with indorsed human bondage," and who took low prices for his corn because he refused to sell to whiskey producers. The same pattern marked Milton's own life. Though deeply thoughtful and learned by the standards of his place and time, he was chiefly a man who obeyed his own strict conscience, planted his banner in the enemy's midst, and defended it with no thought of surrender. He taught his children to do the same.

As a boy in the 1840s, Milton had followed his father's example of piety outside any church. But at nineteen, after much reading and careful deliberation, he submitted to baptism by full immersion in the Church of the United Brethren in Christ. In 1850, in a declaration that determined the preoccupations of his lifetime, he said he had been called to the ministry, and after several years of directed reading, he was ordained. For several years he courted Susan Koerner, a young Brethren woman of German descent, and they married in 1859.

The Brethren Church had been founded by German evangelicals in Maryland only a few decades earlier. It won its first great wave of converts among the early settlers of Ohio, Kentucky, and Indiana, and by Milton's day it was known as a church of the frontier, with the frontier's scorn of social hierarchy and ritual. By 1850 the Brethren claimed some fifty thousand members—not many by Methodist or Presbyterian standards, but young and vigorous. Soon they leaped the Rocky Mountains to establish footholds in California and Oregon. Their aim was the reign of Christ in the New World, and for that to occur, Christian reformers first must sweep away the works of Satan, the worst of which was human slavery.

As a young man in the 1850s, Milton sharpened all his intellectual weapons in the battle between the abolitionists and the "Great Slave Power" of the southern aristocrats. Good was on one side, evil on the other. Between them was a vast, weak-willed crowd susceptible to temptation and unholy compromises. These were the ones you had to watch. Among the worst villains of Milton's early universe were clergymen who cowered from the attack on slavery for fear of losing the weekly offerings of slaveowners and their collaborators. With God's Kingdom on the horizon, the stakes were always high. Daily life was a battle, and people deserved to be trusted only after they proved which side they were on.

Milton discarded the intellectual blinders that many pietists wore. A zealous reader since childhood, he came by most of his learning on his own, through a program of "steady, continued and systematic investigation of subjects. In this he was a rigid disciplinarian." While reading for the ministry, he took courses and taught at Hartsville College in southern Indiana, a small Brethren school that attracted more elementary and secondary youngsters than college students. He never took a degree. Though he became a self-taught theologian and biblical scholar, he liked science and read widely about other faiths. His father apparently had had the same blend of pietism and openmindedness, or so Milton told an early biographer, who said Dan Wright, though "a man of strong convictions," had been "very tolerant of the opinions of others, and very ready to recognize all that was good in any person, religious denomination or political party."

Still, "all that was good" constituted a pretty narrow range, in Milton's view. After the Civil War, as he began his ministry as a circuit-riding preacher—first in Indiana, then as a missionary in the Oregon Territory, then back in the Middle West—he narrowed his sights on the evil that would be his special target for most of his career.

For Wright and other evangelical reformers, the volatile fuel of antislavery was easily siphoned into the movement against Freemasonry and its imitators—"the dark-visaged sister of the now defunct institution of slavery." In our own time, when the men's fraternal lodges have lost much of their clout, it is hard to understand the power they wielded and the fears they inspired in nineteenth-century America. Touring the nation, Alexis de Tocqueville was startled by the "immense assemblage" of Freemasons, Odd Fellows, Red Men, and a hundred others. To members, especially among the growing middle classes of America's cities, the lodges were places of good fellowship, charity, and social uplift. But to Milton Wright and many like-minded Protestants, they were ominous rivals of Christianity itself, elevating vague notions of brotherhood above the need for grace and salvation. The worst, in the opponents' minds, were the Freemasons, the largest lodge and so the most dangerous. Their elaborate rituals mimicked the church's, but were "Christless." An odor of Old World aristocracy wafted from the lodges, of exclusionary social circles, of "ins" versus "outs," of hidden favors and preferences among the initiated and a raw deal for those left out. This was anathema to men who cherished the anti-elite tradition of the American Revolution—despite the fact that Washington, Franklin, and other prominent patriots had been devoted Freemasons. Worst of all about the secret societies was their secrecy itself, which the opponents found intrinsically unchristian and antidemocratic and which, by its very nature, inflamed their imaginations about the danger. A strict ban on membership in any secret society was part of the 1841 Brethren constitution.

Yet after the Civil War, even among the Brethren, the fervor of the antisecrecy, antimasonic movement began to lose its edge. The lodges had spruced up their images and were growing enormously, with millions of members, including many respectable churchgoers in every city and town. To some younger leaders of the United Brethren Church, the issue seemed tired. They disapproved of "lodgery," too, but not so vehemently as their elders, and they feared the issue was hurting the church's efforts to recruit and hold members. In 1869, these young reformers—known within the church as the "Liberals"—proposed a softening of the Church's antisecrecy stand. An antisecrecy majority of conservative "Radicals" prevailed. They wanted no part of a "creed on wheels," rolling with the whims of change, and they feared the Liberals as a rising force that might one day dismantle not just their antisecrecy clause but all the traditional tenets of the church. To help hold them in check, the Radical majority installed their stalwart young spokesman, Milton Wright, as editor of the Brethren newspaper, *The Religious Telescope*. Its offices were in Dayton, Ohio, where Wright moved with his young family in 1869.

His five children—Reuchlin, born in 1861; Lorin, in 1862; Wilbur, in 1867; Orville, in 1871; and Katharine, in 1874—grew up in the shadow of their father's struggle within the church. It went on for nearly forty years.

IN THE EARLY 1890s, Wilbur Wright, then in his early twenties, gave a little talk to a neighborhood gathering. He spoke well, and occasionally entertained friends this way. His younger brother Orville's best friend, Eddie Sines, missed the talk. So the next day Eddie asked Will to tell him the gist of it. It was the time when phonograph recordings were the new craze, and Will replied: "I can let you hear it on the phonograph."

He disappeared into the next room. As Eddie and Orville listened, the sound of a fuzzy phonograph recording began, complete with scratches, hisses, cheers, laughter, applause, and Will's speech, every word of it just short of intelligible. But there was no phonograph. Will faked all the sounds himself.

He was that kind of kid, and that kind of young man—clever, quirky, a little odd, and more likely to think up his own amusements than to follow the crowd. His bow to convention was to be a fine athlete. Among other sports he excelled in gymnastic events, including the horizontal bar. He was an excellent figure skater. And he was a good student.

He went to high school in Richmond, Indiana, where his father was based for several years. In 1884, just short of graduating, he moved back to Dayton with his family. There, for a time, he ran with his older brothers and their crowd in a club called Ten Dayton Boys. When Reuchlin and Lorin left home for college, then jobs and marriage, Will was left as the lone big brother to Orville, who was four years younger, and Katharine, called Katie or Kate, who was seven years younger.

Reuchlin and Lorin went to Hartsville College, the same Brethren school their parents had attended. But Milton and Susan had larger ambitions for Wilbur: They made plans for him to enroll at Yale. In Dayton, he took extra classes in trigonometry and Greek, apparently to enhance his skills for college. His parents hoped he would enter the ministry.

But an accident intruded. In winter, Will often played a game called shinny, a version of ice hockey, on a frozen pond at the Soldiers' Home, a veterans' retreat on the western outskirts of Dayton. During a game in 1886, when Will was nearly nineteen, another player let his stick fly out of his hands. It struck Will in the face hard enough to knock out several teeth. Some weeks later, he became ill with a condition his father described as "nervous palpitations of the heart"—an imprecise term that leaves the facts unclear. He may have suffered from digestive problems, too. Perhaps the deterioration of his health was related to the facial injury, or perhaps there was some other cause.

Something in that series of events drew a line across Will's life. He stopped playing sports. He enrolled in no more college preparatory classes. His father, in a New Year's Eve diary entry at the end of 1886, the year of the injury, recorded that "Wilbur's health was restored." But that seems not to have been true, for Will abandoned his college plans and for several years suffered from some chronic health problem whose nature never was made clear. It apparently made him fear that he would not live for long. Several years later, he reconsidered college, but concluded that any time and money spent on it "might be time and money wasted." If those words mean what they seem to mean, Will spent most of his twenties, the age when most people find their place in life, not at all sure he would live to see his thirties.

WILL BECAME THE SECOND invalid in the Wright home. Three years before his injury, Susan Wright had contracted tuberculosis. She was what her husband called "a declining, not a suffering invalid." Soon she could barely walk across her sitting room without having to pause to catch her breath. Milton's work took him away for much of the year. Reuchlin and Lorin were grown and gone. Orville and Kate were still kids in school. That left Will. So for several years he was his mother's principal caregiver, carrying her up and down the stairs and doing most of the household work. His features resembled hers, and though not much is known about her, it appears that he was like her in personality, too. She was very bright. She had gone to college, a rarity for women of her generation. She was good at math. Her imagination and her hands worked well together; she had a talent for "adapting household tools or utensils to unexpected uses." Her few surviving letters suggest a chilly sense of humor and an evangelical Christian's wary view of the world beyond her front gate.

Milton said later that Will cared for his mother "with a faithfulness and tenderness that cannot but shed happiness on him in life, and comfort him in his last moments. Such devotion of a son has been rarely equaled. And the mother and son were fully able to appreciate each other. Her life was probably lengthened, at least two years, by his skill and assiduity."

To brother Lorin in Kansas, who heard about the family only through occasional letters, it seemed as if the athletic, ambitious Will had dwindled into a housebound idler. "What does Will do?" he asked Kate. "He ought to be doing something. Is he still cook and chambermaid?"

In fact, despite his vague infirmity, Will was busy. Besides caring for his mother, he helped his father with church business. He also embarked on a remarkable project of self-education. Without leaving 7 Hawthorn, he could choose from bookshelves crammed with the classics of ancient Rome and Greece, modern novels, histories of England and France, treatises on mathematics and biology, and sets of *Chambers' Cyclopedia* and the *Encyclopedia Britannica*. A favorite was *Plutarch's Lives*. He likely read Boswell's *Life of Johnson* and Gibbon's *Decline and Fall of the Roman Empire*. His reading "perhaps nearly equaled the advantages of a classical education," his father wrote later. "His knowledge of ancient and modern history, of current events and literature, of ethics and science was only limited by the capacity of his mind and his extraordinary memory."

Traveling in Europe twenty years later, Will might see the name of a village, then recount precise details of a battle fought there in Napoleon's time, solely on the basis of a book he had read in his youth. Many associates later said they were "impressed not only by the range of his reading, but by the fact that no knowledge he had once acquired ever seemed to grow dim." Clearly, a remarkable intelligence was nourished during Will's years of retreat. But he was less like a student in the free-for-all of the seminar room than some solitary young man preparing to take religious orders. He honed his intellect in argument, but only within the cloisters of home and church, and largely, it appears, with his younger brother.

No one can draw a psychological profile on such slim evidence as we have from Wilbur Wright's youth, especially at a distance of more than a century and through the lens of a different time. Still, a few strokes on a sketch pad are possible. He was close to both his parents. Yet with Milton often away and Susan in need of daily care, it is clear that he formed an especially intimate bond with his mother. Indeed, just at the transition from boyhood to manhood, circumstances conspired to make him nurse to his mother, manager of his household, and overseer of his younger brother and sister. All these tasks were identified, of course, with the role of the Victorian mother—that is, with the position of responsibility for the day-to-day welfare of one's family. With his father, too, he was called upon to be more the helper than the one who is helped. Both parents relied on him—Susan for care, Milton for counsel and aid. Given these demanding roles, it is at least not surprising that this boy had more than the usual difficulty in pulling away from his family and launching himself. His own physical frailty, whatever the cause, would have compounded the tendency to stay stalled at home. And a stall of several years, while less able friends ventured out and began to find success, may only have compounded his ambition to do something remarkable. Even if one can only guess at his motives, it is at least certain that the circumstances of this period of his life—his injury and subsequent retreat from a conventional education and career; his mother's illness; his bachelorhood; his self-education—formed a context for the single-minded pursuit of an extraordinary goal.

AFTER EIGHT YEARS as editor of the *Religious Telescope*, Milton Wright had been elected one of five bishops of the United Brethren Church. Four years later, with his enemies, the Liberals, in the ascendency, he lost his bid for reelection and resumed his old life as an itinerant preacher in eastern Indiana, though he continued to lead the Radicals with a stream of pamphlets and editorials aimed at the Liberals. Yet at the churchwide General Conference in 1885, the Liberals won the stunning victory that Wright and his allies had feared. They named a Church Commission that would rewrite the Brethren's "Old Constitution" of 1842 and the even older creed of 1815.

The Liberals' political master stroke was to reelect their chief adversary, Milton Wright, as bishop—but to appoint him as overseer of the Church's West Coast district, where he would be safely out of their business for the better part of each year.

Schism loomed. Wright and his Radicals argued that the Church Commission was an illegal body, that only the General Conference could rewrite the Church's foundational documents. To Milton it was right against rascality. The righteous held to immutable principles and assumed that anyone holding otherwise was a scoundrel and self-seeker. The commission rewrote the documents anyway. Put to a vote of the general membership, now numbering more than two hundred thousand, the revisions were approved by a two-thirds majority. But three-quarters of the Church's members chose to boycott the vote, apparently in tacit approval of Milton Wright. Nonetheless, most delegates to the General Conference of 1889, even most of the Radicals, were by now desperate to make peace and voted to accept the changes.

One bishop dissented—Milton Wright. He and his allies walked out and reorganized as the Church of the United Brethren in Christ (Old Constitution). Ten thousand to fifteen thousand Brethren joined them. Wright and three others were chosen as bishops.

Six weeks later, Susan Wright died at the age of 58.

AT TWENTY-TWO, Wilbur Wright—without an advanced education or job experience—became Milton's chief lieutenant as the older man went about the arduous job of building a new sect from the remnant of the old. While Samuel Langley pondered the results of his toy experiments and whirling-table data, Wilbur helped to manage his father's business affairs and became a key strategist in the Brethren Radicals' court battles over disputed Church property and real estate, especially the largest asset, the publishing operation in Dayton. Will's performance as an advocate was striking. He argued with a mastery of facts, logic, and wit that veteran lawyers would later envy. His writing had the edge of a razor. And he loved to fight, as his notes to the family show. "They try to smile," he gloated after one especially good thrust at the Liberals, "but it evidently hurts."

Will came of age in the Brethren's internal war. The battle against the secret societies and the Liberals strengthened his tendency, well-learned at 7 Hawthorn, to detect wrongdoing at the drop of a hat, and to keep a wary eye on privileged sharpsters operating behind closed doors. As he moved into realms far beyond sectarian politics, the habits of thought he learned under his father remained strong.

But father and son were different, too. Milton was naturally combative, but to him, the struggle with the Liberals was at bottom a matter of faith. To Will it seems to have been chiefly a matter of family duty and even of sport. The bishop's diaries and letters are full of references to Scripture. Affairs of the spirit were at the center of his life. Will's surviving letters contain no such references—none. For him it was right versus wrong, but also and essentially "us" versus "them." If he had ever looked toward a life in the ministry—his parents' hope for him—he now turned away. It's not clear that he even attended services regularly in his twenties, and in his thirties it's clear he did not do so. The bitterness of church politics had shadowed his entire childhood. That may be why he could not find a home among the Brethren or in any other church. He fought in the Brethren's battles with heart and mind. For fifteen years after the schism, he would leap back into the fray whenever Milton needed him. But this was his father's life work, not his.

WHEN HE WAS SEVENTEEN, Orville Wright built a printing press out of a pile of scrap parts that included a folding buggy top, a discarded tombstone, and a pile of firewood. In a few weeks he had it printing a thousand sheets an hour. A pressman from a big Chicago printing house heard about the machine and dropped in to see it. He looked at it from the top and the sides, then crawled underneath. Finally he said: "Well, it works, but I certainly don't see how."

Nothing could have pleased Orville more. He had fallen in love with the complex technology of printing some years earlier, when his father, as chief of the Brethren's publishing arm, brought his sons to work. It was a big operation, filling a four-story building in downtown Dayton, where the Brethren published four newspapers plus hymnals, books, tracts, and stationery, all in a hubbub of fast-handed typesetters and clattering presses. The Wright boys had spent a good deal of time around their maternal grandfather, a carriage-maker who apparently acquainted them with tools and the fundamentals of woodworking. In the publishing house they learned about machines. Orville spent two summers as a printing apprentice, then dropped out of high school before his senior year to launch his own printing business. Soon he started a small weekly newspaper, *The West Side News*, and inveigled his brother Wilbur to serve as editor. A few months later they shut down the *News* and replaced it with a more ambitious daily product, *The Evening Item*. Up against a dozen daily competitors in the Dayton market, the *Item* folded in less than four months. The brothers fell back on printing jobs for merchants in the neighborhood, which gave them a modest but steady livelihood.

Orville had chosen this work. Wilbur had not chosen anything else.

FOUNDED IN 1805 as a center of trade where three creeks joined the Miami River, Dayton had become a city that made things, a "city of a thousand factories," its streets teeming with carriage-makers and wood-benders, machinists and carpenters, engravers and glass-makers, artisans and engineers. They made harrows, stoves, and steam pumps; varnish and motors and medicine; and especially cash registers, the city's leading export. Talk in the shops and the streets led to new ideas: By 1900 Dayton had filed more patents per capita than any other city in the United States. Its sixty thousand people knew machines. They were perhaps especially susceptible to the charms of the bicycle.

The big-wheeled "ordinary" had been around for only a dozen years when it was swept aside by the multiple innovations that composed the "safety": wheels of identical size, close to the ground; a sturdy diamond frame; an endless chain running over pedal-driven sprockets; the pneumatic tire, which ended the machine's deserved reputation as a "boneshaker"; and the reliable coaster brake. That basic design has persisted ever since, but for the addition of handbrakes and the modification of gearing after World War II. For several years after its introduction in 1888, the "safety" cost too much for most Americans. But keen competition brought prices down, and by 1895, hundreds of thousands were being sold each year, despite the onset of a depression.

The automobile followed so closely, and so far surpassed the bicycle in importance, that the bicycle's own heyday has been all but forgotten. But it was a great craze, with millions of people buying, riding, and extolling the virtues of bicycles. The easy pleasure of the two-wheeled cruise—so different from riding a horse, not to mention so much cleaner—was like a narcotic. Bicycle academies flourished. "Wheeleries" were established to rent bicycles to those who couldn't afford to buy. Bicycle accessories became as big a business as bicycles themselves. Hundreds of exhibitors crowded the halls of cycle shows in the biggest cities, and manufacturers sent racing teams around the country to stir the excitement. "It would not be at all strange," the *Detroit Tribune* remarked, "if history came to the conclusion that the perfection of the bicycle was the greatest incident in the nineteenth century."

In the Wright family, Orville was infected first, in the summer of 1892, when he bought a fine new Columbia for $160, a very substantial sum at a time when relatively few workers earned more than $500 per year. He soon entered races and did well. Will bought his own model—used, for $80—several weeks later. Still careful of his health, Will chose not to race, but he did take long rides around town and into the countryside. That fall, without giving up the printing business, the two brothers began to sell and repair bicycles. The business did well enough for them to hand off most of the printing to Orville's friend, Ed Sines, and their brother Lorin, who had returned to Dayton, married his childhood sweetheart, and begun a family. They rented a small storefront in West Dayton, the first of a series they occupied, and worked at their newfound trade.

Will apparently did not care for it much. Some eight years after his injury in the shinny game and his ensuing illness, he finally felt fully well. Though he had suffered an attack of appendicitis the previous fall, he soon had recovered. At the age of twenty-seven, he realized he might resume his interrupted plan—to attain a college degree and pursue a professional career. In the fall of 1894, he considered the possibility seriously enough to write his father, who was traveling, to ask for his opinion and a loan.

> I have been thinking for some time of the advisability of my taking a college course. I have thought about it more or less for a number of years but my health has been such that I was afraid that it might be time and money wasted to do so, but I have felt so much better for a year or so that I have thought more seriously of it and have decided to see what you think of it and would advise. I do not think I am specially fitted for success in any commercial pursuit even if I had the proper personal and business influences to assist me. I might make a living but I doubt whether I would ever do much more than this. Intellectual effort is a pleasure to me and I think I would [be] better fitted for reasonable success in some of the professions than in business. I have always thought I would like to be a teacher. Although there is no hope of attaining such financial success as might be attained in some of the other professions or in commercial pursuits, yet it is an honorable pursuit, the pay is sufficient to live comfortably and happily, and is less subject to uncertainties than almost any other occupation. It would be congenial to my tastes and I think with proper training I could be reasonably successful.

He said he could pay most of his own way by renting and fixing bicycles. But he could not do it "unless you are able and willing to help me some," probably to the tune of six or eight hundred dollars over four years.

Milton did not think twice. "Yes," he replied immediately, "I will help you what I can in a collegiate course. I do not think a commercial life will suit you well."

Neither health nor money was a barrier—but Will never entered college. No available document tells why. Perhaps Milton rethought his finances. Perhaps Will decided a man of twenty-seven ought to pay his own way, yet he never managed to set enough aside. He simply continued to go to the bicycle shop every morning.

SAMUEL LANGLEY SOUGHT SOCIETY among men of learning and distinction; the Wrights spent most of their time within the circle of their family. Names were important to the Wrights. The bishop, an enthusiastic chronicler of his family's genealogy, gave his sons unusual Christian names because he felt "Wright" was common, and needed an ornament. Nicknames proliferated. Lorin was "Fiz." His three eldest children were "Whacks," "Ivettes," and "It," while the youngest, Horace, answered to "Brother," "Mannie," "Boy," "Buster," "Bust," "Buzz," and "Buzzy." Kate called Orville "Little Brother" or "Bubbos" or "Little Bubbo"; the latter two probably had been her attempts to say "brother" as a toddler. To the outsider these endearments might have seemed odd, since Orville was Kate's *big* brother, not her little brother. Actually, it was fitting. Of all the Wright siblings, these two were the closest. But she was the stronger, larger figure of the pair, he the more vulnerable, the more needy of her care and attention.

Wilbur was called "Jullum" or "Ullum," a customer's effort to give a German twist to "William." Though the nickname started in adulthood, for some reason it stuck, at least with Kate. Mostly Wilbur was just "Will," the plainest of all the family names.

Twins, a boy and a girl, had been born between Lorin and Wilbur, but had died as infants. Only Milton ever mentioned them, and then only in fleeting notations in his diary.

Katharine was Katharine to her father (who sometimes infuriated her by misspelling it); Kate or Katie to her friends and brothers, who also created several homemade variants of the German "schwesterchens" ("little sister"), a nod to their German forebears on their mother's side. These were "Schwes," "Schwessie," "Swes," and "Sterchens." Thus she was "sister," Orville "brother." Wilbur was only Will, standing slightly aloof from the sister-brother pair.

Kate had what the Wright men lacked—conventional, straightforward charm. She loved conversation and sparkled in it. Her brothers possessed fine senses of humor, but hers may have been keener and she shared it more generously with people outside the family. She read lots of novels. She loved to go out and battled her father's inclination to keep her in. She was the family's ambassador to neighbors, colleagues, friends, and acquaintances. During her mother's decline, when she was fourteen, Milton had instructed her to prepare to take over as mistress of the house. "Learn all you can about housework. . . . You have a good mind and good heart, and being my only daughter living, you are most of my hope of love and care, if I live to be old. I am especially anxious that you cultivate modest feminine manners and control your temper, for temper is a hard master." She endured the role for a time. Then she enrolled in Oberlin College and returned four years later to become a teacher of Latin at Steele High School. She moved back in with her father and brothers and ran the household when she was not teaching, but her temper was undimmed, her spirit effervescent. "But for you," Milton told her once, "we should feel like we had no home."

At some point she became engaged to be married, but broke it off. The date is not clear. She referred to it in a letter many years later, when her life had become utterly different from any life she might have imagined in her youth, and when she was contemplating marriage again at the age of fifty. After an engagement of two years, "I saw that he really didn't care much about me—not the way I have to be cared about if I am going to care so much." When she suggested "we had better give it up . . . he was evidently relieved." She persuaded herself that "men were not the least interested in me, except as a friend."

As for her brothers, neither Wilbur nor Orville apparently came anywhere close to marriage—certainly not Will. The earliest biography, which was based in part on interviews with Orville and Katharine, said Will "was briefly attracted by one of the girls" at Richmond High School. It is the only such report. According to a close associate of the brothers in later years, "Orv used to say it was up to Will to marry first because he was the older of the two. And Will kept saying he didn't have time for a wife. But I think he was just woman-shy—young women, at least. He would get awfully nervous when young women were around." On the streetcars, "if an older woman sat down beside him, before you knew it they would be talking and if she got off at our stop he'd carry her packages and you'd think he had known her all his life. But if a young woman sat next to him he would begin to fidget and pretty soon he would get up and go stand on the platform until it was time to leave the car." Orville was said to have dated a friend of Katharine's, but nothing came of it. Many years later, Kate told an intimate that "there was a reason why none of us married." But she chose not to elaborate, other than to cite the extraordinary closeness between the three siblings: "Always everything that interested them . . . interested me and they took up all my interests in just the same way . . . I am sure Will and Orv thought as much of me as of themselves when they made any plans." After her engagement was broken, "I couldn't think . . . of my narrow escape without being scared nearly to death." Scared of an unhappy marriage? Or scared of disrupting a bond between siblings that somehow had taken on the inviolability of a marriage? There is no way to know. But clearly the bond was unusual, and terribly strong.

So the two brothers and the sister and their father lived on together. Lorin and his wife, Netta, lived only a few blocks away, and soon had children who became regular and welcome visitors to 7 Hawthorn.

Will worked and read books. Outside the bicycle shop he saw the members of his family and not many others. The Ten Dayton Boys—minus Reuchlin Wright, the eldest brother, who was farming out west—met now only twice a year, at a winter banquet and again on the Fourth of July. All but Will had married. A friend remembered him at one of the summer picnics "putting up the swings for the children, and then standing aloof from the crowd much of the day. In fact, the strongest impression one gets of Wilbur Wright is of a man who lives largely in a world of his own, not because of any feeling of self-sufficiency or superiority, but as a man who naturally lives far above the ordinary plane."

IN 1894, *McClure's* published a long article about a German engineer, Otto Lilienthal, whom the magazine called "the Flying Man." Will knew a little about Lilienthal already; the brothers had published a brief article about him in one of their own papers. The article in *McClure's* told of the German's extraordinary recent progress in making long, flying leaps with artificial wings.

Raised in the suburbs of Berlin, Lilienthal and his younger brother, Gustav, were fascinated by the flight of birds and butterflies. Through their teen years and their schooling as mechanics and engineers, the boys experimented with a series of flying devices. Returning home from the Franco-Prussian War, Otto said to his brother: "Now we shall finish it." After establishing a successful career in mining engineering, with profitable inventions to his credit, Lilienthal began intensive experiments with artificial wings, beginning with cautious hops from a three-foot-high mound in his garden. Now, in the early 1890s, he was astonishing witnesses by sailing through the air for hundreds of feet. He jumped from sand hills and from an enormous artificial hill of his own design, with a shed inside to hold his machines.

Two things were crucial about Lilienthal's achievements. One was the act itself. The second, just as important, was that the act was photographed, and the photographs were reprinted all over the world. The dry-plate negative had been invented only in the 1870s, and it was not until 1890 that a moving object could be captured in a frozen image, without a blur. A photograph of anything in motion was still relatively new in 1894. Published images of Lilienthal hanging by his arms from his birdlike contraption, obviously many feet off the ground, were electrifying.

From later recollections, it's clear that Will read the *McClure's* article and saw the photographs. The author explained that Lilienthal, through observation of birds and exhaustive experiment, had concluded that the key to flight lay in the proper curvature of the wing, and in holding the wing at a proper angle to the wind. He was not flying, Lilienthal explained, but "sailing"—that is, he was "being carried steadily and without danger, under the least possible angle of descent, against a moderate wind, from an elevated point to the plain below." Americans imitating him were calling this "gliding." A glide was a gradual, controlled fall; the more gradual the fall, the closer the experimenter came to such birds as eagles and vultures, which stay aloft for long periods without flapping their wings.

By implication, a human being who learned to stay aloft that way might attach a means of propulsion to his artificial wings and actually fly.

"No one can realize how substantial the air is, until he feels its supporting power beneath him," Lilienthal said. "It inspires confidence at once. . . . I am far from supposing that my wings . . . possess all the delicate and subtle qualities necessary to the perfection of the art of flight. But my researches show that it is well worth while to prosecute the investigations farther."

Two years later, in 1896, newspapers reported the German had fallen to the ground, broken his neck, and died.

The brothers said later that Lilienthal's death reawakened the curiosity about flying they had felt as boys. This is a little odd. It was one thing to be inspired by Lilienthal's publicized successes earlier in the nineties. But his death, if anything, would have discouraged the potential experimenter. Certainly it discouraged others. In any case, there was nothing urgent about their new cu-

"NO ONE CAN REALIZE HOW SUBSTANTIAL THE AIR IS, UNTIL HE FEELS ITS SUPPORTING POWER BENEATH HIM."

Otto Lilienthal experimenting in his monoplane glider, ca. 1895 riosity. Will wrote his letter to the Smithsonian nearly three years after Lilienthal's death. If he and his brother discussed flying machines during that interval, they were merely amusing themselves with a puzzle. Yet in 1899 Will told the Smithsonian of his plan to study the matter closely. The evidence is not conclusive. But it suggests that Will had thought of a way to keep a flying machine balanced in the air and wanted to see whether anyone before him had had the same idea.

THE BIBLIOGRAPHY that Richard Rathbun sent to Dayton was brief, reflecting the opinion of flight experimentation in the scientific community. One book was by Octave Chanute, a prominent civil engineer, entitled *Progress in Flying Machines*. Another was Langley's *Experiments in Aerodynamics*, a report of his whirling-arm work published in 1891. Also on the list were the three volumes of *The Aeronautical Annual*, published from 1895 to 1897. These, too, were collections of scholarly articles. The editor, a wealthy Boston enthusiast named James Means, dedicated the 1897 number "to the memory of those who, intelligently believing in the possibility of mechanical flight, have lived derided, and died in sorrow and obscurity." Will studied the *Annuals* thoroughly, along with the other books on the list and several Smithsonian pamphlets, including a translation of Lilienthal's article, "Practical Experiments in Soaring."

It became clear to him that Lilienthal had been no trickster or kook but an intelligent and industrious worker, cautious and patient. Yet he said the goal of "independent horizontal flight"—not just gliding but actually flying at will, with a motor, independent of the wind—lay not far off. And he tried to tell what only he could tell—what it was like to fly:

> It is a difficult task to convey to one who has never enjoyed aerial flight a clear perception of the exhilarating pleasure of this elastic motion. The elevation above the ground loses its terror, because we have learned by experience what sure dependence may be placed upon the buoyancy of the air. Gradual increase of the extent of these lofty leaps accustoms the eye to look unconcernedly upon the landscape below. To the mountain climber the uncomfortable sensation experienced in trusting his foot into the slippery notch . . . may often tend to lessen the enjoyment of the magnificent scenery. The dizziness caused by this, however, has nothing in common with the sensation experienced by him who trusts himself to the air; for the air demonstrates its buoyancy in not only separating him from the depth below, but also in keeping

> him suspended over it. Resting upon the broad wings of a well-tested flying machine, which, yielding to the least pressure of the body, obeys our directions; surrounded by air and supported only by the wind, a feeling of absolute safety soon overcomes that of danger. . . . The indefinable pleasure . . . experienced in soaring high up in the air, rocking above sunny slopes without jar or noise, accompanied only by the aeolian music issuing from the wires of the apparatus, is well worth the labor.

Will also read the work of an obscure Frenchman named Louis Mouillard, a farmer and poet who lived most of his life in the wilds of Algeria and Egypt. As a maker of gliders, Mouillard was insignificant, but as a missionary of flight, he was, as Will said later, "like a prophet crying in the wilderness, exhorting the world to repent of its unbelief." Indeed, the Frenchman used the rhetoric of the biblical prophet: "O! blind humanity! Open thine eyes and thou shalt see millions of birds and myriads of insects cleaving the atmosphere. . . . Many of them are gliding therein, without losing height, hour after hour, on pulseless wings without fatigue; and after beholding this demonstration . . . thou wilt acknowledge that Aviation is the path to be followed. . . .

"If there be a domineering, tyrant thought, it is the conception that the problem of flight may be solved by man. When once this idea has invaded the brain, it possesses it exclusively. It is then a haunting thought, a walking nightmare, impossible to cast off."

WILL HAD GIVEN UP SPORTS, but any fine athlete retains his instincts after his body has failed him. And here was Mouillard saying that any man who wanted to fly "must thoroughly know his business . . . just as he knows . . . his business as . . . a skater, a bicycle rider, an acrobat—and, in short, as an expert in any gymnastic exercise." Will had mused at length about the minor mystery of how a bicyclist keeps his balance, and here was Lilienthal saying, "The evolution of the flying machine will be similar to that of the bicycle, which was not made in a day, and . . . [the flying machine] will not be either."

Both Mouillard and Lilienthal argued for a pragmatic, step-by-step approach that appealed to Will. Experimenters should get out of "the closet" of mathematical theory, Mouillard said, and into the field, for "I feel well convinced that I never will meet with anybody willing to hazard his life upon the bare dictum of a formula." Progress, Lilienthal agreed, would be a matter of "continual practice." The experimenter need not solve the whole problem at once. Progress would come in tiny steps: "Regardless of the most ingeniously constructed apparatus, the art . . . will have to be acquired just as the child learns to stand and to walk."

To walk, the toddler must balance. Likewise, the skater, the cyclist, and the gymnast must above all keep their balance. "The first obstacle to be overcome . . . is that of stability," Lilienthal urged. Octave Chanute, too, said the key problem was balance, "the most important and difficult of those remaining to be solved."

As Will studied, it became clear how he ought to proceed. It was going to be an athletic endeavor, and its principal goal would be proper balance, as in skating or bicycling or gymnastics. When first mounting the parallel bars or the bicycle seat, you simply tried it, expecting to fall. So it would be with this. In the instant of the fall you might have an idea about how you could have turned your body differently or placed a hand elsewhere. You tried again, fell again, considered, adjusted. In this way he would learn new things. He would attempt to build on others' successes, particularly Lilienthal's, with the aim of answering the question he and Orville apparently had discussed off and on for three years. How could one fly, yet escape Lilienthal's fate?

SOMETIMES WILL RODE HIS BICYCLE to a spot just south of Dayton called the Pinnacles, where strange, pointed outcroppings jutted up from the heights overlooking the Miami River. Buzzards and hawks wheeled over the formations. But they flew so high that even a careful observer had trouble following the movements of their bodies and wings. The small birds of the garden were no more helpful. They dove, darted, and twisted in paroxysms of gymnastic mastery, their movements far too quick and minute to study.

In Dayton, summer is southern, yet the Wrights wore heavy long pants and long sleeves all year round. Even with the windows and doors of the shop open, the atmosphere inside was close at best, stifling at worst. When business was slow, it was natural to step out onto Third Street, and it probably was there, as wagons and bicycles passed back and forth, that Will noticed something in the flight of a pigeon.

From his reading, he had begun to learn a new vocabulary to describe the subtle movements of flight. The "center of gravity" was the point at which a bird—or, in theory, a flying machine—would balance, both fore to aft and side to side. It depended on the distribution of weight. The "center of pressure" was the central point where all the forces pushing the bird upward were concentrated, the fulcrum on which the bird was resting. It depended on the shape, size, and angle of the wings. When the center of gravity and the center of pressure coincided, at least roughly, then the bird had its balance under control.

When the two points got too far off kilter, a plunge to the ground was imminent. Clearly, the birds knew the trick of keeping the two centers in equilibrium. But how they did so was a mystery.

Will now knew what Octave Chanute and Louis Mouillard thought about how birds kept their lateral balance—that is, their balance from side to side. They said a bird draws one or the other of its wings inward, close to the body, thus decreasing the lifting surface on that side. The center of pressure would move toward the opposite side, and the bird would regain its balance. They also said the bird "sometimes rocked its body over toward the high side in order that the increase of weight on that side might help to bring the high wing down."

But one pigeon of West Dayton failed to conform to theory. As Will watched, he saw the pigeon wobble, a quick side-to-side oscillation. "That is," he recalled later, "it tilted so that one wing was elevated above its normal position and the other depressed below its normal position and tilted in the opposite direction." Will pondered the speed of the bird's oscillation. Could it shift its weight rapidly enough to cause so rapid a back-and-forth movement? That seemed unlikely. Besides, Will could see no sign that the pigeon was drawing either wing closer to its body. Something else must account for its wobbling. "The thought came that possibly it had adjusted the tips of its wings . . . so as to present one tip at a positive angle and the other at a negative angle, thus, for the moment, turning itself into an animated windmill . . ." In other words, by turning the leading edge of one wing tip up and the leading edge of the other wing tip down, the pigeon changed the degree of lift on both sides of its body. One wing rose, the other fell; the center of pressure shifted, and the bird was now rolling. When its body had rolled far enough in one direction, the bird then reversed the angles of its wing tips, and began to roll back in the other direction. The pigeon, in this view, wasn't adjusting its center of gravity at all. Nor was it pulling in a wing to move the center of pressure. It was using the air rushing over both wings to move its center of pressure around its center of gravity in perfect control.

How to apply the pigeon's lesson to a man-made glider? It implied a method very different from the one Otto Lilienthal had used. He had shifted his center of gravity by swinging his legs this way and that—an exceedingly difficult and dangerous task, since it sent the center of pressure darting all over with every change in angle and airspeed. Will's idea, instead, was to control the center of pressure by controlling the angles of the wing tips. Apparently he discussed the problem with Orville. Perhaps, they thought, wings could be mounted on metal shafts connected to gears at the center of a glider. Turning the gears would turn the wings, one side angling up, the other down. But they could not imagine such a device being strong enough to do the job, yet light enough to fly.

A new idea occurred to Will in the bicycle shop. He held an empty bicycle tube box, squeezing it, feeling the springy resistance of the stiff paper. He lowered his eyes to the box and saw the shape of a double-wing glider. The box was twisted in the beginning of a spiral—a helical coil, to use an engineer's phrase. If his ideas about the pigeon's wings were right, then he was pressing this flimsy box into the shape of human flight.

WILL BUILT A KITE. Its shape resembled the shape of the cardboard box. It consisted of two rectangular planes made of linen stretched over light wooden frames. Each surface was five feet wide and thirteen inches from front to rear. One plane was superimposed on the other, about thirteen inches apart, connected to each other by uprights or struts. The struts were attached with wire loops, so the upper plane could shift forward and back relative to the lower plane. A horizontal rudder or "elevator" protruded from the center rear strut. In front, Will attached two pairs of cords, with each pair tied to the ends of a stick. The two sticks were his controls.

On a day with a strong breeze, Will took the kite out to a field near Bonebreak Seminary in West Dayton. Two small boys, John and Walter Reiniger, went along. The boys stood with the kite, pulling until the cords were almost

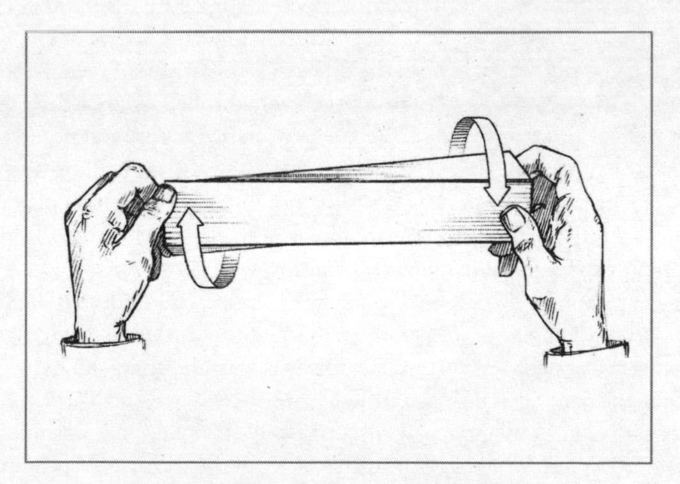

"HE HELD AN EMPTY BICYCLE TUBE BOX."

Wilbur conceives the theory of wing-warping

taut. When the breeze quickened, Will told the boys to give the device a little upward toss. It rose.

By angling the sticks in opposite directions, he could make the kite's surfaces twist, or warp, with one side presenting itself to the wind at a higher angle than the other. The high side rose, the low side dropped, and the kite rolled. By angling the sticks in the same direction, he could change the angle of the elevator, making the kite dive or climb. It worked.

*Chapter Three*

# "Some Practical Experiments"

"I SHALL BE ABLE TO STATE ITS PERFORMANCE WITHIN A SHORT TIME."
The Aerodromics Room in the South Shed of the Smithsonian

SAMUEL LANGLEY ENJOYED few of his duties more than conducting important guests through the exhibit halls and laboratories of the Smithsonian, especially when the matter at hand exhibited his own contributions to science. He had such an opportunity on November 21, 1899, when members of the Board of Ordnance and Fortification arrived at the Smithsonian Castle.

In his youth Langley had possessed the good looks of a leading man of the stage. Now, at 65, his face and physique had grown puffy from many fine meals, and his dark eyes, once intense, were squinty under sagging lids. He always greeted special guests in what an aide called "the 'dim religious light' of the Regents' Room," where dark oils of Smithsonian titans hung under a thirty-foot groin-vaulted ceiling. Here he delivered a lecture on the work completed since the awarding of the aerodrome contract. Under his instructions, "great difficulties" in constructing a new launching apparatus had been "substantially conquered," in part by new flights of No. 5 and No. 6. (It didn't hurt to reinforce board members' recollection that in all the world, he alone possessed the power to send heavier-than-air machines through the air, even if they were unmanned.) The frame of the great aerodrome had advanced far enough that "it is expected to be in condition for its first trial of balancing in free air before the close of the present year." A design for two superposed wings—one wing above the other—had been tried and found less effective than the original design, with one wing forward and one aft. As for the engine, it was being constructed in New York by "the only competent builder in the United States," and "though it has been many months in hand, it is expected here from day to day . . . and I shall be able to state its actual performance I hope within a short time."

Then Langley led his guests out of the Castle, into the secret cloister of the South Shed and up the stairs to the cavernous Aerodromics Room. The unwieldy assemblage of rods and cables clearly constituted something less than a flying machine. Still, Charles Manly thought the members seemed "very much interested and pleased at the progress of the work and at the intricacy of the problems connected with it." More inspection followed at the Eighth Street Wharf, where Langley showed off the new houseboat, the new launching apparatus, and a fully assembled No. 5, a promise of greater things to come.

The military officers on the board must have guessed that Langley had been talking a better battle than he had so far fought. Surely some of them recalled his prediction of trials of a finished aerodrome by now—the end of 1899—and it was clear that he had a long way still to go.

At the moment, actually, Langley was less worried about the work itself than about how to pay for it. With his original allotment of $25,000 running low, he had learned, finally, what no one had made him understand the previous fall—that the BOF never had approved his full request of $50,000, but only half that amount, with no more than an informal promise to consider another $25,000 upon seeing evidence of progress. Taken aback by this revelation, Langley was eager to welcome every board member to the Aerodromics Room, where they might catch the whiff of excitement and promise. He need not have worried so much, for the BOF's reputation—and the War Department's—now lay at the mercy of the aerodrome almost as much as did Langley's own. The board members, eager to see their support for Langley vindicated, approved the additional $25,000 at their next meeting.

Still, it was a close call. By January, the original $25,000 was gone, save for $1,500 reserved to pay Stephen Balzer for his engine. Richard Rathbun found there was not even enough money to cover January paychecks for the aerodrome workers. Langley sent a panicky query to Captain Isaac Newton Lewis, his contact at the BOF, who rushed a check over from the War Department by messenger.

WORKING ON LANGLEY'S engine was threatening to drive Stephen Balzer out of business. By New Year's Day, 1900, the engineer had spent more than twice the total contract price of $1,500. His original deadline had passed nearly a year earlier, and the engine was still far from finished. According to Charles Manly, Balzer now had "practically no means, outside of a small income required for personal living expenses, at his disposal for carrying on the work." Manly urged Langley not to lose faith in Balzer's "integrity of purpose," and to advance him an additional $500. But Manly now felt his own credibility was at risk. He told the struggling engineer: "I have had to practically guarantee this amount myself and further assure the Secretary that the engine will be ready for its official test by the third of February. Now pray don't disappoint me in this matter as I have staked a great deal on your being certain to allow nothing to interfere with the completion of the engine by February 3."

AFTER ONLY A YEAR on the job, Manly knew it was risky to keep the secretary waiting. With each passing month, the Smithsonian staff felt the rising pressure of Langley's desire to make the great aerodrome fly.

He was eternally impatient, as if haunted by a fear that he would fall short of his aims only because underlings and contractors and correspondents were always late. He preferred telegrams to letters, thinking they prompted quick responses—though his correspondents soon learned Langley's telegrams were no more urgent than anyone else's letters. He "was often unfairly impatient with assistants, and would betray irascibility by unduly raising his voice when things did not get on to suit him." One day, John Brashear, his old associate and friend from Pittsburgh, watched Langley huffing and puffing over some delay. Brashear slipped a note to a friend in the room: "Same old nervous driving energy, but gets no further ahead than the easy man."

The portly R. L. Reed, foreman of the aerodromics shop, was a frequent target of Langley's impatience. Cyrus Adler recalled being on hand one morning at about eleven o'clock when Langley summoned Reed. The secretary whipped off a sketchy design for some small device for the aerodrome and directed Reed to make a working model. Reed asked Langley when he wanted it. The secretary, rushing out the door, said, "This afternoon, at two o'clock."

Reed pondered this impossibility for a moment. Then he turned to Adler and remarked: "Do you know why the Secretary never married?"

"No," Adler said. "Why?"

"He would have married one day and expected a grown-up family the next."

The secretary wanted work done not only quickly but to the highest standard of quality. Machines built under his eye should not only work properly but look good. Once, when it was decided that some parts on the aerodrome should be strengthened by brazing—a process that would blend two visually dissimilar surfaces—Langley remarked that it was a shame to spoil the elegant appearance.

To most of the Institution's staff, he was a cold and distant presence, seldom seen and never to be disturbed. He greeted few of the lesser employees by name; he didn't know their names. From assistant secretaries down, staff members learned that if you happened to be walking with the secretary, he wanted you to stay a step or two behind, like a vassal behind a lord. Those who provoked his temper might find themselves face to face with a sputtering, stamping child. Once, rushing out of his office for an appointment, he barked to his valet: "William, my hat!" When the fellow returned a moment later with the secretary's derby, Langley hurled it down the hall and shouted: "I said a *hat!*"

Not only low-level staff feared him. Long after Charles Abbot became one of Langley's successors as secretary, he recalled that as a young scientist in the Smithsonian's astrophysical observatory, hand-picked by Langley himself, even one of his favorites, "I used to have cold shivers down my back whenever he sent for me." Abbot learned to speak only when Langley asked a question. Otherwise the secretary was not listening.

Only one member of the staff could face him down. This was the capable and clever William Karr, the Smithsonian treasurer, master of the arcane ledgers and budgets in which Langley took little interest. Once Langley returned a draft memorandum to Karr and ordered: "Take this and correct it and bring it back." Karr coolly replied: "It's correct already." Usually Langley simply left Karr alone, a practice he would live to regret.

Virtually no one but Langley and his two servants ever saw the inside of his townhouse in the Columbia Heights neighborhood of Washington, with its long view of the Castle through the humid haze. His social life, though robust, was conducted in private sanctums—the elegant homes of his friends and the mahogany lairs of his favorite clubs in Washington (the Cosmos), New York (the Metropolitan), and Boston (St. Botolph's). When traveling, he had only to show one of the complimentary free passes provided to him annually by the Pennsylvania Railroad. (His personal shipping was provided gratis, too, courtesy of the major express companies.) He shunned the rough and tumble of political life in the capital. He detested appearing before Congress, delegating this chore whenever possible to scientist-politicians such as Charles Walcott. And though he had written widely for the popular press, he distrusted and disliked "our friends the reporters." To discourage their curiosity, he ordered his groundskeepers to let the grass grow long around his laboratory sheds, so that passing journalists would think nothing important was going on inside. By his order only three men carried a master key to the Castle's innumerable doors—the assistant in charge of the office, the trusted Richard Rathbun, and himself. The doors to Langley's own offices remained closed to all but his closest aides, and even they were forbidden from a smaller room adjoining his main office, which was sealed off as "a sort of holy place reserved for the Secretary alone."

Though in conversation he was erudite, even charming, he often displayed the awkwardness of a man who worries too much about the impression he is making. He had a sense of humor, but when he told a joke, he would spin on his heel and rush away, as if afraid of the consequences. Even among friends, he had trouble looking people in the eye. He would hold forth as if speaking to the floor.

Yet among his circle of friends, and there were many, he was "not only revered, but loved." They insisted that his forbidding "shell of hauteur" was shyness. All considered themselves lucky to spend time in his company; all spoke of warmth, generosity, and kindness which few others—except children—ever saw. Those close to him were touched by "his longing for friendship, for real affection, and his appreciative gratitude when he found it," said a woman friend who knew him for many years. "I don't think any one could know him well, could have glimpses of his inner life, and not love him." Alexander Graham Bell was a gregarious man with friends around the world, yet Bell's son-in-law believed Bell's closest friend was Langley. The astronomer Charles Abbot overcame his early terror to decide that Langley was "a staunch friend," "a great man with a warm heart, against whom none of us who long served him bear any grudge in our memories." Favors and gifts, small and large, were common. When George Brown Goode's widow made plans to move away from Washington, leaving her fifteen-year-old son in the city to finish school, Langley offered to house the boy. When Langley dined with his elderly friend John Wesley Powell, who had lost his right arm to Confederate fire at the battle of Shiloh, the secretary quietly cut up the food that Powell could not cut for himself. Walcott spoke of "his habit, when he once trusted a man, of implicitly trusting him in all things."

If his traits seemed contradictory—coldness and kindness; the eager search for friendship and the shying away from human contact—the answer may lie in Abbot's simple observation that Langley was "a born aristocrat." His aristocracy was not of family or money, but of learning and culture. He unconsciously divided humanity into the few who merited his attention and the many who did not.

Among those Langley favored, his friendship and loyalty attracted loyalty in return. In the Smithsonian, his aides made Langley's quest their own, fending off visitors and other distractions from his aeronautical work, arranging affairs to forestall his wrath, helping him with the myriad tasks of invention. All through Langley's tenure as secretary, George Brown Goode, Richard Rathbun, Cyrus Adler, Charles Walcott, and the other top men conspired to protect his time from all but the most essential burdens of administration, so that he might spend most of his hours in the Astrophysical Observatory and especially in the South Shed. Partly this was because Langley wanted it this way. But partly it was their own desire to help him. They were all too aware their friend was risking his good name and the Institution's in his aeronautical enterprise. For him and for themselves, they would do all they could to ensure its success.

NO ONE GAVE more energy to Langley's endeavor than Charles Manly, and every sign indicates that Langley recognized the diligence of this remarkable youngster whom he had asked to build the world's first flying machine. Though not yet twenty-five, Manly supervised several simultaneous jobs of construction and testing, all "with the utmost speed." These included building the aerodrome's frame; crafting its wings and rudders; overhauling the No. 5 and No. 6 models for new tests; building a new small houseboat for the models and planning a large one for the great aerodrome; plus creating complex new launching apparatuses. The design and construction of a quarter-scale aerodrome for more testing was soon added to his burdens. Most important, he kept tabs on Balzer's progress in New York. Manly constantly phoned Balzer, wrote letters, sent telegrams, ran up by train for surprise inspections, all with a single aim—an engine of no more than one hundred pounds that could deliver a sustained twelve horsepower. Manly liked Balzer and respected him, and believed his pledge to "strain every point to deliver the engine to you." But Manly's worries rose as each of Balzer's promises gave way to another report that he must make another change in design or materials.

On February 3, 1900, nearly fourteen months after Balzer promised to deliver the engine in ten weeks, Langley's latest deadline arrived. No engine. The next day Balzer promised to ship the machine on February 8. He did not. When Langley decided to build the quarter-scale aerodrome, Manly spoke up for

Balzer again, saying the engineer could build the small engine but only if Langley advanced him another eight hundred dollars. Again, Langley agreed, but now he inserted a forfeit clause. If the small engine was not finished by April 1, 1900, the contract would be void. "I desire to say that I thoroughly understand the urgent need of the engine," Balzer assured Langley, "and will make every effort possible to complete the engine, if possible before the specified time."

In March Langley departed for a vacation in the Caribbean, "very uneasy" about the prospects.

"You must realize," Manly chided Balzer, "that there is a limit to the number of times that a man of the Secretary's experience will allow himself to be given assurance that everything is progressing as well as possible, and then after waiting several months, things are in practically no better shape than they were before, without losing confidence in the person giving him such assurance, and I fear that unless this large engine is completed very soon that it may be the means of the Secretary losing all confidence in my judgment and advice."

THE JAMAICANS CALLED *Cathartes aura*—the turkey vulture—by the nickname "John Crow." Langley found it easier to observe the birds in Jamaica than in Washington simply because they were more common, and he believed the Caribbean variant was "almost as much superior in skill to our buzzard as [it] is to a barnyard fowl." The Jamaicans levied a five-pound fine on anyone who killed one of the beasts. For *Cathartes aura* meant "Golden Purifier"; the vultures cleaned up the muggy island's rotting flesh, and thus were regarded as helpers. Safe from human interference, they wheeled fearlessly through the yards around Langley's hotel. This was a mixed blessing. They came close enough for Langley to see their grotesque red masks, perhaps even to glimpse their habit of excreting on their feathery legs. But their nearness to his balcony also offered "exceptional opportunities for near study" of some of nature's most skilled soaring creatures.

Langley watched the great carrion-eaters by the hour. They flapped their wings only for a moment upon leaving the ground. Once airborne the wings stayed all but motionless, the birds moving "with astonishing smoothness and indeed the extreme of grace." Here was a perfect model for the great aerodrome—a creature able to sustain itself indefinitely with stationary wings, even in a calm. And the bird could turn at will. Langley gazed with bafflement and envy. How? How did the vulture accomplish a turn in the air?

Squinting into the semitropical sun, he glimpsed something like the trick Wilbur Wright had seen pigeons execute in the streets and alleys of West Dayton. In his notes Langley wrote: "I think their movement into the breeze was managed in some way by shifting the angle of the wing, chiefly the angle of advance." The wing would tilt; that side would rise until "the line between the wing tips was nearly vertical." Then, suddenly, the birds would "blow off to the leeward, with the wind in their backs, this final movement being made as quickly and deftly as the turning of one's hand, and occupying but the fraction of a second, so that it was hard to see how, or just when, it was done."

Amending his notes later, he wrote it a second time: There was something in "the angle of the wing." The angle seemed to change on one side . . . the wing on that side rose . . . the bird turned. The movement was so fleeting, barely perceptible, the bird so high overhead, it was hard to see. But there was something to it.

Upon returning to Washington, Langley gave orders that led to a six-month investigation by six members of the Smithsonian staff led by Robert Ridgway, curator of the division of birds. Specimens of "John Crow" were collected (in Cuba, not Jamaica, to keep it legal), measured, weighed, hung up by strings in soaring position, and dissected. The same was done with Washington turkey vultures. The wings were measured separately, to determine their centers of gravity and pressure. To the Smithsonian's astrophysical observatory went an order for twin telephoto cameras, and to the Zoological Park in northwest Washington, an order for the erection of a tall tower with a viewing platform. When cameras and tower were ready, T. W. Smillie, the Smithsonian photographer, went up the ladder to capture simultaneous images of soaring vultures from various angles.

At the same time, the secretary directed Manly to consider ways of changing the angle of the aerodrome's wings in flight, in order to keep the machine balanced. "I have been noting this ability to guide by the slight inflection of the wing, in my studies of the Jamaica buzzard," he said, "and am ready to say that I think, while the quarter-sized working model of the great aerodrome is building, it will be worth while to make some arrangement of the frame or wing-holder which will make it possible to test this idea. . . . I will request you to especially look out for this, as far as you can."

It was the first time Langley had turned his attention from the quest for lightweight engine power.

IN THE UNITED STATES, only one other significant figure besides Langley and Bell stood against the wave of popular doubt about the possibility of human flight. This was Octave Chanute, of Chicago.

In 1900 Chanute turned sixty-eight. He was the son of learned French immigrants who moved from Paris to New Orleans when he was six, then to New York when he was twelve. Raised to be "a full-fledged American," with perhaps a touch of immigrant defensiveness about his claim to America's promises, Chanute went to work at seventeen, holding chains for surveyors on the Hudson River Railroad. Bright and diligent, he rose quickly and participated in the surge of construction that laced the nation with rails in the 1850s and 1860s. By his early forties he was regarded as "the best civil engineer in the West," designing and building bridges throughout the continent's midsection, including the first bridge ever to cross the Missouri River. He built the Union Stockyards in Chicago, one of the great industrial spectacles of the day. He spent ten years as chief engineer of the Erie Railroad in New York, and in 1891 was elected president of the American Society of Civil Engineers. By then he had brought to market a chemical preservative for wooden railroad ties, and it made him a wealthy man.

Chanute was the very picture of the accomplished American professional, eminent among his peers, comfortably fixed in a fine home in a great city, with a happy marriage and two beloved daughters. Plump in figure and round of face he looked like a slightly indignant owl. He had worked hard to gain respect, both for himself and for his still-young profession of civil engineering. If he longed for greater achievement, he also had much to lose. A taste for adventure and a nagging sense of caution mingled in his mind.

Like Langley, Chanute had begun to pursue an interest in flight in a time of low spirits. Since his youth he had thought flight "presented the attraction of an unsolved problem which did not seem as visionary as that of perpetual motion," since "birds gave daily proof that flying could be done." In the early 1870s the city of New York asked him for recommendations on public transit, and his proposal for elevated steam trains dragged him into a nerve-racking political rumble. Worn out, he took a long tour of Europe, where he happened to read about aeronautical experiments conducted by legitimate engineers. Chanute knew something about the startling effects that wind sometimes had on large plane surfaces, such as the roadways of bridges. He was impressed that serious European engineers not only believed the problem of flight was susceptible to engineering solutions, but had conducted research and published papers in reputable journals. He began to collect "such information as was to be found on the subject"—precious little, compared to other engineering fields—and added his own "speculations." Soon his aeronautical studies were interfering with his work. He bundled up his papers, wrapped them in red tape, and vowed to leave them alone until he could take up the subject "without detriment to any duty."

In the late 1880s he unwrapped the bundle and began to add to it again. By this point, he said later, learned debate about human flight had reached a point where "it was no longer considered proof of lunacy to investigate it, and great progress had been made in producing artificial motors approximating those of the birds in relative lightness." Soon, in the upstairs office of his home on Huron Street, models of imagined flying machines hung from every square foot of the ceiling, and stuffed birds stood on every surface. Very early in the morning, "when only the milkman was about," he would drop models of flying machines from his roof. He flew all sorts of kites "to the great admiration of small boys."

Chanute was no impetuous youngster, rushing headlong after his will-of-the-wisp. He had no desire to be seen as a lone kook. Nor did he think one man alone could solve the problem. Instead, he envisioned a worldwide network of engineers collaborating informally, each working on his own plan but sharing ideas and findings. Chanute appointed himself an informal clearinghouse for such efforts, and went to work. He enlisted the help of a clipping service and began to correspond with experimenters in the United States, Europe, Africa, and Australia. Among his correspondents was Louis Mouillard, the dreamy, ne'er-do-well Frenchman in Egypt, author of *The Empire of the Air,* which inspired Wilbur and Orville Wright.

Through several years of reading and correspondence, Chanute became a sort of self-taught professor of aeronautics without portfolio and the acknowledged dean of flight experimenters. At the invitation of a friend, he distilled his findings into a series of articles in *The Railroad and Engineering Journal,* then collected these essays in his *Progress in Flying Machines,* published in 1894.

Where Mouillard, in *Empire of the Air,* had been deliberately inspirational, Chanute used only the soberest language. Yet anyone who skimmed *Progress in Flying Machines* found himself in an outlandish mechanical menagerie. Here were devices shaped like parachutes and umbrellas; birdlike machines with two wings, four wings, ten wings; "an aerial car with paddle wheels revolving in a transverse plane, for the purpose of lifting and propelling." Here was a Belgian shoemaker named De Groof, who in 1874 had constructed "a sort of cross between beating wings and a parachute." Cutting loose from a balloon over London, he had plunged straight to his death. All these and many more had been failures, of course. Yet Chanute told of promising experiments, too—the extraordinary toys of Alphonse Pénaud and others like them, which flew many feet when launched by hand with their rubber-band motors; a French flapping-wing model that had flown for eighty yards on the power of twelve exploding cartridges; and machines with rotating horizontal screws, or propellers. Thomas Edison himself, shortly after presenting the world with his incandescent light in 1879, had fiddled with aerial screws, concluding not that the enterprise was doomed, but simply that "the thing never will be practicable until an engine of 50 horse power can be devised to weigh about 40 pounds." Now, in the 1890s, that sounded plausible, for the internal combustion engine promised to deliver the requisite power without inordinate weight.

Chanute concluded that flapping-wing devices, called "ornithopters," were impractical. Machines with fixed wings, like the rigid wings of a soaring bird of prey, held the most promise. And he believed that experimenters should begin with gliders, as Lilienthal had. But he was interested in every idea and welcomed every thoughtful effort. After all, no one had succeeded, so who could predict which path would lead to success? Only by a broad, concerted campaign, with advances in knowledge shared openly and generously, would the fledgling community of flight experimenters move forward. Embarking on his survey, Chanute had wanted "to satisfy himself whether, with our present mechanical knowledge and appliances, more particularly the light motors recently developed, men might reasonably hope eventually to fly through the air," he said. From what he knew now, "this question can be answered in the affirmative."

FOR YEARS, Chanute dreamed that he might be the one who made the crucial experiments. But it was "only after Lilienthal had shown that such an adventure was feasible that courage was gathered to experiment with full-sized machines carrying a man through the air." He meant courage to try the experiment, not to fly the machines himself. He was too old for that. When he designed his own gliders in 1896, he hired younger men to try them on the lonely dunes at the tip of Lake Michigan, near Chicago. He himself did try a few very short and tentative hops, to get the feel of it, and concluded that "there is no more delightful sensation than that of gliding through the air."

He designed gliders with multiple wings, following the example of the British theorist John Stringfellow, who, in turn, was influenced by triplane designs of the founding genius of aeronautics, George Cayley, a British peer in the era of King George III who had been the first to imagine a flying machine with fixed wings, an engine, and movable surfaces to control the machine's path. Another design Chanute tried was a near-replica of one of Otto Lilienthal's, built and flown by Augustus Herring, an ambitious flight experimenter whom Chanute had hired as an aide. It acted "cranky and uncertain" in the stiff lake breezes. To keep any control at all, the operator was forced to throw his body back and forth "like a tight-rope dancer without a pole."

"With every gust of the wind he would have to shift his weight to keep the machine going straight," said a watching reporter from the *Chicago Tribune*. "The greatest difficulty is right there. The wind shifts so suddenly at times that no one can move fast enough to keep up with it."

A second machine was of Chanute's own design—a high stack of curved, ribbed wings that looked, in an early permutation, "like six pairs of birds superposed," though the men nicknamed it the "Katydid" for its passing resemblance to the longhorned grasshopper of the same name. But the profusion of wings was not its key feature. Chanute had reasoned that if Lilienthal's problem was the operator's inability to shift his weight fast enough to control the wings, then perhaps "the wings should move instead of the man." He attached each wing to a central frame so that it could pivot to stern when struck by a gust, then snap back on a spring. This, he hoped, would be an automatic way of keeping the centers of pressure and gravity together. In a hundred glides, the machine proved "steady, safe and manageable." But its longest distance was only eighty-two feet, and there seemed to be no special magic in the pivoting wings.

A third design did better. It had two plane surfaces—a biplane, or "double-decker" configuration—that Chanute bound together by criss-crossing guy wires in a pattern familiar to anyone who had seen many bridges. Engineers called it a Pratt truss. It was an exceptionally good way of lending strength to the form without undue weight. The biplane design eliminated the need for awkward bracing wires above and below the wings—a nuisance that Langley and Manly were always struggling with. And it flew well. Many glides stretched over two hundred feet, and Herring, hanging from the bottom wing and making small adjustments of his weight to control the glider's path, once flew more than three hundred feet in fourteen seconds. Chanute watched from the sand below. "It was very interesting," he wrote, "to see the aviator on the hillside adjust his machine and himself to the veering wind, then, when poised, take a few running steps forward, sometimes but one step, and raising slightly the front of his apparatus, sail off at once horizontally against the wind; to see him pass with steady motion and ample support 40 or 50 feet above the observer, and then . . . gradually descend to land on the beach several hundred feet away. . . .

"All agreed that the sensation of coasting on the air was delightful. . . . All the faculties are on the alert, and the motion is astonishingly smooth and elastic. The machine responds instantly to the slightest movement of the operator; the air rushes by one's ears; the trees and bushes flit away underneath, and the landing comes all too quickly. Skating, sliding, and bicycling are not to be compared for a moment to aerial conveyance, in which, perhaps, zest is added by the spice of danger."

The double-decker was tested again the following summer, and with greater success. But Chanute no longer hoped to be the inventor who graduated from gliding to an actual, powered flying machine. Herring had stormed off on his own, vowing to put an engine on a biplane glider and fly it to glory. Chanute doubted he would succeed, and Herring soon proved him right. Yes, the two men had glided like Lilienthal, and learned more in a few weeks of trials than in all Chanute's years of quiet study. But the prize of automatic stability seemed no closer than before. "The inconstancy of the wind" had baffled Chanute, and "a flying-machine would be of little future use if it could not operate in a moderate wind." The wind, he had found, "is not a steadily flowing current like that of a river," but "a rolling mass, full of tumultuous whirls and eddies, like those issuing from a chimney. . . . Its effects upon a man-ridden machine must be seen and felt to realize that this is the great obstacle to be overcome in compassing artificial flight. It cannot be avoided, it cannot be temporized with, and it must be coped with and conquered before we can hope to have a practical flying-machine. . . .

"It seems unlikely that a commercial machine will be perfected very soon," he wrote that winter. "It will, in my judgment, be worked out by a process of evolution: one experimenter finding his way a certain distance into the labyrinth, the next penetrating further, and so on, until the very center is reached and success is won."

To experimenters who wished to pick up where he had left off, Chanute recommended that they start their work by testing machines as kites. Then they should board the machines to make "low gliding flights over bare and soft sand hills."

"Any young, quick, and handy man can master a flying-machine almost as soon as a bicycle, but the penalties for mistakes are much more severe. After all, it will be by the cautious, observant man—the man who accepts no risks which he can avoid, perhaps the ultra-timid man—that this hazardous investigation of an art now known only to the birds will be most advanced."

CHANUTE WAS ACCUSTOMED to receiving letters out of the blue from unknown flight enthusiasts. Few were very promising. But a letter that came to him in the middle of May 1900 was unusual. It was signed by Wilbur Wright, of Dayton, Ohio, and began: "For some years I have been afflicted with the belief that flight is possible to man. My disease has increased in severity and I feel that it will soon cost me an increased amount of money if not my life."

Obviously he had read widely in the field, and his language suggested an incisive mind and a confident spirit. "It is possible to fly without motors, but not without knowledge and skill. This I conceive to be fortunate, for man, by reason of his greater intellect, can more reasonably hope to equal birds in knowledge than to equal nature in the perfection of her machinery."

The writer had dissected Lilienthal's work into three parts—scientific principles, methods of experimentation, and the machinery itself. He assumed Lilienthal's principles to be closer to the truth than anyone else's, since he had flown more than anyone else. Therefore his ultimate failure must be attributable to his experimental method and his machine.

In five years, Wright estimated, Lilienthal had spent no more than five hours in the air, and "even the simplest intellectual or acrobatic feats could never be learned with so short practice." A way must be found to lengthen the practice sessions, Wright said, and he had an idea.

He planned to build a tower or derrick 150 feet tall. From this tower he would fly a glider of his own design as a manned kite. "The wind will blow the machine out from the base of the tower and the weight will be sustained partly by the upward pull of the rope and partly by the lift of the wind. . . . The aim will be to eventually practice in a wind capable of sustaining the operator at a height equal to the top of the tower." This experiment would not precisely simulate free flight, he knew, "but if the plan will only enable me to remain in the air for practice by the hour instead of by the second," then perhaps he could acquire the skills that in the end had eluded Lilienthal.

In appearance, the glider would be "very similar to the 'double-deck' machine with which the experiments of yourself and Mr. Herring were conducted in 1896–7." The difference lay in Wright's idea for keeping the machine balanced. His observations of birds, he said, had persuaded him that "they regain their lateral balance, when partly overturned by a gust of wind, by a torsion of the tips of the wings. If the rear edge of the right wing tip is twisted upward and the left downward the bird becomes an animated windmill and instantly begins to turn, a line from its head to its tail being the axis. . . . I think the bird also in general retains its lateral equilibrium, partly by presenting its two wings at different angles to the wind, and partly by drawing in one wing, thus reducing its area. I incline to the belief that the first is the more important and usual method." He would apply this "torsion principle"—the idea of twisting the wingtips to change the angle at which they met the flow of air—in his machine.

Wilbur Wright sounded several notes that won Chanute's approval. "I make no secret of my plans," he said, "for the reason that I believe no financial profit will accrue to the inventor of the first flying machine, and that only those who are willing to give as well as to receive suggestions can hope to link their names with the honor of its discovery. The problem is too great for one man alone and unaided to solve in secret." These sentiments Chanute heartily endorsed. Furthermore, Wright's mention of Lilienthal as his exemplar meant he was a fixed-wing man and a glider man, also like Chanute. He had no confidence in flapping wings, and no interest in trying a powered machine prematurely.

What he asked of the Chicagoan, Wright said, was only "such suggestions as your great knowledge and experience might enable you to give me," including advice on "a suitable locality where I could depend on winds of about 15 miles per hour without rain or too inclement weather," and any available information on Percy Pilcher, the Englishman who had followed Lilienthal to his death in a glider in 1898. If Chanute could advise him whether any such schemes as Wright proposed had been tried and found wanting in the past, that, too, would be appreciated.

Chanute immediately wrote a detailed and warm response, saying he was "quite in sympathy with your proposal to experiment; especially as I believe like yourself that no financial profit is to be expected from such investigations for a long while to come." He sent Wright citations to eight articles that might be helpful and extra copies of two more. Testing gliders as kites was "quite feasible," he said, though he worried that ropes and a tower might compromise Wright's findings and lead to accidents. "I have preferred preliminary learning on a sand hill." Chanute had found steady winds suitable for gliding at Pine Island, Florida, and San Diego, California, but no sand hills at either place. "Perhaps even better locations can be found on the Atlantic coasts of South Carolina or Georgia."

He invited Wright to call on him in Chicago any time, and in the meantime, "I shall be pleased to correspond with you further."

LANGLEY LEARNED UPON his return from Jamaica that despite "experiments upon experiments," Stephen Balzer was no closer to success than before. He was having trouble keeping the engine from overheating. Langley steamed. "The matter is serious." He and Manly could have built a better engine themselves in the time Balzer had wasted. Wouldn't a radial engine do better than a rotary? Couldn't they keep it cool for a few moments with wet rags? All he needed was "the carrying of a man something like a mile in two or three minutes, returning to the start in safety." If a radial could not work, then the very idea of "the gas engine seems to be condemned." The only alternative was to return to steam power, with its cantankerous burners, boilers, condensers, and pumps, and, "I cannot bring myself to contemplate this last possibility." Balzer made yet another promise of quick progress. But Langley could not long defer a decision on "what is to be done if Balzer's promise is broken, as so many of his promises have been."

Finally, Balzer got the engine running reliably enough to conduct a horsepower test. Charles Manly watched over his shoulder. For several minutes the machine roared and sputtered. But the Prony brake, used to measure power, registered no better than four horsepower, fully eight short of the minimum requirement.

Manly was now at his wit's end. For all the disappointments and delays, the young engineer still believed that of all possible solutions, Balzer's offered the best chance of success. Yet Langley's patience was nearly up. Manly committed himself to staying on in New York until some definite result, good or bad, could be obtained.

He worked at Balzer's shop for six weeks. At first things went from bad to worse. Balzer's machinists, unpaid for weeks, walked off the job. In desperation, Manly begged Langley to send two or three trusted Smithsonian machinists to New York, and Langley agreed. Manly advanced money to Balzer out of his own pocket. Just as the engine seemed ready for a better test, "very discouraging and exasperating delays" ensued in the form of utterly unexpected compression leaks, "the most exasperating thing that has come up in the work."

Langley, increasingly concerned for his young charge's physical and emotional well-being, now showed a bit of that warmth valued by his closest friends. He cautioned Manly "not to over-tax your strength by trying to accomplish impossibilities . . . I have no question in my mind whatever, that you are doing all that it is possible for anyone in your position to do, and are working on this as if your own future were at stake with it."

On June 13, Manly watched as the engine raced up to 360 revolutions per minute, with perfect compression, achieving a pull of eight horsepower—finally, he told Langley, a "very encouraging" performance. "It now seems that about all of the faults of construction have been remedied."

Langley was not persuaded. He needed some definite conclusion to these tests, and the engine's failure to reach the required twelve horsepower "was in itself such a conclusion." His summer excursion across the Atlantic was fast approaching. He wished Manly to accompany him, and to scour the European capitals for an engineer to step in where Balzer had failed. "Do not misunderstand me . . . when I ask you, after re-reading our correspondence and remembering that it is impossible that I should continue this work indefinitely, if you can give me positive assurance when the work will be done. I presume that you cannot, and that no man can."

No, Manly conceded, he could give no guarantee "that the engine will be successful within a certain specified *time*. . . . If it were a mere question of money loss instead of the much more serious loss of *time*, I would not hesitate to stand accountable for any financial loss occasioned by continuing the attempts to correct the few faults now remaining in the engine." He insisted the principles of Balzer's design were sound. He would go to Europe if Langley wished. But "the one best thing to do in the interests of the Government" would be to advance Balzer a little more money, send two more Smithsonian machinists to New York, and authorize Balzer, in whom Manly retained "absolute moral confidence," to keep trying.

Langley agreed to grant this last chance. Money was sent. After a final inspection of Balzer's work in New York, the secretary, Manly, and Manly's aide, George Wells, boarded the *Germanic*, bound for Liverpool, on June 27, 1900. "The chief object of your visit," he told Manly, "is to get something that can be absolutely relied on."

DAYTON'S SUMMER CYCLING SEASON remained busy through July. Not until August could Will spare the time to pursue his plans for a glider. Heeding Octave Chanute's suggestion about the Atlantic coast, he finally dashed off notes to weather stations in Myrtle Beach, South Carolina, and the wild and remote Outer Banks of North Carolina, asking for information on the topography and room and board. He planned some "scientific kite flying," he said.

From Myrtle Beach he apparently got no reply. But two helpful letters came back from Kitty Hawk, North Carolina—the first from a weather station man who promised steady winds and a wide beach, the second from one William Tate, whose reply was unexpectedly warm. "You would find here . . . a stretch of sandy land one mile by five with a bare hill in center 80 feet high," Tate said, and "not a tree or bush anywhere to break the evenness of the wind current," which was "always steady, generally from 10 to 20 miles velocity per hour. . . . If you decide to try your machine here & come I will take pleasure in doing all I can for your convenience & success & pleasure, & I assure you you will find a hospitable people when you come among us." Kitty Hawk had no hotel, Tate said, but there was at least a "good place to pitch tents." He advised Will to arrive well before October 15. After that "the autumn generally gets a little rough."

Heeding Tate's warning about the weather, Will worked quickly, assembling parts.

IN A SPARE ROOM of the bicycle shop a new sort of glider materialized. Its design was the product of Will's intensive study and thought, and it drew on his ability to see things in a mental realm where concrete objects and mathematics harmonized. As he said to Kate once, "My imagination pictures things more vividly than my eyes." He saw the glider before he built it. (Orville shared this ability, though the evidence suggests he did not bring it fully to bear on the glider project until later.)

The flying machine problem, Will said, was threefold (a) to build wings of sufficient lift, (b) to build an engine of sufficient power; and (c) to balance and steer the machine in flight. Langley and Manly would have agreed with the formulation, but they were putting their money on the engine as the key to the problem. Will believed the problems of wings and engines to be essentially solved already. The central problem was the "inability to balance and steer. . . . When this one feature has been worked out, the age of flying machines will have arrived. . . ." Balance and steering could be worked out on a glider. Only then would an engine be needed.

Such a glider would carry its own weight and the weight of its operator while remaining stable and safe during its journey through thin air. Even when buffeted by unexpected gusts, it must not pitch up and down like a bucking horse or a roller coaster, and it must not tilt too far over to right or left. In short, it must keep its balance, though it had nothing for support but "the viewless air."

The means of achieving this aim were not so clear. It would be a terribly tricky balancing act. The best precedents were the gliders of Lilienthal and Chanute. But Lilienthal's glider had killed him and Chanute's had driven him to give up in uncertainty. So Will was designing without much to go on. He had his idea about wing-warping, which, even if it worked, would answer only one of his questions—how to keep the glider balanced from side to side. He had Chanute's double-decker plan, which was inherently strong and easily adapted to wing-warping. And he had his small library on aerial navigation. From this, he had to decide on the size of the wing, the shape of the wing as seen from overhead, the curvature (or *camber*) of the wing, the angle at which the wing would meet the wind, and any other structure that would help achieve the goal.

His books and articles contained a good deal of scientific information about the energy of the wind, much of it recorded over 150 years by Europeans trying to make better sails, windmills, and watermills. They had formulated key concepts about the effect that moving fluids (including air) have upon surfaces. One concept was *lift*, the sum of forces that push a surface in the desired direction—up, in the case of a wing. Another was *drift*, later called *drag*, the sum of friction and other forces that slowed the wing's forward motion. A critical fact becomes obvious to anyone who holds a board flat like a wing in a powerful wind: lift and drift vary dramatically according to the angle at which the board meets the wind—the *angle of incidence*, later called the *angle of attack*. If you tilt the board just a little, the wind pushes it both up and back—a little lift and a little drift. As you tilt the board farther back, approaching an angle perpendicular to the wind, lift decreases and drag takes over. The wind simply blows the board backward.

A good glider would maximize lift and minimize drift. Birds had shown Lilienthal and many before him that a curved wing produces more lift than a flat one. But experimenters disagreed about the proper degree of curvature. Lilienthal's own best wing had a curvature of one in twelve—that is, at its highest point, it was one inch high for every twelve inches of width from leading edge to trailing edge. The shape of the wing's curve was a simple arc—a fraction of a circle. Will came to believe that shape would produce too much drift and too much bucking and pitching. So for his own wing, he flattened the curvature to a ratio of one in twenty-three, and he moved the high point much closer to the wing's leading edge. In an effort to maintain "fore-and-aft balance," as Will called it (that is, to control the bucking-and-pitching problem), he decided to affix a horizontal rudder in front of the wings, and he made the rudder movable, so the operator could counterbalance the wind's unpredictable pressure on the wings. To reduce drift, he thought the operator should lie down on the lower wing, his hands reaching forward to control the horizontal rudder.

Then there was the question of how big the wings should be. Here Will turned to algebraic formulas used by Lilienthal and other aerial experimenters. The idea was to calculate how much weight a given wing would support in the air. The formulas factored in the velocity of the air flowing over the wing; the surface area of the wing; and two coefficients. One coefficient attempted to quantify the effect of velocity on air pressure; the other accounted for the effect of the wing's angle on lift. The air pressure coefficient was the work of John Smeaton, a renowned English engineer of the mid-1700s. Lilienthal had relied on Smeaton's coefficient of pressure, as it had come to be called, and he himself had calculated a table of lift coefficients. Since the German had glided farther and more often than anyone, Will relied on Lilienthal's work, and thus on Smeaton's. He estimated the glider's weight at 190 pounds (including his own weight of 140). He figured on a steady Atlantic wind of 15–20 miles per hour. Then he worked the equation. It yielded his result: he would need a glider with a total lifting surface of about two hundred square feet, distributed between two wings and the forward elevator. He decided on wings some seventeen and a half feet from tip to tip, five feet from leading edge to trailing edge. But the Dayton lumber yards had no spruce to make such long spars. He couldn't find even common pine that long. Chanute sent him the address of a Chicago yard that could fill his order, but there wasn't time. So Will resigned himself to picking up what he could find in the East.

Through the last days of August and early September, he worked alone in dreadful heat. Milton was away on church business. Orville was apparently not much help. He was devoting his free hours to a new mandolin. "We are getting even with the neighborhood at last for the noise they have made on pianos," Kate informed her father. "He sits around and picks that thing until I can hardly stay in the house."

Kate's own summer had been lonesome, punctuated only by a spur-of-the-moment excursion to Chicago to visit friends and see the sights. Back in Dayton on August 19, her twenty-sixth birthday, she "added one more year to my already advanced age" with only her preoccupied brothers for company, plus a bust of Sir Walter Scott, which they presented to her as their birthday gift. Now Orville announced that he, too, was about to leave her. He had decided to follow Will to North Carolina once the glider was assembled there. She pulled her own trunk out of storage for Will and cleaned out Milton's trunk for Orv.

Will went about the house in a silent fog of concentration. Suddenly he realized he had not yet told his father about his plan.

> I am intending to start in a few days for a trip to the coast of North Carolina in the vicinity of Roanoke Island, for the purpose of making some experiments with a flying machine. It is my belief that flight is possible and, while I am taking up the investigation for pleasure rather than profit, I think there is a slight possibility of achieving fame and fortune from it. It is almost the only great problem which has not been pursued by a multitude of investigators, and therefore carried to a point where

further progress is very difficult. I am certain I can reach a point much in advance of any previous workers in this field even if complete success is not attained just at present. At any rate, I shall have an outing of several weeks and see a part of the world I have never before visited.

Milton's reaction has not survived, but he seems to have expressed some alarm. Will promised he would be careful. Kate tucked a jar of jam in his trunk.

AS WILL PREPARED to travel far from civilization, Langley and Manly did the opposite. From Liverpool, they went straight to London, where they presented their compliments to Sir Hiram Stevens Maxim. Besides Langley and Lilienthal, no figure stood larger in the chronicle of those who had pursued the goal of powered flight.

Born in Maine in 1840, Maxim had taught himself coach-building and engineering, then found success in New York's newborn electrical industry in the 1870s and 1880s. Visiting the Paris Exhibition in 1881, he is said to have met a fellow engineer who advised Maxim that if he wanted to make his fortune, he should "invent something that will enable these Europeans to cut each others' throats with greater facility." On that advice, Maxim developed a gun that would fire six hundred rounds per minute—the first truly automatic machine gun. When the weapon was adopted by the armies of Great Britain, Germany, Italy, and Russia, Maxim had his fortune, and the leisure to pursue his fascination with flight. He convinced himself of the theoretical possibility of artificial flight, then focused on the problem of a proper engine. Like Langley, he chose not to focus on balance and steering. "It is neither necessary nor practical to imitate the bird too closely, because screw propellers have been found to be very efficient," he wrote in 1892. "Without doubt, the motor is the chief thing to be considered."

At his rented estate in Kent, Maxim constructed a flying machine that weighed four tons. Its wings—one set above and one below—stretched 107 feet from tip to tip. The craft held two 180-horsepower steam engines, one for each of two eighteen-foot propellers, and space for a crew of three men. Maxim laid a track eighteen hundred feet long across his fields. One rail ran below, to carry the craft on its takeoff run. Another rail ran along above the first, with a guard to keep the machine from rising more than a few inches. Maxim just wanted to see if he could get it off the ground; he had no way to control the craft in the air. On July 31, 1894, at a cost of some twenty-thousand pounds, the machine accomplished Maxim's aim. Engines roaring, it reached a speed of forty-two miles per hour, heaved itself off the track, and promptly smashed the upper guardrail. Maxim cut the engines and bumped to earth, his aerial experiments at an end. He asked the United States to grant him a patent, but his application was denied because the machine lacked a balloon to provide lift, and so "was not, in the eye of the law, a useful invention." Nonetheless, Maxim had been recognized ever since as a pioneer of the heavier-than-air school of flight enthusiasts, and his prestige, like Langley's, lent the field a little more respectability.

Hearing Langley tell of his troubles, the imperious Maxim dismissed Stephen Balzer's rotary design and advised Langley to see Comte Albert de Dion, the French pioneer of the internal combustion engine. With Georges-Thadee Bouton, the count had built steam- and gasoline-powered engines for tricycles, quadricycles, and now automobiles. Maxim believed de Dion had a suitably light engine that could meet the need for twelve horsepower.

But in Paris, de Dion could tell Langley nothing more than how much time he had wasted. In near-perfect English, the count dismissed rotary engines, which he said he had tried and abandoned. He also spoke of difficulties inherent in building a radial engine of sufficient lightness. Yes, he was thinking of building a flying machine engine of his own, but no, he "would not care to give any drawings for work that is incessantly being modified." And no, he had "no engine to sell." De Dion advised Langley that his own successes had been won only "at the cost of incessant experiment in detail and of numberless failures." "As for my work," Langley said, "like any other inventor's work, he believes there is nothing for it but time and patience."

So Maxim and de Dion lost their chance to power the first man into the air—but they also effectively killed Stephen Balzer's last chance. Large reputations weighed heavily with Langley, and here were two premier inventors in England and France who said Balzer's plan was no good. Both men had reinforced the secretary's instinct to stick with what had been done before—the very instinct he was following in the design of the aerodrome's frame and wings. He had barely left de Dion's shop when he drafted instructions to Manly, who had had no luck finding an engine-builder in either Germany or Belgium. "In view of the advice of Mr. Maxim against a rotary engine and the strongly confirmatory opinion of the Comte de Dion . . . I decide as follows: . . . further work on the original Balzer engine is to be discontinued when the experiments in actual construction are finished."

Manly sailed for New York, went directly to Balzer's shop, and tested the engine. In June it had made eight horsepower. Now it barely registered six. Manly told Balzer to box everything and ship the lot to the Smithsonian. Balzer was finished.

In Paris, Langley stewed and grumbled. He blamed not only the hapless Balzer himself, but the steadfast Manly, too, for sticking up for Balzer so long. "I should be disposed to begin afresh," he wrote Richard Rathbun, "rather than spend more time and money in any modification of this unhappy Balzer affair, for if it has not wrecked the whole of my efforts to make an effective flying machine threatens to do so as we draw near the end of the money."

Rathbun knew how to soothe his superior. "I regret that you have not been able to secure more instruction about light engines," the assistant secretary wrote from Washington, "as that would make your work here so much easier, but still, if you do it all yourself, the honor will be the greater, and I for one have entire confidence in your ultimate success."

IN SEPTEMBER 1900, the nation was sweltering in a weather pattern driven by a hurricane over the western Atlantic. As Will arrived by train at Old Point, Virginia, and crossed Hampton Roads to Norfolk by steamer, the storm was entering the Caribbean Sea on a line that would take it across Cuba toward Galveston, Texas, where it would kill six thousand people. In Norfolk the immense weather pattern drove temperatures close to one hundred degrees. Will, dressed in his usual business woolens, searched the lumber yards in vain for spruce. He had to settle for sixteen-foot lengths of white pine—three feet shorter than he needed—which he sent on to nearby Elizabeth City, North Carolina. This was the closest mainland town to the Outer Banks. But he could find no one there who had even heard of a village called Kitty Hawk, let alone someone who could tell him how to get there.

In a steaming hotel room, Will wrote a letter calculated to calm his father about his safety. "I supposed you knew that I was studying up the flying question with a view to making some practical experiments." The region offered the wind he needed and safer conditions than any place close to Dayton. "In order to obtain support from the air it is necessary, with wings of reasonable size, to move through it at the rate of fifteen or twenty miles per hour. . . . If the wind blows with proper speed support can be obtained without movement with reference to the ground. It is safer to practice in a wind provided this is not too much broken up into eddies and sudden gusts by hills, trees &c." Moreover, it was "much cheaper to go to a distant point where practice may be constant than to choose a nearer spot where three days out of four might be wasted.

"I have no intention of risking injury to any great extent, and have no expectation of being hurt. I will be careful, and will not attempt new experiments in dangerous situations. I think the danger much less than in most athletic games."

Finally Will found a malodorous fisherman named Israel Perry who agreed to carry him, his trunk, and his supplies to his destination. Night was coming on as Perry, who struck Will as "a little uneasy," piloted his schooner out of the protected Pasquotank River into the broad, choppy waters of Albemarle Sound. They spent the night struggling against a sudden gale that tore two sails from Perry's half-rotted masts and tossed waves over the stern. Will believed that "Israel had been so long a stranger to the touch of water upon his skin that it affected him very much."

BILL TATE, forty years old, was a fisherman, a commissioner of Currituck County, and the best-educated citizen of Kitty Hawk, population about sixty. He also acted as assistant to the postmaster—his wife, Addie. Tate was the son of a shipwrecked Scotsman. This made him not at all unusual among the residents, many of whom were descended from people who had made lives on the Outer Banks only because Atlantic storms had stranded them there. The Tates lived in Kitty Hawk's finest home—a two-story frame house, unpainted and unplastered, with no books, no pictures on the wall, and little furniture. The rest of the village was a collection of wind-bleached shacks. The weather station stood about a mile away. It was "about as desolate a region as exists near civilization," a visitor said.

Tate had replied to the kite man's letter on August 18 but had heard nothing more from Ohio. Now, three weeks later, he opened his door to find a formal young man, neatly dressed but haggard, who introduced himself as Wilbur Wright, "to whom you wrote concerning this section." The visitor told Tate he had had nothing to eat for forty-eight hours but one jar of jelly. Addie Tate fed him eggs and ham and insisted he stay with them.

For two weeks, the Tates, their two young daughters, and a passing parade of fishermen, Coast Guard lifesavers, and their families watched as the stranger worked in the yard in front of the Tates' house. The product of his labor was a two-wing kite large enough to carry a man. Gently curving ribs gave the wings a shape like a bird's wing. They were covered in white French sateen, a finer fabric than most Kitty Hawkers had ever seen. A rectangular surface of the same material stuck out in front. Slim wooden uprights and criss-crossing wires gave the structure a light, springy strength.

On September 28, a second Mr. Wright arrived at Kitty Hawk. This one wore a mustache and was a little more fastidious in his carriage and grooming. He joined his brother at the Tates'. The two outlanders tinkered for another day or two, then said their machine was ready for a test.

• • •

ON WEDNESDAY, October 3, Will and Orv emerged from their tent after breakfast to find the wind rippling their sleeves and blowing a hollow chorus in their ears. Whoever left a drill on the ground for an hour would come back to find it half-covered with sand, a tiny drift at its edge. If they leaned over to tie a shoe, they could feel the sting of invisible grains on their hands. For old kite flyers like the Wright brothers, it was a good day.

They walked out on the dune. Bill Tate went along to help. Out near the lifesaving station, the brothers stood at either end of the glider, gripping the uprights to keep it steady. Turning it to face the wind, they felt its weight evaporate. They attached lines and let the thing rise as a kite. Another test or two showed that by pulling a string tied to the horizontal rudder at the front, they could ease the glider up and down as they pleased. This was enough to make Will want to try a manned test. He stepped into the eighteen-inch slot they had left for an operator in the lower wing. With Orville at one tip and Tate at the other, each holding the kite strings, the three men ran along together. Their shoes sank in the sand, but they developed enough momentum for Will's feet to leave the ground. Orv and Bill kept running, hauling the lines behind them, and Will rose to a height of about eight feet. When he tried the bar in front that controlled the rudder's angle, the machine pitched and rose.

"THEY FELT ITS WEIGHT EVAPORATE."

The Wrights' first glider, flown as a kite at Kitty Hawk, 1900

"Lemme down! Lemme down! Lemme down!" he cried. Orv and Tate slowed to let the glider come down. Later Will would only say: "I promised Father that I would take care of myself."

TO ANYONE WHO YEARNS TO FLY, the pull of a multiline kite on its strings is mesmerizing. Though not aloft himself, the operator feels the sensations of flight through the lines—the tension of wing against wind and the remarkable power to turn the kite right or left or up or down by tugs on the strings. Both Wrights knew something of this feeling from boyhood. Now, with the largest kite they had ever flown, they lingered on the beach whenever the wind was good, squinting upward as the sun passed overhead and fell toward Albemarle Sound. The kite had lessons to teach, and the Wrights were patient students. They developed a system for measuring its lift and drift at various angles to the wind. With a grocer's spring scale they measured the pounds of pull on the rope, and they measured the angle of the rope from the vertical, probably using a straight edge with a protractor head, a common tool they likely would have used in their shop at home. With a little high school trigonometry, they could arrive at figures for lift (the pull straight up) and drift (the backward pull). The numbers were perplexing. The kite possessed lift, but significantly less than Lilienthal's tables had predicted, and not enough to support the weight of an operator in a moderate wind. The brothers gave up Will's scheme of flying the glider from an improvised tower and began to accept the inevitable. If Will wanted to ride the machine, he would have to run down a slope and glide forward through the air, like Lilienthal. And it would take a strong wind to lift him.

Bill Tate began to rush through his daily work in two or three hours so that he could hang around with the brothers for the rest of each day. Kids and fishermen walked over to watch as time permitted. They asked the obvious questions: What were they up to? What was this really all about? As the Ohio men answered, little by little it became clear that they were not just interested in experiments with kites, but in inventing a flying machine.

To the Kitty Hawkers, the Wrights' kite would have looked nothing like a bird or a kite or anything familiar at all, except perhaps an oversized crate with its sides cut out. The Ohioans called the cloth platform in front a horizontal rudder, but the Kitty Hawkers knew boats and this looked like no rudder they had ever seen. Yet as the crate hovered overhead, bucking a bit with every gust, the experimenters would tug on a string in front. Then Tate and the others saw that indeed the little contraption in front did act as a rudder, though it moved the crate up or down instead of side to side. When the brothers pulled other lines, the whole contraption contorted slightly, as if an invisible hand were squeezing it out of its proper shape. Then one side or the other would descend a foot or two, like a shoulder dropping. One day a gust caught the contraption as it sat on the sand and flipped it up and over in a crashing cartwheel, cracking spars and ribs. But the brothers went at the thing with their tools, and three days later they had it back in the air. Sometimes they loaded heavy chains on the lower surface to see if it would lift the weight. It did.

One day Bill Tate's eight-year-old nephew Tom was standing around. He was a chatty kid who had been telling the visitors tall tales; they liked him. The brothers asked if he cared to ride. Tom said he would, and a moment later he was lying on his belly up in the air.

THE WRIGHT BROTHERS had spent their lives in Dayton and a few small towns elsewhere in the Midwest. They had been to Chicago once. They had never seen an ocean, nor any place where the ground was not clothed in midwestern oaks, maples, crops, and gardens. This blank wilderness was as exotic to them as Antarctica.

A writer of the day called the Outer Banks "the strip of beach that encircles the whole North Carolina coast like a sort of front porch rail, sometimes a mile or two out, sometimes, as at Cape Hatteras, far out of sight at sea." Kitty Hawk stood at one of the widest points in the rail, but even there it was only a mile from the ocean to Albemarle Sound. Water dominated the view at nearly every point. A few years before the Wrights arrived, a visiting naturalist counted ten thousand Canada geese passing over his head in a single afternoon, and "that flight represented only a small part of the myriads frequenting the sound both south and north." He watched many more tens of thousands of gray-brown ghost crabs crawl ashore after sunset to feed on clams.

Winds from the Atlantic brought the sour scent of salt and the shore. From the west, across the sound, came the tang of bay leaves and pine. Along the sound there were woods where bears wandered and wild hogs rooted. Toward the ocean, there was only sand. It stretched in barren plains and rose in drifted humps.

"The sand!" Orville told Kate, "the sand is the greatest thing in Kitty Hawk, and soon will be the only thing. The site of our tent was formerly a fertile valley, cultivated by some ancient Kitty Hawker. Now only a few rotten limbs, the top most branches of trees that then grew in this valley, protrude from the sand. The sea has washed and the wind has blown millions of loads of sand up in heaps along the coast, completely covering houses and forest. Mr.

Tate is now tearing down the nearest house to our camp to save it from the sand." Orville said their nephew Milton could play all day and not have to wash his hands before dinner, as "you can't get dirty. Not enough to raise the least bit of color could be collected under a finger nail."

He found time to write two long letters to Kate—not enough to suit her, but more than Will, who apparently wrote no one at home once his brother had arrived. Orville's mind wandered to business details at home, but Will said nothing about the shop. Orville took in the sunsets—"the prettiest I have ever seen . . . deep blue clouds of various shapes fringed with gold"—but Will was often asleep on his cot before the sun reached the horizon. When the wind died and the glider sat idle, Will went out on the dunes to watch hawks, eagles, buzzards, and ospreys. He climbed dunes to get better views of the birds in flight from in front or behind, not just from below. He wanted to see the angles at which they held their wings, and which weather seemed best suited to soaring. Orville wrote travel notes to Kate, telling her stories of Kitty Hawk's poverty-stricken cows and horses and hogs; of the mockingbird who nested in the live oak by their tent; of the domestic habits and local celebrities of Kitty Hawk. Will recorded field notes:

> No bird soars in a calm. . . .
>
> If a buzzard be soaring to leeward of the observer, at a distance of a thousand feet, and a height of about one hundred feet, the cross section of its wings will be a mere line when the bird is moving from the observer but when it moves toward him the wings will appear broad. This would indicate that its wings are always inclined upward, which seems contrary to reason. . . .
>
> Buzzards find it difficult to advance in the face of a wind blowing more than thirty miles per hour. Their soaring speed cannot be far from thirty miles.

The ocean was only a few hundred yards from their tent. But there is no record that either brother set foot in it.

By mid-October, everyone in the village was on speaking terms with the two "Mr. Wrights." There was an old New Yorker named Cogswell who had moved to the Outer Banks for his health and married a sister of Addie Tate's. He told the brothers if they didn't leave soon Bill Tate would die from the excitement.

They wanted another try at manned glides before leaving. They were much surer of their controls than they had been on the first day, when

Will had gotten scared in the high winds. On Saturday, October 20, with Tate along to assist, they loaded the glider on a wagon and went four miles down the island to the tallest of three big dunes called the Kill Devil Hills. It offered a long smooth slope, entirely bare of vegetation, running northeast toward the surf. Near the top, facing a hard wind, Will stepped into the operator's slot.

Otto Lilienthal and Octave Chanute had warned that any glider with wings longer than twelve feet was inherently dangerous, since it could not be controlled by shifting one's weight. Will had never made a free flight in a glider. Yet on this day he chose to defy the world's only authorities on the basis of only his own calculations and preliminary experiments. It was a characteristic moment.

The three men ran down the slope and the wind took the craft. Will held the horizontal rudder steady, and the machine skimmed the slope, Orv and Tate panting along at the wingtips. The wing-twisting mechanism was tied off, so if one tip or the other rose, Orv or Tate simply pushed it back down to level. They rushed back up the hill and did it again and again, about a dozen times in all. In the longest, Will skimmed along the curve of the hill for nearly four-hundred feet. He found he could ease the craft down to the sand ever so gently and easily; the tracks in the sand were forty feet long.

At the end of the day they hauled the glider back to the top of the hill, hoisted it, ran a few steps and hurled it into the breeze. The machine wafted out toward the Atlantic, dipped, and thunked into the sand. They left it there.

Will had hoped for hours of practice in the air and wound up with roughly two minutes. Yet "we were very much pleased with the general results of the trip, for setting out as we did, with almost revolutionary theories on many points, and an entirely untried form of machine, we considered it quite a point to be able to return without having our pet theories completely knocked in the head by the hard logic of experience, and our own brains dashed out in the bargain."

On October 23, five weeks after Will's arrival, the brothers departed. Bill Tate went down to the sand hills, lifted the glider onto a cart, and hauled it back to his house. Addie Tate took her shears to it. She snipped the French sateen from the Wrights' ribs and spars, washed it, cut it to size, and stitched new dresses for her girls.

LANGLEY WAS STILL poking around Europe for other prospects. He had instructed Manly to do his best to create a workable radial engine from the remains of Stephen Balzer's disaster, but Langley was not yet ready to entrust all his hopes for success to his young engineer's untested skill. He still hoped that the Old World's finest minds had a solution to the challenge.

Near the end of his stay in Paris, he went to the nearby Château de Meudon, where Napoleon Bonaparte had founded a balloon school and factory more than a century earlier. Lately the place was home to the Third Republic's Military Balloon Park. Here Langley peered at a gigantic hangar, watched a balloon rise from the ground, and met several leaders in the French campaign for lighter-than-air supremacy. It was a path that was leading from the balloon to the zeppelin—a stately, elegant, yet slow and ponderous means of flight. One of its enthusiasts was Henri Deutsch de la Meurthe, an oil millionnaire who, the previous spring, had established a prize of one hundred thousand francs for the first man to make an aerial round trip from the outskirts of Paris to the Eiffel Tower, a distance of some seven miles. Langley also was introduced to an immaculate sprig of a young man, odd and elegant and very much the center of attention. This was the Brazilian Alberto Santos-Dumont, a young pioneer balloonist on the verge of great fame. Santos-Dumont invited Langley to visit him the next day at Suresnes to see the cigar-shaped balloon he was building for an attempt at the Deutsch Prize. Langley found him "discouraged at his failures" so far. Examining Santos-Dumont's engine with intense interest and a newfound sense of expertise, Langley was quite sure, he told Rathbun later, "that, with the experience I have acquired at Washington, I could make the thing go, substituting . . . machinery at once more powerful and light, which I believe I could do." But Langley was not interested in gently floating beneath a directed gas-bag. He wanted speed, and power.

BY LETTER, Langley now learned that Manly, in Washington, was feeling hopeful. He had been conducting new tests and now felt entirely convinced that the rotary design was, as Langley had suspected, a doomed enterprise. He had swiftly converted Balzer's handiwork to a radial design, with stationary instead of rotating cylinders, cooled by temporary water jackets. Immediately he got "very encouraging" results—"750 revolutions per minute with the horsepower varying between twelve and sixteen." This promised to meet Langley's standard and then some.

With only a short time left to find an alternative engine in Europe, Langley insisted that Manly be definite. Manly must tell him "either . . . that he will build something in Washington which will do the work, or that I had better abandon hope of his doing so, and attempt to buy on this side with all the difficulties in the way." He must "say in substance 'I can build here in reasonable time' or 'I give it up.' "

Richard Rathbun, on the other side of the Atlantic, met with Manly. The young engineer committed himself. Rathbun cabled back to their chief: "Manly can build here in reasonable time."

## Chapter Four

# "Truth and Error Intimately Mixed"

"THE WRIGHTS NEEDED NO ASSISTANCE AND WANTED NO GUESTS."

Octave Chanute, Orville, Edward Huffaker and Wilbur (standing) at the Wrights' camp, 1901

AT THE BEGINNING, in 1899 and 1900, Will's letters always referred to "my . . . ideas," "my principles," "my plan."

"I have been afflicted with the belief that flight is possible . . ."

"I am about to begin a systematic study of the subject . . ."

"I wish to use spruce . . ."

"I have my machine nearly finished."

But in the autumn of 1900, shortly after returning to Dayton, he reported to Octave Chanute that "my brother and myself spent a vacation of several weeks at Kitty Hawk, North Carolina, experimenting with a soaring machine." Will continued to handle the correspondence with Chanute and the rest of the world. But the pronoun changed.

"We began experiments . . ."

"The machine we used . . ."

"We soon found . . ."

"We also found . . ."

"Our calculations . . ."

"Our reasons . . ."

"Can you give us any advice . . ."

One can mark the day when acquaintances become friends or lovers, but it remains a mystery exactly how and when these brothers and business partners became partners in the pursuit of flight. So does the question of what cost in pride, if any, Will paid as the change occurred.

Many years later, when neither his brother nor his sister were alive to comment, Orville told a biographer that he "often chided Wilbur for talking down to him as if he were still a 'kid'; and [for] his habit of saying or writing 'I' when he meant 'we.' "

The hint that Will tried to hog the credit is doubtful. On the contrary, whenever asked, he took pains to say the two always had formed an indivisible team, as in this widely published remark: "From the time we were little children my brother Orville and myself lived together, worked together and, in fact, thought together. We usually owned all of our toys in common, talked over our thoughts and aspirations so that nearly everything that was done in our lives has been the result of conversations, suggestions and discussions between us." In fact, the scant surviving evidence from the brothers' childhood suggests it was otherwise—that Wilbur and Orville, like any two brothers four years apart, ran in different circles, especially when the gulf of adolescence stood between them.

Will used the word "I" in discussions of the project only in 1899 and 1900, when there was nothing yet to hog the credit for. In those months, "I" could only have meant what it seemed to mean—that the project was Will's alone. He thought of it; wanted it; did the reading and thinking and work, or most of it. Certainly he and Orville discussed the project. And it must have been in those discussions, and in the camaraderie of the vacation to the Outer Banks, that Orville began to take a strong interest, and finally, step by step, to become a full partner.

Perhaps when this offbeat hobby became something much more than a hobby, Orville felt defensive about his own claim to the original idea, and his brother knew it and wanted to ease any strain. From long experience in the Brethren battles, the Wright family had the habit of solidarity. One suspects there was an unspoken rule: It must never seem to be Wilbur versus Orville. The family ethic was the Wrights versus the world, and the proper division of credit meant nothing compared to what bound them together.

The bonds were extraordinary. At one level they were the same that unite any close siblings—those of common experience and a shared way of looking at things. Scattered through their letters and Katharine's is a family code of mangled words and phrases—"Great big sing!" "Sooccess!" "Them is fine!" These were funny things said by small children—the Wrights themselves, or their nieces and nephews—and recycled to fit new occasions down through the years. They were the threads of small events and mishaps, of walks to school and long Sundays, that formed the weave of a shared life.

Another bond was intellectual. The brothers shared a deep curiosity, a love of discovery, and a yen to solve problems. Of course, many families produce two children who think rather alike. Few produce two children who think together, though to outsiders, the brothers' way of thinking looked less like cooperation than combustion. They argued for sport, their voices rising, relishing the debate as they might have relished a tennis match, whacking assertions and evidence back and forth. "I love to scrap with Orv," Will said once. "Orv is such a good scrapper." From the fracas, ideas emerged. They enjoyed the fight. But the fight was also a method, and the tougher the fight the better the results. Will once analyzed it for a friend:

> No truth is without some mixture of error, and no error so false but that it possesses some elements of truth. If a man is in too big a hurry to give up an error he is liable to give up some truth with it, and in accepting the arguments of the other man he is sure to get some error with it. Honest argument is merely a process of mutually picking the beams and motes out of each others eyes so both can see clearly. Men become wise just as they become rich, more by what they *save* than by what they receive. After I get hold of a truth I hate to lose it again, and I like to sift all the truth out before I give up an error.

Their niece, Ivonette, remembered the sound of their voices in the parlor on Sunday afternoons. "One of them would make a statement about something very important, and then there would be a long pause. Then the other would say, 'That's not so.' 'Tis too.' 'T'isn't either.' " And "at the end . . . Wilbur would say, 'Well, I think you were right. You certainly were right about that.' And [Orville] would say, 'No, I think you were right.' "

There were also bonds of purpose and of character. The brothers disagreed all the time about technical matters, and occasionally about business tactics, but never about what they wanted to accomplish or what was right or wrong. Their judgments of people seem to have been identical. In hundreds of letters they exchanged, there is scarcely a single overt expression of affection. Yet the letters also leave no doubt that each put the other first in his loyalties, that each trusted the other utterly and understood him completely.

Outsiders often found the two men virtually indistinguishable. Their handwriting was similar. They were about the same height (Will was taller, at five feet, nine and a half inches); same weight (140 pounds); same blue-gray eyes. One stranger said if you closed your eyes you couldn't tell which one was talking. They finished each others' sentences. When one finished making a point, the other often said, "That's right."

But that was all outside the family. Inside it, they were regarded as very different.

First, their alliances within the family differed. With Reuchlin and Lorin out of the house, Will was the big brother, while Orville and Kate, once the little kids of the family, remained accomplices and confidants, perhaps the two closest of all the children. Of the five, Will was the one closest to their father.

In the house on Hawthorn Street and the shop on West Third, the two brothers often moved about as if within separate force fields. Will's aloneness was more obvious. He was the austere intellectual, always composed and cool. He vanished into silent solitude, his eyes focused on interior scenes. You had to work to get his attention. Every day at lunch, Carrie Kayler, the servant girl, would watch him come through the kitchen door, hang his cap on a peg, cross the kitchen, and remove one cracker from the jar on the counter. After lunch, he would go out through the same door without his cap. A moment later—every day—he would step back in with a wry smile, remove his cap from the peg, and go out again. If not working he seldom was without a book, though once, at a gathering, the family was astonished to learn he had somehow mastered the harmonica without anyone hearing him practice.

Orville was preoccupied in a different way. He was the bustling man in a hurry. Straight as a ruler, his collar points sharp and crisp, he bounced from task to task, intent on the errand of the moment and talking in bursts. "Excitable," said his father, who remembered that as a boy, Orville liked to "punctuate" the tales he told his mother by "placing his hands on the seat of a chair" and throwing his heels up behind him—"the natural language of exhilaration."

Will had a devastating dry wit, but there was more fun in Orville, though he preferred the types of fun that grow tiresome—teasing and practical jokes. One of Will's letters to his father contains an unexplained reference to Orville's "peculiar spells." This may have referred to times when Orville stubbornly insisted on having his own way—perhaps, in turn, to resist Wilbur's own implacable will. There was a fussiness in Orville that was absent in Will. The younger brother cared far more about his appearance and clothes. Orv's pants were always pressed; Will's bagged and sagged. Will could charm and impress a group in a speech but was taciturn in one-on-one encounters with strangers. Orville was incapable of addressing a group, but in one-on-one exchanges, friends and strangers alike found him warm and delightful. Both enjoyed children, and were good with them, though Orv was a little more patient with the nieces and nephews. "Orville never seemed to tire of playing with us," Ivonette remembered. "If he ran out of games he would make candy. . . . Wilbur would amuse us in an equally wholehearted way, but not for long. If we happened to be sitting on his lap, he would straighten out his long legs and we would slide off. That was a signal to us to find something else to do."

Both wrote clearly and well, though Orville detested writing while Will enjoyed it and wrote far more often. Family members thought Orville was perhaps the better mathematician.

The chemistry between them was complex and probably irrecoverable. "Wilbur was good at starting things," Ivonette remembered. But "he'd get discouraged, and Orville would just be bubbling over. He'd want to go on with it, you know—'If you do so and so, it will do so and so.' " But Orville was too much the tinkerer, always seeing a new way to improve something, and never finishing it. Will would rein him in and say, "Quit inventing. Let's go with what we've got." Once, when Orville was doing an assembly by himself, Kate told Will that "Orv seems to be getting along pretty well with putting the machine together, though, of course, we all understand that a machine can never really be finished!"

Both undoubtedly possessed intellectual gifts of a very high order. But Orville's intelligence tended to be purely practical, while Will was both practical and wise. He saw more, and thought more broadly. It appears that he had the superior imagination of the two, and extraordinary powers of concentration.

Once the brothers began to talk about the problem of flight in earnest, their ideas became more or less indissoluble. But it is impossible to imagine Orville, bright as he was, supplying the driving force that started their work and kept it going from the back room of a store in Ohio to conferences with capitalists, presidents, and kings. Will did that. He was the leader, from the beginning to the end.

THE DAYTON WRIGHTS saw little of Reuchlin, the eldest son, and his young family. Reuchlin had married Lou Billheimer, the daughter of United Brethren missionaries, and moved to Kansas, where he struggled to make a living as a farmer. Relations were a little strained, for Reuch and his father had clashed in the past, and the Dayton Wrights did not care for the Billheimer clan. Reuch and Lou had three children—Helen Margaret, Herbert, and Bertha Ellwyn (after a daughter, Caroline, died). During their Christmas visit to Dayton in 1900, Will, Orv, and Kate had seen a good deal of the children, and silently disapproved as the sisters bullied their little brother, and especially as Reuch and Lou showed a tendency to take the girls' side, since "Herbert was the one who raised the least strenuous kick." Will's heart went out to Herbert, "a bright manly little fellow," and "a little quieter in his disposition than most children." But he held his tongue, not wanting to interfere.

Then Will learned that Reuch and Lou planned to take Herbert out of school and start him early in a business career. Will was appalled, and decided he must say something on the youngster's behalf. He sent a letter, which does not survive, expressing his concern. Lou, apparently misreading much of what Will had said, fired back a hurt letter of her own, saying that not only Herbert but all the Wright males could use a strong push in the direction of ambition and success.

Responding, Will chose his words carefully. He pointed out the discrepancy between raising a child on the Golden Rule and then pushing him into the marketplace to make his living.

> When I learned that you intended to put him into business early I could not help feeling that in teaching him to prefer others to himself you were giving him a very poor training for the life work you had chosen for him.
>
> In business it is the aggressive man who continually has his eye on his own interest who succeeds. Business is merely a form of warfare in which each combatant strives to get the business away from his competitors and at the same time keep them from getting what he already has. No man has ever been successful in business who was not aggressive, self-assertive and even a little bit selfish perhaps. There is nothing reprehensible in an aggressive disposition, so long as it is not carried to excess, for such men make the world and its affairs move. I agree that a college training is wasted on a man who expects to follow commercial pursuits. Neither will putting a boy, who has not the aggressive business instinct, to work early, make a successful business man of him.

Will conceded the truth of Lou's dig about the shortcomings of the Wright men. "I entirely agree that the boys of the Wright family are all lacking in determination and push," he said. "That is the very reason that none of us have been or will be more than ordinary business men. . . . Not one of us has as yet made particular use of the talents in which he excells other men. That is why our success has been only moderate. We ought not to have been business men." For that very reason, it would be unwise to push Herbert into business, for he was cut from the same cloth. The boy had the gifts to accomplish something "really great," Will said, but he must be steered in a direction that matched his talents.

> There is always [a] danger that a person of his disposition will, if left to depend upon himself, retire into the first corner he falls into and remain there all his life struggling for a bare existence, (unless some earthquake throws him out into a more favorable location), when if put on the right path with proper special equipment he would advance far. Many men are better fitted for improving chances offered them than in turning up the chances themselves. . . . If left to himself he will not find out what he would like to be until his chance to attain his wish is past.

This letter may be as close to autobiography as Wilbur Wright ever came. In the concern for his nephew, one can trace his understanding of what had happened in his own life—the missed chances for college and a vocation in science or teaching; the unintended drift into business; the misplaced talents seeking a proper outlet.

A single word provided a clue that at thirty-four, he had not lost hope in his own chances: "Not one of us has as *yet* made particular use of the talents in which he excells other men."

IN JUNE 1901, Octave Chanute asked the Wrights if he might pay them a visit. The household went into an uproar of preparation. Prominent churchmen had dined at the Wrights' table, and the occasional Dayton dignitary, but no one like Chanute, an international leader in his field, a friend of such men as Samuel Langley.

In a warm conversation that lasted far into the evening, the Wrights described their experiments and plans in detail. Chanute advanced a proposal. He had been thinking of organizing "a sort of tournament" where aeronautical experimenters might share ideas as they competed in gliding. Such an event could be held at the Wrights' camp at Kitty Hawk later that year. He proposed to join them there and invite two other men—Edward Huffaker, Langley's former assistant, who was building a new glider for Chanute; and George Spratt, a young Pennsylvanian who was building his own glider. Chanute said all of them, sharing ideas, could make faster progress together than on their own. He had few hopes for his own new glider, but the Wrights might "extract instruction from its failure."

The Wrights needed no assistance and wanted no guests. They had little enough time to work on their own ideas, and none to spare for the ideas of others. Will tried gently to parry Chanute's proposal, but the older man failed to take the hint.

Still, this unexpected inconvenience paled in comparison to the compliment of such a figure as Octave Chanute paying close attention to their experiments. Will could not resist the temptation to crow a little to Reuchlin. He recited Chanute's accomplishments as an engineer and as "the leading authority of the world on aeronautics," then reported that he had been "very much astonished at our views and methods and results, and after studying the matter over night said that he had reached the conclusion that we would probably reach results before he did."

Several days later, the Wrights received a gift from Chanute in the mail—a hand-held brass instrument, about the size of a compass, called a clinometer, which could be used to measure the angle of a sand dune, the descent of a glider, or the angle of their tether rope. It would dramatically improve the precision of their measurements. "We are of course delighted with so beautiful a little instrument," Will replied, "and our pleasure in it is very much increased by the fact that it is to us a token of your friendly interest in us and our experiments. We find it very convenient and quite accurate."

The brothers had brought on a new man to help at the shop—Charlie Taylor, a mechanic who had married into the family of their landlord. They asked him to watch the business while they were gone, and they got away earlier than expected, on the seventh of July.

AS THE WRIGHTS APPROACHED the Atlantic, Samuel Langley was in the South Pacific. He had left Washington a few weeks earlier, before the worst of the tidewater humidity crawled up the Potomac. In Chicago he spoke with Octave Chanute, then went on to the West Coast, where he boarded a ship for Tahiti.

Langley was satisfied now that Charles Manly could handle the aerodrome work unsupervised for a few weeks. And he would not deny himself the diversions and relaxation of a summer journey. On his trips he could enjoy being the secretary of the Smithsonian Institution without the burdens of the job itself. Crews on passenger trains and great steamships accorded him all the little privileges and comforts due a man of his high station. He could escape the political headaches of Washington and give his mind over to science, the arts, and the contemplation of nature.

He could look back over the winter of 1900–01 with a mixture of frustration and satisfaction. He and Manly had done a good deal of hiring. The aerodrome staff now included seven machinists and three carpenters. This meant faster progress, but also a monthly payroll of more than eight hundred dollars, a figure that included neither Manly's salary nor the ever-growing costs of the engine, the giant houseboat (and all its associated fees for dockage, towing, and upkeep), not to mention the flying machine itself. At this rate his appropriation would not last for long.

For his expenditures over the last year, he could record steady improvement in the great aerodrome's engine. Manly had converted Stephen Balzer's handiwork from a rotary to a radial design that was achieving eighteen horsepower in bench tests. That was much better than Balzer's best. But the engine still lacked the power they thought necessary to get the aerodrome into the air.

To settle questions of the aerodrome's design and construction, Langley had ordered Manly to build a quarter-scale model that would match the great aerodrome down to its own tiny, custom-made gasoline engine. If it could keep its balance, then the full-sized version should be able to do so, too. For weeks that spring Manly raced to get the quarter-scale ready for a trial before Langley's trip, despite smashed cylinders, ruptured propellers, and a bent frame. Then, on June 18, 1901, on the same remote stretch of the Potomac where Langley had flown his original models in 1896, the quarter-scale made several straight-line flights in a barely detectable breeze of two miles per hour. The engine overheated each time, cutting the flights to a few hundred feet. Nonetheless, Manly recorded, they "showed conclusively that the balancing of the aerodrome was correct, at least as far as motion in a straight line and in a quiet atmosphere was concerned."

Even before this, Manly had gained enough confidence in his and Langley's handiwork to make a critical personal decision. He and his patron had spoken several times about "the question of who will be the aeronaut to accompany the large aerodrome in its coming field trials." Langley had pondered the problem with a good deal of anxiety, knowing the moral burden that would fall on him if the aerodrome met with disaster. Manly had said he wanted the job, though the secretary thought he was "not anxious" for it. And Langley hesitated to give it to him, believing that "engineering knowledge . . . is of less consequence than a general clear head and ability for quick decision and action united with a most important quality implying the absence of nervousness which for want of a better word I will call unperturbability, a quality which prevents a man from losing his head at the critical moment." He considered one of Charles Walcott's young men at the Geological Survey. But Manly now put himself on record, perhaps at Langley's request. "I am now, as I have been at all times during the past two years, ready to occupy this position . . . I fully recognize the danger to which I should be subjected in such experiments in free flight and I desire to assure you that in accepting such a position I do so entirely at my own suggestion and of my own free will and accord, and I beg that you will permit me to express to you my great appreciation of the very kind interest which you have shown in pointing out . . . the personal danger to which I would be subjected.

"I shall make no attempt to here express my appreciation of your unceasing kindness to me ever since my first connection with the work, for such a task would be impossible."

Langley agreed. With that resolved, and the design of the aerodrome settled, they could look forward to the perfection of the engine over the coming winter, then trials in 1902. In the meantime, for a few weeks in the South Seas, the secretary could think about something else.

LANGLEY HAD READ ABOUT the fire walkers of the Pacific islands—aboriginal priests who uttered incantations that allowed them to walk barefoot across red-hot stones without burning themselves. Reliable witnesses had given accounts of such ceremonies in Hawaii, Tahiti, and elsewhere around the world. Langley had read the great Scottish anthropologist James Fraser's account of them in *The Golden Bough*, and just recently, a scholarly description by the scientist Andrew Lang. In Tahiti, "respectable eyewitnesses and sharers in the trial" told Langley the thing really had been done. His curiosity was aroused. "I could not doubt that if all these were verified by my own observation, it would mean nothing less to me than a departure from the customary order of nature." There were many who said his aerodrome was an attempt to violate that order. So this would be "something very well worth seeing indeed."

Thus it happened that the elderly Boston astrophysicist in frock coat and starched collar, descendent of Puritan divines, met the elderly Tahitian Papa-Ita, one of the last priests of Raiatea and "the finest looking native that I have seen; tall, dignified in bearing, with unusually intelligent features," and "dressed with garlands of flowers." Langley offered to cover the cost of a fire walk ceremony, but the priest said through an interpreter that would not be necessary. He planned to make a fire walk on July 17, the day before Langley's ship was to sail for home.

At a place near the beach, within the sound of breakers on the barrier reef, the secretary observed the preparations closely. He watched men lay firewood at the bottom of a trench twenty-one feet long, nine feet wide, and two feet deep. They set the logs on fire, then covered the blaze with three or four layers of stones, each one weighing, he estimated, forty to eighty pounds. Four hours later, when the ceremony was about to occur and a crowd had gathered, Langley observed that the lowest of the three layers of stones had become literally red-hot, some of the stones "splitting with loud reports." The stones of the top layer—which were "certainly not red-hot"—were pushed away with long poles. It was the next layer down—the middle layer, not directly in the fire, though "no doubt very hot"—that Papa-Ita, after uttering spells, proceeded to walk across several times. Other Tahitians followed, mostly in bare feet, then a number of Europeans, mostly in shoes. (Langley himself demurred.) None was burned, though tongues of fire could be seen among the glowing stones at the bottom of the trench.

But "the crucial question," Langley said, "was, How hot was the upper part of this upper layer on which the feet were to rest an instant in passing?"

He received Papa-Ita's permission to have one stone removed for an immediate test. He chose one that everyone had stepped on. Its lower end had rested in what appeared to be the hottest part of the fire. "I could think of no ready thermometric method that could give an absolutely trustworthy answer." So to arrive at a rough estimate, he had the stone dropped in a large bucket half-filled with water. The water roared into a violent boil. When the boiling stopped, Langley measured the water remaining in the bucket, to determine how much had evaporated. By this method he made a rough estimate that the stone's *mean* temperature was about twelve hundred degrees Fahrenheit—nearly hot enough to glow red, and certainly hot enough to burn skin.

Langley broke off a chunk of the stone and took it back to Washington. There, Smithsonian scientists solved the puzzle. The stone turned out to be a highly porous form of basalt. You could hold a piece of it in your hand, heat the other end indefinitely with a blow-torch, yet your hand would not be burned. The substance was "an exceedingly bad conductor of heat." The stones got burning hot on the bottom, but not on the top.

In view of the recent interest the fire walk had aroused among scholars, Langley decided to report his findings. In a letter to the British journal *Nature*, he described the scene, his improvised field test, and his ultimate discovery that the properties of the stone, not magic, explained this apparent violation of the "order of nature." In conclusion he said: "I have reason to believe that I saw a very favorable specimen of a fire walk. It was a sight well worth seeing. It was a most clever and interesting piece of savage magic, but from the evidence

I have just given I am obliged to say (almost regretfully) that it was not a miracle."

"Almost regretfully"—that was the other side of Langley, not the scientist but the romantic and aesthete, the open-minded seeker reaching for the unknown. Henry Adams, who often went to Langley for tutorials on science, said the secretary "nourished a scientific passion for doubt." But at a deeper layer he also harbored an unfulfilled passion for belief, a yearning for worlds beyond physical nature. He delighted in fairy tales, dabbled in psychic research, and spent years staring in wonder at the fabulous landscape of the sun. As a scientist he was duty-bound to quash tales of miracles and magic. Yet in a part of his mind he had wanted the upper surface of that stone to be fiery hot. It was the part of his mind where his longing for flight burned brightest.

Adams said Langley had "the physicist's heinous fault of professing to know nothing between flashes of intense perception." That is, he would claim certain knowledge only of that which could be directly observed. All else was speculation and hypothesis. Yet Langley gave the speculative realms their due—even more, perhaps, than Adams realized. "Rigidly denying himself the amusement of philosophy, which consists chiefly in suggesting unintelligible answers to insoluble problems, he still knew the problems, and liked to wander past them in a courteous temper." His flirtation with the "miracle" of the fire walk apparently crystallized certain philosophical thoughts, for he wrote a short essay on the subject soon afterward, entitled "The Laws of Nature."

What, he asked, would constitute a miracle? Langley recalled the argument of the philosopher David Hume, who, a century and a half earlier, had defined miracles as violations of the laws of nature—a proposition that Hume found inherently illogical. Langley said Hume's proof against the existence of miracles remained compelling. But he believed that now, in 1901, the argument had become antiquated and irrelevant. Hume's century, the eighteenth, "was satisfied with itself and its knowledge of the infinite, and content in its happy belief that it knew nearly everything that was really worth knowing," and "this 'nearly everything' which it thought it knew about the universe, it called the 'laws of nature.' "

The nineteenth century had brought a vast change in the common perception of the universe. "It is, that the more we know, the more we recognize our ignorance, and the more we have a sense of the mystery of the universe and the limitations of our knowledge." The "laws of nature" turned out to be "little else than man's hypotheses about nature." Each reported miracle was like the fire walk—a phenomenon that upon close observation required an adjustment, an expansion, of man's understanding of nature. In fact, the revelations of the astronomers' telescopes and the biologists' microscopes suggested that the expansion of knowledge had barely begun. Miracles were to be found not in violations of nature, but in the vast, unknown regions of nature itself.

Langley had now spent more than fifteen years testing "hypotheses about nature" as it related to the sustenance of plane surfaces in thin air. At the beginning of his experiments, it would have been liberating and exhilarating to grasp the idea that no fixed "laws" prevented human flight, as Newton had argued. Now, it seemed, Langley was coming to an older man's humility. He was nearly seventy. Waiting impatiently for Charles Manly to complete an engine that would defy Newton, Langley may have suspected that the rewards for his labors so far were terribly small compared to what he still had to learn.

His experiments on the aerodromes seemed exhaustive and endless. But how much of the evidence could he count on? He had accumulated volumes of data. But "it may be a just inquiry as to what constitutes observation and, above all, who judges the evidence." Science changed, but "human nature remains very much the same, and always has a good conceit of itself." He recalled the best chemists of the 1700s, who had been certain that all combustible substances were permeated by an element called phlogiston. The best evidence left no doubt of phlogiston's existence. Yet phlogiston had turned out to be utterly imaginary. Now, in a new century, that same, fallible human nature was judging the evidence adduced by experiments in new realms.

"There is a great deal of this 'human nature' even in the best type of scientific man," Langley warned. "We of this twentieth century share it, with our predecessors, on whom we look pityingly, as our successors will look on us."

THE LAWS OF NATURE ruled supreme in the Outer Banks of North Carolina. One could make a case that this slim arc of islands simply is not a place for permanent human habitation, for nature generally has its way there, despite the best efforts of man. The islands change constantly. Tides, currents, and waves push against each other, scraping and gouging the beaches as if they were bystanders at the edge of a brawl. Beach becomes marsh; marsh becomes beach. Sediment drops offshore to form immense shoals, their bulk rising like a whale lolling in the shallows. The waves wash the sand up on the beaches, where the wind scrapes up the particles and hurls them inland. There they join the shifting dunes or bury buildings and roads.

Some changes happen slowly, through the action of wave upon wave and the piling of one sand particle on another. Some happen in a few explosive moments. A hurricane can dump so much water in the sounds between the islands and the mainland that when the storm surge passes and the water washes back toward the ocean, it cuts new channels, making two islands where a day before there was only one, and spreading in vast, shallow ponds in the valleys among the dunes.

As the brothers bustled about in Elizabeth City in July 1901, buying groceries and lumber, a massive storm battered the islands. At the weather station at Kitty Hawk, the anemometer raced upward to a reading of ninety-three miles per hour, then the cups tore away in the wind. The wind smashed the remains of the Wrights' glider, which lay in the sand where Addie Tate had left it shorn of its fabric. It was still raining when the Ohioans' hired wagon creaked into the village the next day. As they unloaded their sodden crates of lumber and supplies, Bill Tate and the others said it had been the worst storm anyone on the Banks could remember.

After one night under Tate's roof, the brothers set out for the Kill Devil Hills four miles to the south, hauling their lumber and supplies and groceries. Near the base of the dune they called the Big Hill, they pitched their tent, drilled a well, and set to work raising a shed big enough to protect their new machine.

A WEEK LATER, Edward Chalmers Huffaker arrived in Kitty Hawk, lugging the boxed-up parts of Octave Chanute's new glider. Huffaker was a farm boy with a master's degree in physics. He had turned down the University of Virginia's offer of a scholarship to pursue a doctorate and instead had become a civil engineer. Popular articles by Langley and Hiram Maxim in the early 1890s had induced him to make his own flight experiments and Langley had been sufficiently impressed that he offered Huffaker a job at the Smithsonian, had assigned him an office, and put him to work on aerodromes.

In manners and temperament, the secretary and the scion of Chuckey City, Tennessee, were less than a perfect match. Once, on a weekly inspection tour with Cyrus Adler, Langley peered into rooms where scientists and clerks attended to business in stiff coats or cutaways, their shirt collars well-starched, their posture formal, their manner industrious. When Langley reached Huffaker's doorway, he saw his new assistant slumped in his chair, boots planted on his desk, thumbing through his reading and aiming tobacco juice at a distant spittoon. Langley shuddered and moved on, Adler recalled, then muttered, "Well, Huffaker is as God made him."

In fact, both Langley and Adler believed Huffaker's mind to be "very good and original." In Tennessee Huffaker had built hand-held models. In Washington he had experimented with strands of China silk, setting them loose to float upward in the baked air over Smithsonian Park. His experiments led to an extensive article on thermal air currents. Langley said Huffaker's observations qualified him as "one of the most acute observers of this class of phenomena whom I have known." In the small fraternity of flight experimenters in 1901, Huffaker's claim to expertise was stronger than most. He had Chanute's say-so on the Wrights' promise as flight experimenters, but he could justly consider himself very much their senior in experience.

Yet when Huffaker walked into the little camp on the sand flats, his quiet hosts, though cordial, seemed neither eager for his help nor in any need of it. They were in the midst of installing a gasoline stove in the workshop they had just built, a substantial shed sixteen feet across and twenty-five feet long, with large, awning-type doors that swung up at either end to permit the easy storage of their machine in rough weather. The machine was still in parts, but Huffaker could see it would be a good deal larger than any glider he knew about. He inspected the neophytes' work with admiration. The spars and ribs were sound and smooth. The cotton muslin surfaces, precisely stitched, felt taut and strong to the touch. The next day, Huffaker and Wilbur Wright walked the eight miles to Kitty Hawk and back for supplies and mail, Huffaker listening as the Ohio man described his experiments of the previous year in detail. "Their work is done with a good deal of accuracy and celerity," Huffaker recorded in the diary that Chanute had asked him to keep. It was "first-class in every respect." When the first wing was finished, Huffaker found "the completed surface is strong and well made, the workmanship being almost perfect."

THE KILL DEVIL HILLS stood only some fifty miles to the southeast of the Great Dismal Swamp. When the wind blew off the Atlantic, the air was clear and fresh. But northwest winds were nearly as frequent, and when they came in the summer months they brought mosquitoes in fearful numbers. Shortly after Huffaker's arrival, the wind shifted, and "in a mighty cloud almost darkening the sun," the insects descended on the camp. Typhoid fever had been more pleasant, Orv told Kate. "The sand and grass and trees and hills and everything was fairly covered with them. They chewed us clean through our underwear and 'socks,' lumps began swelling up all over my body like hens' eggs."

The men tried to escape by going to bed, though it was only 6:00 P.M. They wrapped themselves in blankets, leaving just their noses sticking out. But "here nature's complicity in the conspiracy against us became evident." The wind, blowing hard until now, dropped dead. The heat of "our blankets then became unbearable." But, of course, when they stripped off the covers "the mosquitos would swoop down upon us in vast multitudes." After "ten hours in a state of hopeless desperation," they dragged themselves into the early light to try to work on the machine, but soon were forced to retreat again. The battle continued for several days. The humans were repeatedly routed.

On July 25, Chanute's second recruit, George Spratt, joined them. That night, the campers spread out across the dunes to gather sun-bleached driftwood stumps to make a smoke ring around the camp. It was ingenious but not very effective. Each man had a choice as to how to spend the night—slapping insects or choking on wood smoke. Spratt hauled his cot back and forth all night long, unable to bear either choice for more than a few minutes at a time, and in the morning he "announced that that was the most miserable night he had ever passed through." In fact, the veterans told him, it had been the best night in several. With great reluctance he braved another night. The next day the wind shifted, pushing the insects back toward their inland swamps, and the outlanders at last could proceed with their work.

Eager to entertain his sister, Orville narrated the ordeal in a detailed chronicle of some 750 words. Will, concentrating on the experiments ahead, told the same tale to his father in just ten words: "The mosquitos have been almost unbearable for the last week." Again, he sought to reassure the bishop that his sons were not risking life and limb: "We expect to be careful to avoid real risks. We do not think there is any real danger of serious injury in the experiments we make. We will not venture on thin ice until we are certain that it will much more than bear us."

The Wrights worked steadily on the assembly of the glider. George Spratt pitched in. Huffaker did not. He strolled on the beach, gathered shells for his children, and went to Kitty Hawk for more food. While the Wrights and Spratt worked, Huffaker talked . . . and talked . . . and talked, mostly on the topic of character-building. It did not take long for the brothers to realize that Chanute had foisted a first-rate boor upon them. Huffaker had displayed poor enough manners in the prim cloisters of Samuel Langley's Smithsonian. One can only imagine the personal habits he exhibited in a wilderness camp far from any modern convenience or running water. But it was his carelessness and slothfulness that really annoyed the Wrights. He left delicate stop watches and anemometers lying around in the sand, and he used the Wrights' box camera for a stool.

"He is intelligent and has good ideas," Will conceded to his father, "but little execution." George Spratt, on the other hand, "we like . . . very well. He is not lazy."

• • •

WHEN AMERICANS in the late nineteenth and early twentieth centuries talked about would-be inventors of flying machines, the name Darius Green often entered the conversation. He was the fictional creation of the Massachusetts novelist and poet John Townsend Trowbridge, whose comic epic, "Darius Green and His Flying-Machine," had been a staple of public readings and popular anthologies since its first appearance in 1869. Had he written his poem thirty years later, Townsend might have used George Spratt as his model.

Darius was a Yankee farm lad, ambitious but not too bright, who reasoned:

> "The birds can fly an' why can't I?
> Must we give in," he says with a grin,
> "That the bluebird an' phoebe
> Are smarter'n we be?"

So, working in secret "in the loft above the shed," Darius gathered "wax and hammer and buckles and screws and all such things as geniuses use," and concocted an elaborate pair of wings and a tail. On the Fourth of July, he would leap from the loft and "astonish the nation an' all creation by flyin' over the celebration!" But all came to smash in the muck of the barnyard, to which he plummeted "in a wonderful whirl of tangled strings, broken braces and broken springs." Darius Green was a rebuke to all those who hoped to rise above their proper place. "This is the moral," Townsend scolded. "Stick to your sphere."

In appearance, George Spratt was Darius Green grown up—a dreamy young Yankee farmer who neglected his chores to tinker with odd contraptions and scribble arcane calculations in the pursuit of flight. The son of a prosperous physician in the town of Coatesville, Pennsylvania, near Philadelphia, he took a medical degree himself in 1895. But his plans to enter practice fell apart during a long spell of rheumatic fever, leading to a chronic heart problem, and in any case, "to be a physician . . . was never my first choice." Deeply ambitious "to be of direct service in promoting the plans of the Creator," he tried other kinds of work but failed at each. His father induced him to take over the management of the family farm. But Spratt considered this worse than medicine, "and I began to think that life was a burden indeed with no redeeming qualities, and that . . . I was . . . forgotten by the powers that design paths for men." Still frail from his illness and "not good for a day's work in any line," he lived with his parents, taking "nothing . . . but food and clothing and [asking] for nothing more for I know I have been a disappointment to them."

One day he flew a kite to pass the time, and "the sight of it captivated me so that I recognized in it my mission." He considered a future devoted to such a pursuit, "and I finally said why not stake my whole self upon it." He dared to hope that his destiny was to discover the principles of flight and apply them to flying machines. He had "no burning desire" to fly himself, "but I would love dearly to solve my share of the problem and like to claim the whole problem as my share." Finding "more pleasure being with nature than with people," he spent much of his time engrossed in the observation of birds, flying beetles, and grasshoppers. Indeed, "I have been almost forced to believe that birds and insects have come to me and performed for me as they never do for others, answering my questions with all the voice they could command."

When Spratt began experimenting with a model glider, he met the response that greeted most Darius Greens. His father ridiculed him, "saying if the Lord intended man to fly he would have given him wings." When Spratt gave the flight enthusiast's standard response—that God had not given man fins either, "yet [he] crosses the Atlantic in a week and takes a good part of a city with him"—then Dr. Spratt, Sr., would get "vexed" and retort that "there is no parallel in the cases, he wasn't talking of fishes & boats etc. etc."

In 1898, when Spratt was twenty-eight years old, he wrote to Octave Chanute to explain his theories and ask if they had merit. When Chanute said, "I believe you are on the right path as regards flying machines," Spratt was thrilled, and a warm correspondence ensued. Like Wilbur Wright, he was delighted to find a respected older figure who approved of his preoccupation and with whom he could discuss his ideas.

Yet Spratt was hindered by lack of wind in his Pennsylvania valley, too little money, the distractions of farm work, ignorance of engineering, indifferent mechanical skills, and "something that is a remarkably close second to laziness if it is not the genuine article." Indeed, his letters suggest he would have made quicker progress had he spent more time actually experimenting and less time agonizing over the choices he had made and the shortcomings that bedeviled him. Nevertheless, the Wrights took to him immediately. They enjoyed his thoughtful talk about flight and respected his knowledge of the woods and fields. He lacked their self-confidence. But they would have liked his odd combination of ambition and humility, which was not unlike their own. And Will presumably recognized a kindred spirit, a man for whom the quest for flight had become the fount of identity.

• • •

THE GLIDER OF A YEAR BEFORE had fallen well short of Will's expectations for lift. So the brothers had built a new machine nearly twice as large as the old. The wings spread twenty-two feet from tip to tip and were seven feet across from the leading to the trailing edge, making a fatter rectangle than the 1900 wings. Mimicking Lilienthal, they increased the curvature to a ratio of one in twelve, though they retained the elliptical shape of the first design. The glider weighed only ninety-eight pounds, but even with one man on each corner, the march up the dune had them sweating and panting. Sand slipped into their leather shoes and infiltrated their woolen stockings. Sand gave way under every footstep. When the angle was right, the wind filled the wings and took the weight. But in the valleys the machine was dead weight, and if the wind came at a bad angle, it bucked and pulled.

When they made the top, they were rewarded with vistas of stark beauty. To the west lay the broad reach of Albemarle Sound, and a quarter-mile to the east, the white-capped Atlantic, where crowds of gulls and pelicans wheeled and dove. On the high dune itself there was nothing to see but the sand—no trees, no vegetation of any kind, only black swallows rising and falling. From a distance of twenty yards, the surface of the hill seemed to glow, for the wind blew the sand into a filmy sheen. If the men sat to rest, grains of sand collected in their hair and behind their ears. By evening they felt as if their skin had been brushed with chalky powder.

At the top of the hill, before each test, they would pause to catch their breath and calculate. One man would raise a hand-held Richard anemometer above his head to check the speed and direction of the wind. Another readied a stopwatch to time the glide. A tape measure was kept handy for measuring the distance. They used the clinometer from Chanute to measure the slope of the dune, and they took care to find a slope that would match the usual slope of the glider's own path. By doing this, they could fly along the contour of the dune, practicing their skills just a few feet off the sand. That way, mistakes were likely to cause only bumps and scrapes, not broken bones or worse. There was a good deal of waiting for the wind to blow hard enough and in the right direction. When they had recovered their breath and the wind was right, they were ready for another test.

Will got in the slot in the middle of the glider. Orville and one of the assistants—Spratt or Huffaker or Bill Tate or Bill's half-brother, Dan—would grab a strut at either end. Then, at Will's say-so, all three would run. As the men at the ends grunted and strained, struggling for traction in the confoundingly soft sand, Will would lean forward on the frame, gripping the bars that controlled the rudder and hooking his ankles on the frame behind him. The assistants ran and hauled, straining for a little speed. Then the burden suddenly lightened as the wind and the glider met.

To Will, lying prone in the operator's cradle, it was a moment of confusion, a mixture of bracing physical buoyancy with furious mental figuring about the effects his movements might have on the glider's behavior. "To the person who had never attempted to control an uncontrollable flying machine in the air, this may seem somewhat strange," he said later, "but the operator on the machine is so busy manipulating his rudder and looking for a soft place to alight that his ideas of what actually happens are very hazy."

His first flights were puzzling.

Orville and George Spratt stood at the leading corners, grasping the uprights. At Will's signal, the three set off in the usual lumbering sprint. Then came the release . . . and a shuddering thunk into the sand. They tried again with only a little better result; the machine, which they thought would rise far more readily than the glider of the year before, nosed downward and hit the sand perhaps fifty feet down the hill from the starting point. Hoist, run, release . . . *thunk*—twenty feet. Will, trying to counteract the glider's obvious desire to nose down, wriggled backward a couple of inches to bring the center of gravity farther toward the rear. With his hands, he pushed the horizontal rudder to a sharp upward angle. With each try he moved farther back until he was sprawled across the wing, his hands barely able to control the angle of the rudder. In this awkward and impractical position, on the seventh try, he found the glider at last consenting to stay in the air.

Huffaker and Spratt looked on in delight as the machine rose and dipped for several hundred feet down the slope. Spratt had not seen anything like it. "Long flight, very good," Huffaker jotted in his diary. They tumbled down the hill in pursuit, then hauled the machine back up. The next try produced a similar glide, with much rising and falling.

The ninth try brought a very bad moment. With Will still stretching to keep his weight near the back while gripping the control bar at the front, the machine refused to hug the sand at a safe altitude. The nose rose and rose. The glider, now twenty feet off the ground and pointed upward, was coming to a dead stop. A flash went through Will's and Orville's minds simultaneously—the glider was behaving exactly as Lilienthal's had just before its fatal crash. Later, the glider's behavior would be called a stall.

Orville cried out. Will shoved the rudder but with no effect. Will inched forward, trying to bring the nose down, and suddenly the glider *was* coming down—yet not with its nose first, but flat, like a parachute. It landed softly enough that Will was not even shaken. A few moments later the same thing happened, and on the final try of the day, the glider actually reversed direction, sliding backward to crunch into the sand on its tail.

The next day was Sunday, so there were no experiments. Huffaker was bubbling about the results thus far, but to the Wrights, Will's flights would have been profoundly unsettling. A glider that was supposed to rise much more easily than their 1900 machine, and stay just as well balanced, was careering up and down, essentially out of the operator's control. Huffaker could admire all he liked; he hadn't been in the air, and his enthusiasm merely displayed his ignorance. "We find from Mr. Huffaker that he and Chanute have greater hopes of our machine than we have ventured to hold ourselves," Will told Milton. "In fact they seem to have little doubt that we will solve the problem if we have not already done so, so far as the machine itself is concerned . . . Mr. Huffaker remarked that he would not be surprised to see history made in the next six weeks. Our own opinion is not so flattering. He is astonished at our mechanical facility, and as he has attributed his own failures to the lack of this, he thinks the problem solved when these difficulties are overcome, while we expect to find further difficulties of a theoretical nature which must be met by new mechanical designs."

The only performance they were happy about was Will's flat descent out of "the very fix Lilienthal got into when he was killed," Orv said. (If this was meant to reassure his sister, who was not yet acquainted with the fine points of gliding, Orville could hardly have chosen his words less carefully. "His machine dropped head first to the ground and his neck was broken."

On Monday morning the glider performed a strange new trick, nosing downward unaccountably. The Wrights went to work on the horizontal rudder, reducing its size, but it made no difference. Orville took a turn in the operator's position, perhaps for the first time. His flight was steadier than Will's attempts of that day but still disappointing. Testing the machine as a kite only compounded the mystery. "It seems as if the tendency is to turn down as the speed increases," Huffaker recorded—a phenomenon that utterly confounded the Wrights' carefully calculated expectations.

They were back at the gate of the maze, bewildered by the problem of balance. In their new technical vocabulary, the problem had to do with two abstract concepts that helped experimenters discuss how to keep their artificial birds from plunging to the ground. One was the center of gravity, the point at which a wing or a glider would balance, both side to side and fore to aft. The other was the center of pressure, the point at which all the forces pushing the wing upward were concentrated. When a glider bucked and pitched in the wind, its center of pressure was racing back and forth between the front and the rear of the glider. "The balancing of a gliding or flying machine is very simple in theory," Will wrote a few weeks later. "It merely consists in causing the center of pressure to coincide with the center of gravity. But in actual practice there seems to be an almost boundless incompatibility of temper which prevents their remaining peaceable together for a single instant, so that the operator, who in this case acts as peacemaker, often suffers injury to himself while attempting to bring them together." The brothers were now suspecting that the relationship between the centers of gravity and pressure differed markedly from one wing to the next, depending on their sizes and shapes.

It was that mischievous center of pressure that troubled Will and Orv most as they talked all that Sunday and tried to glide again on Monday and Tuesday. To stay safe and progress beyond Lilienthal's work, they had to establish fore-and-aft balance without shifting their center of gravity. That meant controlling the center of pressure. In this, their plan had been simple and effective. They were building their gliders with a control surface in addition to the wings—the horizontal rudder in front. With the rudder in a neutral position, parallel to the horizon, the glider was balanced so the center of gravity was slightly forward of the center of pressure. This gave the glider a slight tendency to nose down, which kept its speed up and prevented it from stalling. When the desired flying speed was reached, the pilot would adjust the rudder so that it met the wind at a slightly positive angle. This would bring the center of pressure forward under the center of gravity. In this happy state of equilibrium, a glider could proceed through the air smoothly and safely. And by shifting the angle of the forward rudder—moving the center of pressure in relation to the center of gravity—the operator could induce the glider to pitch up or down, to rise higher into the sky or return to the ground.

In the glider of 1900, the center of pressure had remained pleasingly tame, its fore-and-aft balance reasonably stable. But in this new machine, the center of pressure was running wild. Why?

The problem, they reasoned, must lie in the curvature of the wings. Aside from the increase in size, that was the critical difference between their wings of 1901 and their wings of 1900. They had changed the curvature to be closer to Lilienthal's design. But the new glider had less lift and was harder to control. Could Lilienthal, the greatest of all the flight experimenters, be wrong?

After more attempts on Tuesday, July 30, Will sat down to think things through, recording pros and cons with a pencil in his little palm-sized diary. On the positive side of the ledger, the glider was proving itself to be strong and resilient despite being "very severely used in some forty landings." And, so far,

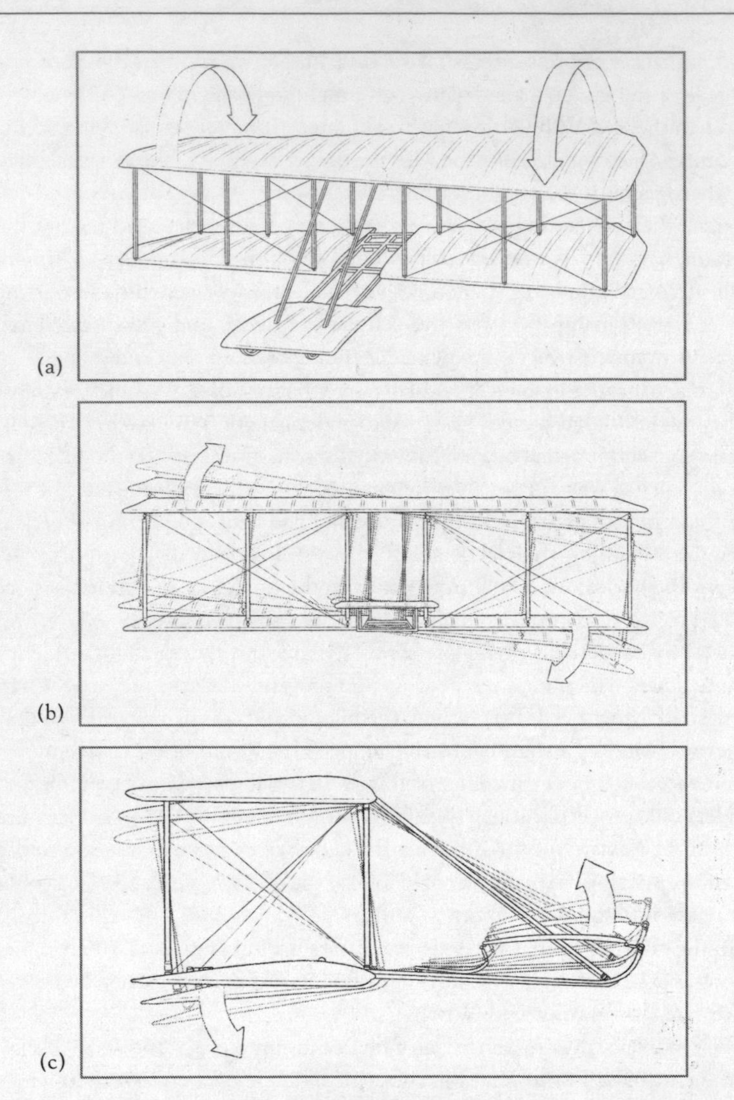


By pulling cables that warped, or twisted, the wings of their 1901 glider (a) the Wrights began to control their lateral balance, later called roll control (b). The horizontal rudder (elevator) in front controlled fore-and-aft balance, later called pitch control (c). When they added a tail in 1902, they perfected their ability to turn right or left, later called yaw control, thus achieving control in three axes—roll, pitch, and yaw.

it was proving to be safe, despite the great increase in wing surface over the 1900 glider. Also, the first attempts at wing-warping had rendered "all that could be desired" in maintaining the glider's lateral balance.

The list of negatives was longer: the shortage of lift, "not much over ⅓ that indicated by the Lilienthal tables"; the poor fore-and-aft stability as compared to the 1900 glider; the high degree of drift, which reduced the glider's speed. It was sluggish and uncontrollable—on the whole, a disappointing machine that was leading them to doubt the very ground on which their experiments stood. "The Wrights have no high regard for the accuracy of the Lilienthal tables," Huffaker noted.

The brothers hauled the glider into the shade of the shed and went to work. Their campmates looked on with increasing admiration. Abandoning Lilienthal's tables, they used their instincts and reduced the curve of the wing. To Huffaker the Wrights' swift and sure remodeling job may have been a goad, for he roused himself to get Chanute's glider ready for testing—only to have the cardboard tubing of its framework crumple under a drenching rain. The Wrights snapped a picture of the soggy mess and later included a print among several they sent to Spratt as souvenirs of the outing. At the time, Will confided, "I took it as a joke on Huffaker, but afterward it struck me that the joke was rather on Mr. Chanute, as the whole loss was his. If you ever feel that you have not got much to show for your work and money expended, get out this picture and you will feel encouraged."

Spratt volunteered to catch the boat to Elizabeth City for provisions. He returned with a cornucopia—ham, bacon, butter, coffee, fruit, maple syrup, even some ice. The delivery was well-timed, for Octave Chanute walked into camp the next day. If he was very disappointed with the disintegration of his own machine he didn't say so. He had made the long trip chiefly to see the Wrights in action.

The old engineer watched as the brothers made final adjustments to the remodeled glider. The curve of the new wing was now shallower, the leading edge smoother. Kiting tests showed they now had a glider that was much closer to their original intentions for the season. Its lift was still less than hoped for, but adequate. If all went well, the brothers at last could test the wing-warping mechanism to see if their most cherished innovation would work with a man aboard.

After a week of reconstruction and kiting, they were ready for a manned trial. On August 8 the wind was raking the hills at speeds approaching twenty-five miles per hour. But they were eager for trials, and they pulled the glider out of its shed. In glide after glide, Will swooped above the slope for two hundred, three hundred feet and more, easily controlling the fore-and-aft balance of the machine, thanks to the remodeled wing. He got so he could skim along just a foot or so above the sand, making tiny adjustments in the horizontal rudder to follow the graceful curves of the slope. As his skills and confidence increased, he ventured higher. "The control of the machine seemed so good," Will said later, "that we then felt no apprehension in sailing boldly forth. And thereafter we made glide after glide, sometimes following the ground closely, and sometimes sailing high in the air."

Now it was not only Chanute and Huffaker but the brothers, too, who believed a breakthrough might be at hand. At last they could test the wing-warping mechanism in a machine with a lot of lift and good fore-and-aft control. If it worked, they would have the first glider that could be controlled in the air without the dangerous reliance on weight-shifting that had led to the deaths of Otto Lilienthal and Percy Pilcher.

Until now, the wing-warping cables had been tied down tight. Now they made the cables ready for use, to be controlled by the operator's feet.

The experiments that followed were utterly confounding. On some glides, Will would rock the wing-warping control right or left and nothing would happen at all. On others, the same motion would produce a sharp turn. Two ground-skimming glides ended in near-disaster. On the first, the left wing suddenly dipped; Will shifted his weight to the right to bring the glider level. But he neglected the horizontal rudder, and the glider plunged into the sand with a shuddering jolt. Will's body hurtled forward face-first into the rudder, splintering pieces of wood and banging his nose and one eye. A similar landing sent him tumbling head over heels.

Will and Orv had come to take it for granted that wing-warping was the key to side-to-side control. It had been central to their hopes since the day in the bicycle shop when Will had twisted the cardboard box. It simply stood to reason—if one increased the angle of the wing on the right and decreased the angle on the left, the machine should rise on the right and turn to the left in a banking motion, like a bicycle.

But this machine was doing nothing of the sort, at least not dependably. On some glides it behaved in precisely the contrary way. In his diary Will jotted: "Upturned wing seems to fall behind, but at first rises." In those ten words lay deep disappointment and puzzlement. It was "a very unlooked for result," Will conceded to Chanute, "and one which completely upsets our theories as to the causes which produce the turning to right or left."

"We found that if we jerked the warping [control] back and forth rapidly the machine would make its way down the hill," Will said later, "but if we persisted in the movement long enough to determine its real effect the machine quickly acquired such a peculiar feeling of instability that we were compelled to instantly seek the ground." Their understanding of the whole enterprise was suddenly turned upside down, for the side of the glider with greater lift was hitting the ground first. Chanute urged them not to despair. They had achieved more than anyone else, he said—more than Chanute himself, more than Percy Pilcher and even than Lilienthal. "Yet we saw that the calculations upon which all flying machines had been based were unreliable," Will said later, "and that all were simply groping in the dark. Having set out with absolute faith in the existing scientific data, we were driven to doubt one thing after another. . . . Truth and error were everywhere so intimately mixed as to be undistinguishable." Apparently Lilienthal had been wrong about the proper curvature of wings. He had been wrong about lift as well; there was something wrong with his lift and drift tables. Wings simply did not behave as the German had said.

They tried a few more glides, but with no better results. Will caught a cold. Spratt left. With rain approaching from the south, they watched a buzzard sail across the flats for more than a mile without once flapping its wings—a perfect display of soaring that could only have stoked their frustration.

Now Huffaker left, too. "He looked rather sheepish on departure," Will told Spratt later, "which I attributed at the time to the fact that he was still wearing the same shirt he put on the week after his arrival in camp."

The brothers, too, had had enough. They broke down the camp and departed.

On the long train ride home, Will said later, "We doubted that we would ever resume our experiments. . . . When we looked at the time and money we had expended, and considered the progress made and the distance yet to go, we considered our experiments a failure." Someday people would fly, he remarked to his brother, but the two of them would not live to see it.

SECRETARY LANGLEY was out of town when a letter arrived at the Smithsonian from Isaac Newton Lewis, recorder of the Board of Ordnance and Fortification, inquiring about "the present status of the work on your aerodrome," and in particular, "when you expect to complete the construction of this apparatus." Charles Manly, following up on his own, learned not only that the board would meet the following week, but that its "personnel . . . had almost entirely changed since the appropriations for the work were made and that a number of the present members had been lately inquiring quite frequently regarding the progress of the work."

The slow schedule was the fault of the engine, Langley replied, which "has now turned out to be the formidable part of the problem." After the "great delays" occasioned by "the failure of the engine builder," Stephen Balzer, and the dashed hopes of "a most thorough and complete search both in this country and in Europe for a competent builder," Langley had been "forced greatly against my wish to undertake the construction of the engine with the very limited facilities at my disposal in the Institution shops." A suitable engine had been finished, "but unfortunately of not quite large enough size to furnish the full power which I deemed essential to have at the disposal of the aeronaut at the first trials in free flight." Construction of another engine had been started, only to meet another "serious delay of several months in obtaining the very special materials necessary for the engine."

Langley predicted—though he did not quite promise—a trial by the end of the year. About the time the board was asking where the aerodrome was, Langley learned that the last of his fifty thousand dollars had now been spent. He did not ask the War Department for more. He had other sources to tide him over. Ten years earlier, Bell had given him five thousand dollars to use for research as he wished. This money was still available, as was another five thousand dollars left to the Smithsonian by a Washington physician and friend, Dr. Jerome Henry Kidder. These monies were pooled in a Bell and Kidder Fund, and designated for aerodromics.

The day after Langley submitted his report to the Board of Ordnance and Fortification, Manly tested a new set of water jackets on the engine. They leaked badly into the cylinder head. That meant new castings, new cylinders, new steel shells, and new water jackets. The winter of 1901–02 began, and there was no trial of the aerodrome.

*Chapter Five*

# "The Possibility of Exactness"

"MYSTERIES EVAPORATED."

The Wrights' lift balance. Inside the wind tunnel, the air flow struck the experimental wing shapes (top, held vertical). The dial, bottom left, recorded the degree of lift.

OCTAVE CHANUTE arrived home in Chicago deeply impressed by what he had seen on the dunes. When Wilbur Wright confessed in a letter that he was discouraged about the strange behavior of the glider when its wings were twisted, Chanute hastened to reassure him. "I think you have performed quite an achievement."

In fact, the old engineer understood neither the nature nor the depth of the Wrights' problem. With Chanute's own money, his old correspondent in Algeria, Louis Mouillard, had built a glider that allowed the operator to increase the resistance on one side by altering the angle of the wing. This was meant to make the machine skid into a turn on that side, like a canoe whose pilot is holding his paddle firmly in the flow of water. To Chanute, the Wrights' twisting mechanism appeared to aim at the same effect. He did not understand that the brothers had quite a different motion in mind—not the skidding turn of a canoe, but the banking turn of a bicycle. With more lift on the left and less on the right, the Wrights expected the left side of their glider to rise, pushing the machine into a banking turn to the right, pivoting around the low right-hand wing. The fact that it sometimes did just the opposite—turning in the direction of the *higher* wing—was what had them so confounded on the train back to Dayton.

But at the Kill Devil Hills, Chanute had not, like the Wrights, seen a glider failing to fulfill carefully calculated expectations. He had seen the largest glider ever constructed soaring down a slope for hundreds of feet under reasonably good control. To his eyes, whatever problems might remain, it was an enormous achievement.

Chanute had taken his own pictures of this spectacle, and the Wrights had taken more. Such images alone, he well knew, would ignite new interest in progress toward his great goal of manned flight—and he was not at all indifferent to having such progress associated with his own name. So Chanute talked up the Wrights among his colleagues in the Western Society of Engineers, based in Chicago. As the organization's current president, Chanute had a large say about the choice of guest speaker at the society's monthly meetings. The very next date—the evening of Wednesday, September 18—was open. Chanute conveyed the invitation to Dayton and urged Will to get his photographs ready quickly for conversion into lantern slides, "the more the better." He left little room for demurral.

Will was a little staggered by the sudden prospect of addressing a prestigious professional society, and a little irritated with Chanute for dropping this fait accompli upon him with less than three weeks to prepare. He never had addressed an audience outside the Church, and given his confusion over what had happened at Kitty Hawk, he hardly knew what to say. He could look like a fool.

His sister would have none of this. "Will was about to refuse but I nagged him into going. He will get acquainted with some scientific men and it may do him a lot of good." That settled, Will told Orville to hurry with the photographs, and to Chanute he replied, "After your kindness in interesting yourself in obtaining an opportunity to address this society, for me, I hardly see how to refuse, although the time set is too short for the preparation of anything elaborate or highly finished. If a brief paper of rather informal nature, with the enclosed pictures, will be sufficient to interest the members for a short time, I shall be glad to respond if desired."

He had hardly sent this off when another note came from Chanute, asking: "May they make it 'Ladies' night'?"

One can hear the resigned sigh and the shiver of nerves in Will's response: "As to the presence of ladies, it is not my province to dictate, moreover I will already be as badly scared as it is possible for man to be, so that the presence of ladies will make little difference to me, provided I am not expected to appear in full dress, &c." Much earlier than he had intended, he was going to tell the world about his secret project.

Whatever Chanute's motives in wangling the invitation, it had the marvelous effect of demolishing any thoughts of quitting. The brothers plunged back into their work. With Charlie Taylor, they installed a new shop engine to power new tools that would aid in the construction of new gliders. Will pored over his penciled records from the recent trip. Orville developed glass photographic slides in the darkened summer kitchen behind 7 Hawthorn.

"We don't hear anything but flying machine . . . from morning till night," Kate declared. "I'll be glad when school begins so I can escape."

AS WILL PREPARED HIS REMARKS, the September issue of the popular magazine *McClure's* arrived by mail in subscribers' homes across the nation. Among the articles were a serial installment from Rudyard Kipling's epic *Kim* ("*A low murmur of horror went up from the coolies*"); a report from the leader of the ill-fated Baldwin-Ziegler polar expedition ("*To solve the mystery that lies hidden at the North Pole has been for many years the cherished ambition of my life*"); a memoir by the stage actress Clara Morris ("*The warm pressure of Mr. Barrett's hand, his brightening eye, gave me such an impression of sincerity*"); an account of the spectacular growth of American cities based on the 1900 census ("*The face of this good old world of ours never before saw such changes as those which we Americans have beheld, during the century just closed, here upon our Western continent*") and an essay on the possibility of human flight by the distinguished scientist Simon Newcomb, professor of mathematics and astronomy at Johns Hopkins University, friend to Alexander Graham Bell, Samuel Langley, and all the leading American scientists, past president of the American Astronomical Society, the American Mathematical Society, and the American Association for the Advancement of Science. Newcomb had calculated constants in the movements of planets and stars and measured the speed of light. America had no more august scientist to consider the claims of the flying-machine men.

The question that the *McClure's* editors put to Professor Newcomb—"Is the Airship Coming?"—plainly irritated him. It encapsulated his countrymen's glib assurance that they lived in an age of mechanical miracles—that any machine that could be imagined could be invented. To Newcomb and his academic brethren, the fascination with inventions was more tiresome evidence of technology's supremacy over pure science in America. While German and English scientists probed the deep secrets of the universe, Americans had been dazzled by a parade of ingenious gadgets—the telephone, the phonograph, electric lighting, the cinema, the wireless, the automobile. Now a new crowd of tinkerers, aping that self-promoting mechanic, Thomas Edison, were saying that a flying machine loomed on the horizon.

Newcomb was the man to put such types in their place. Like Langley, his contemporary, with whom he shared a prickly friendship characterized by sharp scientific disagreements, Newcomb had made a tortuous climb to eminence. Born in Nova Scotia, he started his career as a schoolteacher, then became a "computer"—a human calculator—on the staff of the *American Ephemeris*, the great nautical almanac of the stars compiled at Harvard, where he took a degree at the Lawrence Scientific School. He became chairman of mathematics and astronomy at Johns Hopkins and director of the *Ephemeris*. From these heights he reigned as perhaps the nation's preeminent mathematician and astronomer. He was a stern character. As a hobby he composed severe essays on political economy, all taking the strictest laissez-faire view of what was best for his adopted land. "The basis of Professor Newcomb's character is intellectual and moral honesty pushed to its highest degree," an admirer said. "He loves truth and detests shams."

The spectacle made by clever mechanisms in recent years scarcely implied that putative flying machines, too, would come to pass, Newcomb warned. "The very common and optimistic reply to objections, 'We have seen many wonders, therefore nothing is impossible,' is not a sound inference from experience when applied to a wonder long sought and never found." To his friends Langley and Bell, he brandished a startling rebuke: "No builder of air castles for the amusement and benefit of humanity could have failed to include a flying machine among the productions of his imagination. The desire to fly like a bird is inborn in our race, and we can no more be expected to abandon the idea than the ancient mathematician could have been expected to give up the problem of squaring the circle." But, "We cannot conclude that because the genius of the nineteenth century has opened up such wonders as it has, therefore the twentieth is to give us the airship."

If a great deal more could be learned about the nature of gravity, perhaps man could learn to overcome it. But so profound a mystery might lie entirely beyond human understanding; it was silly hubris to assume otherwise merely because Bell had found a way to transmit speech via copper wire and Edison had caused a carbon filament to glow.

Newcomb enumerated other reasons to doubt the boosters of flying machines. What would such a thing be good for? "It would, of course, be very pleasant for a Bostonian who wished to visit New York to take out his wings from the corner of his vestibule, mount them, and fly to the Metropolis. But it is hardly conceivable that he would get there any more quickly or cheaply than he now does by rail." And who would risk it? An oceangoing ship whose engine broke down could float safely until a repair was made. But any mechanical accident on a flying machine would be fatal.

Finally Newcomb lowered the boom on Langley's great project. He granted that it was possible to calculate the proper size of a wing to support an artificial bird, and the speed it must travel to stay aloft, and thus the horsepower needed to propel it. But here one met the classical problem, dating to Newton. How large would such a machine have to be?

Newcomb posited two flying machines, one exactly twice the scale of the other in all its dimensions. The wing surface of the larger machine would increase as the square of the wing surface of the smaller craft. But the *volume* of an object—and thus its *weight*—increases as the *cube* of its dimensions. The square of two is four. The cube of two is eight. Thus the larger machine would have *twice* the wing space of the small machine but *four times* the weight—a burden that would require a very powerful engine made of unimaginably light material.

Nature presented its own demonstration of the problem, Newcomb said. The most numerous flyers by far were the smallest—the insects. Given current materials, he said, "The first successful flyer will be the handiwork of a watchmaker, and will carry nothing heavier than an insect. When this is constructed"—one can hear the old professor chortle through his beard at his joke—"we shall be able to see whether one a little larger is possible."

ON THE EVENING OF September 18, 1901, sixty or seventy members of the Western Society of Engineers and their wives rustled into their meeting room and took their seats. Most of the men, like Chanute, were civil engineers. They were highly skilled and practical men who built the solid, heavy works upon which America's new urban civilization depended—railroads, bridges, sewers, streets, and towers of commerce. Most, too, were Chicagoans. Some had earned their professional stripes in the rebuilding of the city after the Great Fire of 1871. Members were accustomed to such topics as "Engineering

Problems in Cement Manufacture" and "Proposed Specifications for Steel Railroad Bridges." So the program for the evening, announced as "Late Gliding Experiments," was not the usual fare. It promised more in the way of light amusement than practical enlightenment.

Octave Chanute, round and dignified in business dress, rose to introduce his obscure guest. As usual, he was cautious, almost apologetic, about raising the topic of aerial navigation among serious students of technology. "Those who ventured, in spite of the odium attached to that study, to look into it at all became very soon satisfied that the great obstacle in the way was the lack of a motor sufficiently light to sustain its weight and that of an aeroplane, upon the air." But with progress under way toward light, powerful engines, such as those that Samuel Langley was known to have built, "there is now some hope that, for limited purposes at least, man will eventually be able to fly through the air.

"There is, however, before that can be carried out—before a motor can be applied to a flying machine—an important problem to solve—that of safety, or that of stability." Two brothers from Dayton, Ohio, were making progress in this area, Chanute declared, and "these gentleman have been bold enough to attempt some things which neither Lilienthal nor Pilcher nor myself dared to do." He himself had watched them glide on the Atlantic seashore only a month ago. "I thought it would be interesting to the members of this society to be the first to learn of the results accomplished, and therefore, I have the honor of presenting to you Mr. Wilbur Wright."

Will stepped forward, pale and somber, clutching his sheaf of painstakingly composed pages. The lights were put out except for a shaded lamp for him to read by; this and the beam of the stereopticon used for his photographic exhibits threw shadows across his severe features and bald pate. Since receiving Chanute's invitation, he had given every available hour to thought and writing. "We asked him whether it was to be witty or scientific," Kate said, "and he said he thought it would be pathetic before he got through with it!" He had taken the train up from Dayton that morning, disembarking in time to search out Chanute's home on Huron Street. Will was amused to find his host's upstairs office in a state of extraordinary clutter, worse than Bishop Wright's.

Will had not worn such fine clothes in a long time, perhaps ever. They were all Orville's—shirt, collar, cuffs, cufflinks and overcoat. "We discovered that clothes do make the man," Kate reported, "for you never saw Will look so 'swell.' " He never had delivered a scientific presentation. He never had addressed an audience outside a church. Yet he quickly proved himself fully prepared to make his debut in the world of serious men.

At seventy-five hundred words, the talk was twice the length the engineers were accustomed to. But it flowed well, with colorful strands of insight and wit. The logic was sure and compelling, the explication plain. Will crystallized the arguments of the glider-first school of experimentation—Lilienthal, Chanute, and Pilcher—and made it plain, without a hint of boasting, that if anyone was betting on who would solve the problem first, they would be reckless not to choose the Wrights.

"The difficulties which obstruct the pathway to success in flying machine construction," he began, "are of three general classes." The first had to do with the construction of wings that could sustain the weight of machine and man; the second with the generation and application of power; the third with "the balancing and steering of the machine after it is actually in flight." The first two problems—wings and engines—were more or less solved, he said, or would be soon. The real issue was control—how to keep one's balance in the sky; how to go where one wanted and return to earth at a place of one's choosing. "When this one feature has been worked out the age of flying machines will have arrived."

In going over his data and the photographs and thinking the whole thing through, Will had thrown off the pessimism he felt leaving Kitty Hawk. He was back in the mode of a practical scientist, ready to build upon what he now saw as useful lessons, not disappointments. He led the engineers through each of the brothers' achievements and conclusions—the importance of actual practice in the air, as opposed to theorizing; the notion that balance depended on control of the center of pressure; the decision to borrow and modify Chanute's double-decker truss design, and to place the operator in a prone position to reduce resistance to the wind; the decision against a tail and in favor of a horizontal rudder plus wing-warping to control the dancing center of pressure. He asserted the advantage of approaching the problem gradually, with gliders, before attaching an engine, rather than starting with a powered craft, which had to be right from the start to be right at all, just as Langley had recognized. "The problems of land and water travel were solved in the 19th century because it was possible to begin with small achievements and gradually work up to our present success," he said. "The flying problem was left over to the 20th century, because in this case the art must be highly developed before any flight of any considerable duration at all can be obtained." The alternative was the gliding or soaring machine; whatever principle held the soaring bird aloft, it would work equally well with a glider, if a man could just catch the trick of the thing.

He explained the measurements of lift and drift that he and his brother had made on an actual glider in flight—the first ever made by any experimenter—and the "astonishing" implications: Lilienthal had been wrong. All the existing data about lift and drift had to be recalculated. And he and his brother meant to do the work.

Engineers who built rail trestles and sewers might not have followed the speaker through every obscure detail about centers of pressure and coefficients of lift, but they could see the pictures on the screen—a huge, cloth-and-wood contraption like a box kite, unmistakably in flight and carrying a man, *this* man. Plainly the fellow had done what he said ought to be done, and with remarkable success. He was an American Lilienthal—Chanute had said so—yet he, unlike the German, had not yet died.

In so many words, Will dismissed the work of Lilienthal and of Hiram Maxim. The world-renowned inventor of the machine gun had constructed wings and engines powerful enough to get an eight-thousand-pound aeroplane off the ground in 1893, Will noted. Yet Maxim had given up immediately upon realizing that he had no way of controlling its flight. So much, too, for Samuel Pierpont Langley's ten-year pursuit of an aeronautical engine. And so much for Professor Simon Newcomb, whose measure Will apparently took during his long days of work on the paper in Dayton, and the other masters of formulas and theory who proved the impossibility of flight at their desks, with pencils.

> There are two ways of learning how to ride a fractious horse. One is to get on him and learn by actual practice how each motion and trick may be best met; the other is to sit on a fence and watch the beast a while, and then retire to the house and at leisure figure out the best way of overcoming his jumps and kicks. The latter system is the safest; but the former, on the whole, turns out the larger proportion of good riders. It is very much the same in learning to ride a flying machine; if you are looking for perfect safety, you will do well to sit on a fence and watch the birds; but if you really wish to learn, you must mount a machine and become acquainted with its tricks by actual trial.

WILL'S RELATIONSHIP with Octave Chanute—until now mostly a scientific exchange through occasional letters—was deepening into personal friendship. That fall Will began to write to the older man with increasing frequency, delighted to have found a friend so eager for each new detail of the brothers' experiments. Many of his letters stretched to several thousand words. When Will apologized for writing at such length, Chanute replied with warmth:

"I am amused with your apology for writing long letters, as I find them always too brief."

Will also felt a certain degree of pressure from Chanute, and a degree of frustration with him. It nettled him that Chanute could not yet bring himself to accept what now seemed obvious to the Wrights—that Otto Lilienthal's hallowed calculations of lift and drift were in error. "His faith in Lilienthal's tables is beginning to waver," Will wrote George Spratt, "though it dies hard."

Yet Will could not help but feel warmed by Chanute's rising esteem. At the meeting in Chicago, among distinguished figures of his profession, the older man had treated his much younger guest as a respected peer. In their private conversations he had been very much the eager mentor, urging Will to revise his presentation ("a devilish good paper," Chanute called it) for publication in the society's professional journal—a badge of prestige that would certify the brothers as serious investigators. Chanute urged Will to throw off whatever doubts had snared him at Kitty Hawk, and practically begged the brothers to accept his financial aid.

Bemused, Will shrugged off the offer. This was, after all, only a hobby. The Wrights might not have wives and children to support, but they could hardly abandon their livelihood for the sake of an intellectual diversion, however fascinating. Indeed, the brothers, feeling ready at last to expand their bicycle business, were preparing for several years of intensified labor in that line. "If we did not feel that the time spent in [aeronautical] work was a dead loss in a financial sense," Will said, "we would be unable to resist the temptation to devote more time than our business will stand."

In spite of Will's pleasant reception in Chicago, the brothers were not at all sure what to do next. Will had described the obstacles with admirable clarity, but the obstacles remained. Their glider was "a fractious horse" indeed, and a dangerous one, and they had little idea how to tame it.

Only one task needed to be done right away: Will had to put his talk to the engineers in shape for publication. Here Orville voiced a caution. Will had come down awfully hard on established authorities, especially Lilienthal. Now these rebukes would be committed to print. And who was Wilbur Wright? An unknown, without a degree or even much experience. Could they really be so sure of their own beliefs about puzzles they only had begun to probe? If they were going to challenge the world-famous Lilienthal—and by implication John Smeaton, whose estimates of air pressure Lilienthal had used—they ought at least to try a few experiments to confirm their hunches.

Will saw the wisdom of it. So, as the streets of West Dayton turned golden-brown with the coming of autumn in 1902, the neighbors were treated to a minor diversion—the younger Wright boys, not so young any more, taking turns on one of their black bicycles, pedaling swiftly up and down the streets and stealing frequent glances at a contraption in front at the level of the handlebars—a bicycle wheel mounted horizontally and free to spin. The brothers had affixed two metal plates on the rim, ninety degrees apart. On the left, in the nine-o'clock position, was a small, flat, metal plate set perpendicular to the air flow. In front, at the twelve-o'clock position, was a slightly larger plate, curved in an arc to match Lilienthal's one-in-twelve curvature. When the bicycle was in motion, the air flowing over the curved plate would generate lift and turn the wheel clockwise. The air pressure against the flat plate would push the wheel counter-clockwise. The angle of incidence of the "wing" was carefully set so that it would generate just enough lift to offset the air pressure on the flat plate. If Lilienthal was right, the wheel would remain stationary. If he was wrong, the wheel would rotate. And if he was wrong in the way the Wrights suspected—that is, if his values for lift were too high—then the curved plate would generate too little lift and the wheel would rotate counter-clockwise.

With the rig in place, one or the other of the Wrights mounted the bike and started pedaling. As he glanced up and down, one eye out for horses, carts, and bicycles, the other on the experiment, the horizontal wheel turned counter-clockwise in the rushing current of air.

More attempts brought the same results. The brothers increased the angle of incidence of the curved plate until it generated enough lift to balance against the flat plate. The angle was nowhere near what Lilienthal would have predicted.

In these slight movements of a bicycle wheel, the Wrights read the auguries of a revolution. They had suspected something was wrong in Lilienthal's tables. The wheel proved it.

Still, it was a pretty crude experiment. It had to be checked. With gathering excitement, they tore the ends off a discarded wooden starch box and fitted a fan to one of the open ends. This made a little wind tunnel of the sort they had encountered in their readings. They constructed a little wind vane that, like their bicycle rig, balanced the air pressure on a flat plate against the lift generated by wing surfaces. They placed it inside the box and switched on the fan. The wobbling vane confirmed the bicycle tests.

Working fast, they checked and rechecked results, comparing their figures to all they knew about Lilienthal's glides. Only three weeks after his speech in Chicago, Will wrote Chanute: "I am now absolutely certain that Lilienthal's table is very seriously in error, but that the error is not so great as I had previously estimated." The fog was thinning. He was discerning shapes in the mist. "The results obtained, with the rough apparatus used, were so interesting in their nature, and gave evidence of such possibility of exactness . . . that we decided to construct an apparatus specially . . . for [testing] surfaces of different curvatures and different relative lengths and breadths."

They had come upon an extraordinary tool. New possibilities beckoned. A bigger and better wind tunnel would be Kitty Hawk in a box, with a perfect wind and no risk to life and limb. Instead of trudging up a giant sand hill, waiting for the wind to arrive at the proper direction and speed, guessing at angles and vectors, and laboriously tinkering with wings as long as a living room, they could insert a six-inch piece of metal in the wind tunnel, throw a switch, write a number on a page, and move on. Every test was another dot in the sketch they now could create—a sketch of a wing that really would fly.

"We have been experimenting . . . with an apparatus for measuring the pressure of air on variously curved surfaces at different angles," Will informed his father on October 24, "and have decided to prepare a table which we are certain will be much more accurate than that of Lilienthal."

WITH THE BISHOP AWAY on a long trip, Kate found the house on Hawthorn Street nearly as quiet in the fall as it had been in August, when her brothers were away. Every morning she caught the Third Avenue streetcar for Steele High, where she taught in the same room as the year before, the one overlooking the river, though "the pupils . . . are not so nice as they were last year." After school, she went for bicycle rides. She saw to her father's mail and business affairs and supervised Carrie Kayler in the housework. In the evening, she read. The only break in the routine came when her nieces and nephews dropped in, though the children were going through "not so pretty" phases that fall. Still, they touched her. "Ivonette and Milton were invited to a birthday party last evening and Uncle Orv, happening to go by the house, found baby Sister in tears because she couldn't go, too. She would have been a sweet little guest. Uncle Orv consoled her for a few moments by producing some candy and having 'a party' on his lap."

Her brothers walked in at suppertime, ate and argued, then walked out again, heading back to the shop. The house had been through a little flurry when Will went to Chicago, then dull peace returned. Now, "The boys are working every night on their 'scientific' (?) investigations," she told her father. "Poor Will says he'll have to eat crow for, after all, he has discovered by accurate tests that Lillienthal's tables are not far off." The weather was fine. A relative of the next-door neighbors had died. A letter of Kate's had been discovered at the post office without a stamp. A friend of Bishop Wright's spent the night. Apart from these events, she said, "I can't think of any news."

THE WALK FROM 7 Hawthorn to the bicycle shop took five minutes—one block west to South Williams, a block north to West Third, half a block farther west and across the street to 1127 West Third, between Frank Hale's grocery and Fetters & Shank, "Undertakers & Embalmers; Coffins, Caskets & Robes of any Style or Quality Furnished at Reasonable Prices."

Above the door at 1127 was a sign for "Wright Cycle Company." A sign above that, at the peak, said "C. H. Webbert." That was the building's owner, the Wrights' landlord and friend, Charlie Taylor's uncle-in-law, a prosperous plumber. Inside the door was a clean-swept but sparsely furnished room with a gas stove and a cabinet full of bicycle accessories and supplies—tires, wheel rims, spokes, toe and trouser clips, foot pumps, 20th Century and Aladdin bicycle lamps; leather bicycle saddles; and jars of grease. The odors of rubber tires, gluepots, and oil mingled in the air. Through a door at the back of the room, visitors glimpsed bicycles and machinery in a warren of storage and work rooms. In those rooms, rough tables were littered with scraps of wood and sheet metal and abandoned pieces of flying toys contrived for the nieces and nephews out of paper, bamboo, and cork. Spruce shavings crackled underfoot. Racks of tools for wood-working and metal-working hung on the walls. Shelves were lined with jars of cotter pins, nuts, and nails. Overhead, long leather belts ran along the ceiling, connecting the single-cylinder natural-gas shop engine to the big power tools: a twenty-inch Barnes drill press; a fourteen-inch Putnam lathe; a bench grinder.

There was a workroom upstairs, too, where the brothers did much of their experimenting while Charlie Taylor worked on bicycles downstairs. The brothers had installed two bells to minimize interruptions. The first rang when the front door opened, the second when a customer removed the air pump from the wall—the most frequent reason for customer visits in a day of delicate tube tires. If the first bell rang, followed quickly by the second, the brothers could stay at their work; the customer only "wanted air." If only one bell rang, then one or the other had to go downstairs and attend to business.

But November was the heart of cycling's off-season. No more than four or five customers happened in each day, so even during business hours, the Wrights could accomplish a good deal of aeronautical work. As the days grew shorter and the wind turned cold, Wilbur and Orville went at their investigation with greater intensity than ever. With the starch-box wind tunnel, they had intended only to check their figures to support Will's claim. But in doing so they realized they had created the means by which they could discover everything Lilienthal had died trying to find out. "We had taken up aeronautics merely as a sport," Orville said. "We reluctantly entered upon the scientific side of it. But we soon found the work so fascinating that we were drawn into it deeper and deeper."

If Lilienthal's tables could not be trusted, then they must write tables of their own. And they would test not just one wing shape, as the German had, but shapes of all sorts. As Orville put it, there was "a multitude of variations [for] the pressures on squares are different from those on rectangles, circles, triangles, or ellipses; arched surfaces differ from planes, and vary among themselves according to the depth of curvature; true arcs differ from parabolas, and the latter differ among themselves; thick surfaces differ from thin, and surfaces thicker in one place than another vary in pressure when the positions of maximum thickness are different; some surfaces are most efficient at one angle, others at other angles. The shape of the edge also makes a difference, so that thousands of combinations are possible in so simple a thing as a wing."

They constructed a wind tunnel substantially larger than their improvisation with the starch box. The new device was a wooden crate six feet long and sixteen inches square. It stood on four legs at waist height, with a fan and steel hood mounted at one end. To hold their model wing surfaces, they fashioned bicycle spokes and worn hacksaw blades into two delicate structures they called balances. They were simple but powerful mechanical computers—tiny analogs of the forces at play on the wings. With trigonometry, the Wrights could use them to analyze the performance of any wing shape they could imagine.

It appears that Will did most of the close work of fashioning the model wings. They varied in length and width, but most had exactly the same wing area, six square inches, so their performance could be easily compared. Nearly all were snipped from a sheet of steel one thirty-second of an inch thick—easier to shape than wood—and pounded into the desired testing shape. Will could cut a shape and hammer the curve he wanted in fifteen minutes. Then he would build up the leading edge to the thickness he wanted with solder and wax.

Orville, meanwhile, perfected the arrangements for the wind tunnel itself. The fan was mounted on the axle from the brothers' old grinder. This was connected to the shop engine, which would drive the fan at some 4,000 revolutions per minute to create an experimental wind of 20 to 25 miles per hour.

They constructed the balances together. One they called the lift balance, the other the drift balance. Each was equipped with a dial and needle to register results. Experimental wing shapes were mounted not horizontally, like a bird's wing, but vertically, so the "lift" could be measured from side to side. When they turned on the fan, wind rushed through a metal honeycomb—to keep the airflow straight—and struck the wing. This caused the entire balance to swing gently to one side, like the action of an opening gate. The dial recorded how far the gate swung. Correcting for drift and gravity, the brothers could judge the performance of each wing shape at a particular angle in a minute or two. On the lift balance, for instance, they would affix the model wing to the balance; adjust it to the desired angle; turn on the fan; adjust the balance to correct for drift; look through the window on top of the tunnel to read the number indicated by the needle; shut off the fan; record the result; then change the wing's angle and test again. When the full range of angles had been tried, a new wing would be attached and the cycle of lift tests would start over. Later, they tested the same wing shapes on the drift balance. Both balances measured the forces on the wings *directly*, eliminating the need to rely on Smeaton's coefficient of air pressure. They found they could run dozens of tests in a single day. In four weeks they tried more than a hundred surfaces, running each through its paces at angles from zero, past thirty degrees, all the way to ninety degrees, gleaning more information in a couple of days than in two summers at Kitty Hawk. Through the window on top of the tunnel, they watched the needle respond to the wind blowing on surfaces that were flat and sharply curved, square-tipped and round-tipped. They tried tandem wing arrangements, like Langley's; double-deckers like their own gliders; even triple-deckers. They tried surfaces of various aspect ratios—the ratio of the wing's length to its width, or "span" to "chord"—from perfect squares to long, narrow rectangles. The tests revealed astonishing variety and unpredictability in the way surfaces responded to the flow of moving air. Their progress delighted Will's anxiously waiting correspondent in Chicago. "It is perfectly marvelous to me how quickly you get results with your testing machine . . . ," Chanute said. "You are evidently better equipped to test the endless variety of curved surfaces than anybody has ever been."

The elderly engineer added a caution that showed he still had not taken the full measure of his young friends: "I hope that you will . . . tabulate the results as you go along, for the temptation to omit this is always strong." This must have provoked smiles in Dayton. No one in the world would be less tempted than Wilbur and Orville Wright to omit the recording of such critical data.

A few days before Thanksgiving, the brothers began a series of exacting tests upon the forty-eight surfaces that had emerged as the most promising from the first round of experiments. Each surface was run through a battery of trials with the lift balance, at fourteen angles from zero to forty-five degrees; then the drift balance was substituted, and again the surface was put through its paces.

Mysteries evaporated. They learned that Otto Lilienthal's table had been correct insofar as it went, for the single wing type he tested with his whirling arm. It was not Lilienthal who committed the error that made their gliders fly so contrary to expectation. That distinction now belonged to John Smeaton. Much to the relief of Chanute, who had published Lilienthal's results far and wide, Will informed him that "the Lilienthal table has risen very much in my estimation since we began our present series of experiments for determining lift." Will admitted obliquely to his own mistake: "It will not do at all to attempt to apply Lilienthal's table indiscriminately to surfaces of different aspect or curvature; but for a surface as near as possible like that described in his book the table is probably as near as correct as it is possible to make it with the methods he used."

Comparing Lilienthal's data with Langley's the brothers concluded that Langley's key findings about lift had been "little better than guess-work," Orv said later. Lilienthal had gotten some things wrong, Will told Chanute, but the German's *Birdflight as the Basis of Aviation* towered above Langley's *Experiments in Aerodynamics*. Lilienthal's errors, Will said, "are so small compared with his truth, that his book must be considered an extraordinary one to be the work of a single man. "The 'monumental work' of a great American," he added in a reference that offered a glimpse of his private opinion of the Smithsonian secretary, "will not bear a moment's comparison with it."

Column by column, row by row, they filled in their tables. Almost as an afterthought, it seemed, the brothers found their attention drawn to one particular shape. In their series of forty-eight, it was model wing number 12. It was a narrow rectangle with a six-inch span and a one-inch chord—the distance from the leading edge to the trailing edge. Its curvature was one in sixteen. In a postscript to Chanute, Will said "#12 has the highest dynamic efficiency of all the surfaces shown."

With that, the brothers pushed the wind tunnel into a corner. It was the middle of December 1901—time to celebrate Christmas at 7 Hawthorn, then to begin making bicycles for sale in the spring.

At this display of Wrightian self-discipline Chanute practically suffered an impatience-induced stroke. "You have done a great work," he declared, "and have advanced knowledge greatly. Your charts carry conviction to my mind . . . I very much regret in the interest of science that you have reached a stopping place, for further experimenting on your part promises important results." Granted, the Wrights could not yet expect a financial return for their labors. "If however some rich man should give you $10,000 a year to go on, to connect his name with progress, would you do so? I happen to know Carnegie. Would you like for me to write to him?"

To Will, this dumbfounding offer was perhaps a bittersweet taste of the life he might have chosen had he been able to choose again. "Of course nothing would give me greater pleasure than to devote my entire time to scientific investigations," he replied two days before Christmas 1901, "and a salary of ten or twenty thousand a year would be no insuperable objection, but I think it possible that Andrew is too hardheaded a Scotchman to become interested in such a visionary pursuit as flying.

> But to discuss the matter more seriously, I will say that several times in the years that are past I have had thoughts of a scientific career, but the lack of a suitable opening, and the knowledge that I had no special preparation in any particular line, kept me from entertaining the idea very seriously. I do not think it would be wise for me to accept help in carrying our present investigations further, unless it was with the intention of cutting loose from business entirely and taking up a different line of lifework. There are limits to the neglect that business will endure, and a little pay for the time spent in neglecting it would only increase the neglect, without bringing in enough to offset the damage resulting from a wrecked business. So, while I would give serious consideration to a chance to enter upon a new line of work, I would not think it wise to make outside work too pronounced a feature of a business life. Pay for such outside work would tend to increase the danger. The kindness of your offers to assist, however, is very much appreciated by us.

IN THE MIDDLE OF WINTER, the brothers sketched a glider design for the coming summer of 1902. Every dimension was new. None was based on the guesswork of the past. The aspect ratio of the wings (the ratio of span to chord, or length to width) was the same as wing model No. 12—six to one. The wing's curvature would be much shallower than the 1901 design—one to twenty-four. This, they were now convinced, was the true curve of flight, promising both powerful lift and a well-behaved center of pressure. They smoothed the stubby rectangle of the rudder into a leaf shape. They revamped the wing-warping system, switching from the difficult, foot-activated assembly of 1901 to a hip cradle. This was far better suited to the pilot's instinctive tendency to shift his weight toward the high wing when the glider tipped. Now, when the wings tipped to one side or the other, the pilot would automatically warp the wings in the proper way as he instinctively shifted his weight. They made the elevator control more instinctive, too. To make the earlier gliders pitch downward, the operator had to pull *up* on the controls. A new mechanism allowed him to roll his wrists in the desired direction.

The year before, in the waning days of the summer of 1901, the glider's frightening sideways skids had pushed the brothers to the verge of quitting. Now an idea arose. The skids occurred—not always, only sometimes—when the operator warped the wings to attempt a turn. The very notion of wing-warping was to increase the lift on one side of the glider while decreasing lift on the other side, thus inducing a banking roll toward the side with less lift. Perhaps, they reasoned, the glider's nose swung, or "yawed," to one side in the midst of the banking maneuver because the upward-twisted tip offered too much resistance to the wind. The wing on that side would rise but simultaneously slow down, and the glider would slither out of control.

The problem was exceedingly difficult to hold in the mind and examine. A jumble of invisible forces were at play.

But a simple solution occurred to the brothers. It was a fixed vertical tail. Whenever the glider began to yaw into a skid, the tail would offer counterbalancing resistance to the wind, thus evening out the pressure on both sides of the glider and allowing the warped wing to bank into the desired turn. In 1900 and 1901 the brothers had scorned the idea of a tail, believing that wing-warping alone constituted a full answer to the challenge of turning. Now, for the first time since Will's 1899 kite, they designed a tail, a two-surface affair with a combined area of nearly twelve square feet, attached to the trailing edge of the wings.

"The matter of lateral stability and steering is one of exceeding complexity," Will told Chanute in February, "but I now have hopes that we have a solution."

*Chapter Six*

# "A Thousand Glides"

"A FEAT OF CONTROL NO OTHER EXPERIMENTER HAD EVEN TRIED"
Wilbur banking to the right in the 1902 glider at the Kill Devil Hills

SHORTLY AFTER WILL'S ADDRESS appeared in the *Journal of the Western Society of Engineers*, a brief summary appeared in the widely circulated *Scientific American*. In a show of local pride, the *Dayton Daily News* published the address as well, under the headline "Modern Flying Machine a Dayton Product."

"It is conceded," the paper advised readers, "that aerial navigation sooner or later will be the modern mode of transportation. Like the [auto]mobile it will be slow in developing but will, it is predicted, be the coming method of rapid transit." But when the editors reprinted a photograph showing the glider in motion at Kitty Hawk, with Will at the controls, they placed it so that the machine appeared to be flying straight up, like a rocket. Probably no one in town recognized the error except the Wrights themselves.

These were the first public notices of the brothers' project. But they ignited little interest outside the tiny circle of serious experimenters. When people talked about "aerial locomotion," they were talking about massive, steerable gas-bags, and the glamorous figure of the predicted age of air travel was the twenty-eight-year-old Brazilian whom Langley had met in 1900, and who had flown again and again over the rooftops of Paris.

ALBERTO SANTOS-DUMONT was the youngest son of one of Brazil's richest coffee growers. As a child he had played with kites, propeller toys, and silk-paper balloons, and on the family's heavily mechanized plantation he had developed a knack for machinery. He read Jules Verne's fictional stories of airships and true accounts of the French brothers Montgolfier, who had invented the hot-air balloon in the era of the French Revolution. "In the long, sun-bathed Brazilian afternoons . . . I would lie in the shade of the

"THE GLAMOROUS FIGURE OF THE PREDICTED AGE OF AIR TRAVEL"

Alberto Santos-Dumont, ca. 1901, with the motor of his Airship No. I verandah and gaze into the fair sky of Brazil, where the birds fly so high and soar with such ease on their great outstretched wings, where the clouds mount so gaily in the pure light of day, and you have only to raise your eyes to fall in love with space and freedom. So, musing on the exploration of the vast aerial ocean, I, too, devised air-ships and flying machines in my imagination."

As a boy, he and his friends played a catch-you-off-your-guard game like "Simon Says." It was called "Pigeon Flies." They would sit around a table. The leader of the round would say, "Pigeon flies!" at which the others would raise a finger to signify "yes." The leader would continue: "Crow flies!" "Bee flies!" "Hawk flies!"—then try to cross the others up by saying, "Dog flies!" or "Pig flies!" If you raised a finger at this, you lost the round and paid a forfeit. Whenever a leader inserted, "Man flies!" Santos-Dumont recalled, "I would always

"MAN FLIES!"

Santos-Dumont's Airship No. V circling the Eiffel Tower

lift my finger very high, as a sign of absolute conviction, and I refused with energy to pay the forfeit."

In 1891, his father, partially paralyzed by a neurological injury and in search of medical care, took the family to Europe. Alberto, then eighteen, fell in love with Paris. He soon returned for good with a not-so-small fortune bequeathed by his father. Free to do whatever he wished, he engaged a tutor to instruct him in the sciences and became an accomplished aerial balloonist. Like all balloonists, he flew at the mercy of the winds, unable to steer or control his speed. He was pursuing a secret goal—to invent and fly a lighter-than-air craft propelled by an engine and capable of being steered.

He built his first ship in 1898 and crashed it twice in the Bois de Boulogne. More study followed, and more construction. The next year he made many short flights over Paris in dirigible—that is, steerable—gas-filled airships, and Paris society anointed him as a favorite and a fashion-plate, *"le petit Santos,"* an exotic Latin bird to be petted and wondered at.

When the industrialist Henri Deutsch de la Meurthe offered one hundred thousand francs to the first man to fly an airship from the Aéro-Club de France to the Eiffel Tower and back in under half an hour, a journey of seven miles, Santos-Dumont made a spectacular series of attempts. The first ended in a chestnut tree on the estate of the Rothschild family, the second on the roof of the Trocadero Hotel. On October 19, 1901, Santos made his third attempt. He reached the tower in only nine minutes, but lost precious time when his engine broke down. Crawling outside his little basket and along the keel without a safety harness, he fixed the engine while thousands of Parisians watched from far below. He missed the prize by forty seconds. When the Aéro-Club officials denied him the award, Parisians nearly rebelled. The club relented. Santos gave most of the money to the poor, and became one of the most famous men in the world.

Among the thousands of letters he received was one from a childhood friend in Brazil: " 'Man flies!' old fellow! You were right to raise your finger, and . . . you were right not to pay the forfeit; it is M. Deutsch who has paid it in your stead. Bravo! You well deserve the 100,000 franc prize. They play the old game now more than ever at home, but the name has been changed and the rules modified—since October 19, 1901. They call it now 'Man flies!' and he who does not raise his finger at the word pays his forfeit."

In the following spring of 1902, Santos-Dumont was an honored guest in Washington, where the Brazilian ambassador presented him at the White House for an overpowering presidential handshake. Theodore Roosevelt, an accomplished horseman and well-known adventurer, said he would be delighted to go aloft in Santos-Dumont's machine. "And I believe he would," the Brazilian told newsmen afterward. "He received me very pleasantly. He is very clever."

On the other side of Smithsonian Park, Santos-Dumont spent an afternoon with Samuel Langley, and afterward offered remarks on the secretary's heavier-than-air machines. "They are beautiful. One is being made—large enough to carry people—from a model which he made some time ago. He expects to be able to give it a trial in about three months. He is your foremost man in this kind of work. There is no question of that."

NEARING HIS SEVENTIETH YEAR, Bishop Milton Wright looked forward to a peaceful retirement among his children, his grandchildren, and his books. He could take satisfaction in a life well spent in the service of Christ, from the victory over slavery down to his defiant leadership of the Radical Brethren in the battle against compromise with freemasonry. He was securely fixed—the sale of some valuable farm property had supplemented his salary—and cherished by his family. But he remained a warrior in the army of the Lord. A quarter-century earlier he had written: "The salvation of one's self and of his fellows demands a struggle. It is a choice between heroic warfare, with assurance of solid fruits in victory, on one hand, and ease, desolation, and death, on the other. Hence the warfare must come; and, though it cost ever so much sacrifice of ease, friendship, and earth's possessions, it must go on." In the years since then, nothing had changed his mind. The gray-bearded bishop remained on watch for the enemy, and so did his children, who had inherited the warrior ethic.

The enemy now raised his head in the person of Millard Fillmore Keiter, a sharp-faced Brethren minister who had assumed Milton's duties as publishing agent in 1893. When Milton heard Keiter was doctoring church accounts, he assigned Wilbur and Lorin Wright, who was trained as a bookkeeper, to examine the records. "There is something rotten somewhere," Will reported. But no official audit was taken until after Keiter, still under suspicion, slipped out of his post in 1901. The audit showed an unexplained deficit of some $1,480—more than enough to provoke Milton's wrath, especially when reports of more defalcations accumulated. The bishop expected the Church to mete out swift justice. But when its Publishing Board convened in February 1902 to review the matter, a four-bishop majority outvoted Wright and his allies. Despite obvious evidence that Keiter had done wrong, the prevailing bishops wanted the matter buried without ceremony and Keiter given a free pass. With the ruckus of the 1889 schism still ringing in their ears, they wanted no more scandal, especially any that might dampen collections.

In the Wrights' eyes, the lone snake had spawned a viper's brood, and within the Church itself. The entire family was stunned by what Will called the "inconceivable, incomprehensible and incredible" actions that allowed Keiter to skate. "Some members must have felt that since they could not join the Masons they would make the Church a similar institution so far as 'defending a brother right or wrong' is concerned." To Milton it was the battle of slavery revisited, with cowardly churchmen keeping their mouths shut in order to keep their parishioners' purses open.

So the "Old Constitution" wing of the Brethren plunged back into internal warfare. When Bishop Wright was excluded from the pages of the Church newspaper, he struck back with his own petitions and pamphlets and reported Millard Keiter to the civil authorities. Keiter was arrested for forgery. This was "going to law" against a Christian brother, a violation of Brethren discipline. When Keiter was acquitted on technicalities, he saw a chance to silence his aged nemesis, and filed his own disciplinary charges against Milton for "agitating controversy." A decision on Wright's conduct was postponed, but a war of editorials and pamphlets raged all through the spring and summer of 1902. The secular press, supplied with information from the Keiter forces, spread the embarrassing story of a United Brethren bishop under fire for actions "contrary to the teaching of the church." Milton, losing friends and respect, lost weight and sleep as well, his bowels roiling from nervous tension. "Feel loss of sleep and cares considerably," he confided to his diary.

In public the old man conceded nothing. He knew how his stand appeared. Even a close ally in the fight against "the Keiterites" said the elder Wright could seem "imperious, harsh, and unsympathetic." A fellow bishop would later excoriate Wright as the victim of his own ego, which "esteems itself as being immaculate, without spot or blemish, stainless, without taint of evil or sin, pure and perfect." But it was not self-regard that drove Milton's behavior so much as the habits of a lifetime. By inheritance, training, and long experience, he was built to call out wrongdoers, then to smite them.

The bishop's children, built in the same fashion, were his main support. Kate shored up her father's morale with encouragement—"Will and Orv seem to think that it is turning out just right"—and shared indignation at the "Keiterites." "Not one of that crowd knows what common honor, honesty or gratitude means." Orville helped with such chores as typing and printing. But it was Will who stepped in as his father's chief counselor—indeed, as the behind-the-scenes leader of the bishop's whole campaign. He determined political tactics, advised Milton on legal maneuvers, studied Church law, and prepared his father's defense against the disciplinary charges.

Although Will told Milton that "my chief regret is that the strain and worry which you have borne for fifteen years past shows no sign of being removed," he apparently gave no thought to a compromise that would spare his father the ordeal. He framed the argument that Milton would advance in the coming battle. "The question of whether officials shall rob the church, and trustees deceive the church for fear of injuring collections, must be settled now for all time . . . to cheat the people by lying reports is more dishonest than Keiter's stealing, and so far as church interests are concerned, the penalty will be greater."

As the spring of 1902 turned to summer and the busiest days of the bicycle-selling season passed, Will juggled two essential matters—the Keiter affair and the construction of a new glider. To prepare for Milton's disciplinary trial, he went repeatedly to Huntington, Indiana, where the trial would be held at the Church's new college. At home, he stole time for the glider from "attending to some matters for my father which have occupied much of my time and attention recently." For a time he was unsure whether Orville and he could get away at all that summer. "If we go to Kitty Hawk it will probably be some time between Aug. 15th and Sept. 15th," he told Chanute. "It is a pity that the hills near Chicago are not smooth bare slopes."

Chanute offered to help with the tedious computing needed to complete the Wrights' lift-and-drift tables. He came to Dayton in early July and went away with instructions on how to do the tabulations. Early in August Will took four days for another trip to Huntington, after which he scribbled a searing tract against "Keiterites" and their "condoners."

"When my father and myself came to examine the charges carefully, we at once saw that the whole thing was a mere sham. There never was any real intention of bringing the case to trial. . . . The real purpose was to harass the accused."

He finished the essay on August 15, submitting it for publication as a tract by the Church, then returned full-time to the glider. Careful planning was essential, because the brothers were putting together what woodworkers call a kit—a collection of parts that are constructed in one place for assembly in another. All the parts they needed had to be planned correctly in advance, and none forgotten. Once they reached Kitty Hawk, it would be too late to buy or order anything left behind.

They could not make the curved wingtips and ribs themselves. This was work for specialists at a local firm that made parts for the carriage industry. They had the equipment needed for steaming strips of ash, then bending the pliant wood to the required curvature. The wingtips probably came off the rack in the form of four standard carriage bows. The ribs had to be specially ordered. The Wrights would have handed over sketches with precise dimensions, all based on the wind tunnel data.

They planned to reuse the uprights from their old glider, but everything else had to be new. Most parts they could make themselves—spars for the leading and trailing edges of the wings; outriggers and cross-pieces to form the forward rudder and the tail; belly skids for under the wings. For these pieces they bought spruce from one of the local lumber yards. For its weight, it was strong and resilient, good for withstanding repeated jolts. They ordered the spruce cut up into pieces of roughly the right length and shape. Then they went at the pieces with draw knives and spokeshaves, rounding the corners to preserve the wood's essential strength while reducing weight and wind resistance. When this was done, the pieces were ready to be drilled and notched to make holes for screws and mortises for joining. Then the brothers brushed all the wood parts with several coats of varnish, to protect against the moisture-laden North Carolina air, which could destroy the glider as thoroughly as a catastrophic crash if it penetrated the precisely bent wood.

Now the wooden skeleton of the wings could be assembled. In place of screws or nuts and bolts, the brothers used waxed linen cord, a wonderful, all-purpose twine which conveniently stuck to itself and the wood, making it easy to tie tight lashes and knots. In a jolting landing, the lashed joints gave a little, then snapped back, minimizing the possibility of broken joints.

Next came the skin, made from yard upon yard of Pride of the West white muslin. This was the trickiest part of the entire job, and it depended entirely on the sewing skills that Susan Wright had taught her sons. Kate watched, aghast, as her brothers pushed furniture out of the way and filled the first floor of the house with ribs and spars and endless yards of linen. "Will spins the sewing machine around by the hour while Orv squats around marking the places to sew. There is no place in the house to live."

They scissored the fabric into strips, then machine-sewed the pieces back together so the threads would run "on the bias," at a forty-five-degree angle to the ribs. Thus each thread acted as a tiny cross-brace, helping to hold the wing together under the pressures of flight and landing. With painstaking measuring, stretching, and sewing, they created a long, snug pocket for each rib, to keep the fabric anchored and to preserve the precise curvature of the wing when it was subjected to the forces of lift. Then, inch by inch, starting at the trailing edge of the wings, they slipped the tight-fitting cloth skin over the wooden skeleton. The tips of the wings were covered separately, and required an artist's touch at folding, tucking, and stitching.

Bishop Wright's trial was scheduled for the last week of August, just when the brothers meant to leave for North Carolina. As they hurried from task to task, applying finishing touches and packing supplies, they stewed over their father's crisis and worried about his health. Will hesitated to schedule a departure date. He wanted to be nearby if Milton needed him again in Huntington.

"They really ought to get away for a while," Kate told the bishop. "Will is thin and nervous and so is Orv. They will be all right when they get down in the sand where the salt breezes blow etc. They insist that, if you aren't well enough to stay out on your trip you must come down with them. They think that life at Kitty-Hawk cures all ills you know."

Finally, perhaps with Milton's urging, the brothers decided to go. It may have been the first time in his life that Wilbur put his own concerns ahead of his family's. Kate shooed them onto the train on Monday, August 25. She regretted the loss of Will's counsel, yet she was genuinely worried about the summer's effects on his well-being. "I am sorry that Will is not here," she told Milton, "but it was the best thing in the world for him to go away. He was completely unnerved."

"The boys" were barely gone when she began to brood about their safety. She had only her own imagination to form a picture of their annual doings on the distant Atlantic dunes, and "I always feel a little uneasy about that trip." To the experimenters themselves, she offered a brave face. "Dad seems worried over your flying business this year. The habit of worry is strong on him. I am not much alarmed . . . but don't run any risks. We've been worried enough for one year."

IN THE SOUTH SHED BEHIND the Smithsonian Castle, Charles Manly labored day after day and often long into the night to solve a conundrum. From the beginning, the engine had come first. The entire enterprise was built on the notion that "the success or failure of the first flight will depend to a large extent on the power available." By 1902 Manly had all but completed the construction of a superb internal combustion engine, surely one of the finest in the world. Also essentially complete was the most promising machine ever built to accomplish the goal of manned, powered flight. The airframe, Manly thought, "gave an appearance of grace and strength to which photographs failed to do justice," with wings of more than a thousand square feet.

The problem was to bring the magnificent engine and the magnificent airframe into harmony. "If one is building bridges . . . ," Manly said, "it would be criminal negligence to fail to provide a sufficient 'factor of safety' "—that is, extra strength and extra power to ensure success. But he and his chief saw no room for a "factor of safety" in a flying machine. Greater power required increased engine weight. A heavier engine required reinforcements in the airframe, which in turn added weight, which in turn required more horsepower, which in turn required improvements in the engine, and so on.

To destroy the shibboleths of Isaac Newton and his friend Simon Newcomb, to build his "swift Camilla" to "skim along the main," Langley needed "an engine of unprecedented lightness." Thus, "weight is the enemy." But weakness was an enemy, too, and lighter meant weaker. Any reduction in the engine's weight made it more fragile and prone to breakdown, and less capable of bearing the weight of the whole. Any reduction in the airframe's weight made it more vulnerable to the stresses of wind, lift, and engine vibration. Writing about his unmanned models of the 1890s, Langley had said: "Everything in the work has got to be so light as to be on the edge of breaking down and disaster, and when the breakdown comes all we can do is to find what is the weakest part and make that part stronger; and in this way work went on, week by week and month by month, constantly altering the form of construction so as to strengthen the weakest parts." It was just the same with the great aerodrome, but with a more complex and delicate engine, and infinitely more at stake.

Time after time, delays in the South Shed arose from the struggle to reconcile the quarrelsome trio of weight, strength, and power.

Manly: *"To determine more accurately what mode of construction would give the greatest stiffness and strength for a minimum weight, it was decided to make up some test pieces of different forms before making up complete ribs."*

Langley: "Mr. *Manly adds that the entire vibration could be practically cured by counterweighting the pistons at a cost in weight of not over 40 lbs., presumably less."*

Manly: *"Although this wing was a great improvement in every way over any of the previous constructions, it was felt that it was too weak for the large aerodrome."*

Langley: *"Although an engine may develop sufficient power for the allotted weight, yet it is not at all certain that it will be suitable for use on a machine which is necessarily as light as one for traversing the air."*

Manly: *"Great difficulties . . . were experienced in these tests of the engine in the aerodrome frame before the shafts, bearings, propellers, and, in fact, the frame itself were all properly co-ordinated so that confidence could be felt that all the parts would stand the strains which were likely to come on them when the aerodrome was in flight."*

Langley: *"These engines were to be of nearly double the power first estimated and of little more weight, but this increased power and the strain caused by it demanded a renewal of the frame . . . and the following sixteen months were spent in such a reconstruction simultaneously with the work on the engines."*

Of course, the troublesome calculus of weight, strength, and power was not the only source of delay. Defective parts broke. Workmen quit. Orders were late. Estimates were mistaken.

As "the work lengthened out in all directions far above the estimate of the time required," the pressure on Langley and Manly mounted. The BOF's fifty thousand dollars had been exhausted. The additional ten thousand dollars from Alexander Graham Bell and Jerome Kidder was spent. Langley was forced to draw from yet another Smithsonian reserve fund, and it was dwindling as quickly as his credibility with the War Department. Worse, the secretary had become the target of a whispering campaign among his colleagues in the scientific fraternity.

All of this—the engine, the delays, the money, the politics—put a premium on getting the aerodrome into the air at the earliest possible moment. There was simply no time to develop the ideas about balance and control that Langley had conceived while watching Jamaica's "John Crow"—the idea for altering the angles of the wings. He and Manly discussed the idea in "many conferences," Manly admitted later. But Manly never followed through on Langley's "instructions and suggestions." The reason he gave later was "the extreme pressure already on [me] which had for its object, not the production of a flying machine which would embody all of the control which we wished it to have, but which would be burdened only with such devices and arrangements as would enable it to transport a human being, and thus demonstrate the practicability of human flight." They resigned themselves to relying on the cruciform Pénaud tail in the rear, which the operator could manipulate with a wheel.

To solve the puzzles of balance, steering, and landing, they would need more time, and thus more money. So Langley and Manly focused all their work and all their hope on the goal of one "successful flight of a few miles." Once that was achieved, Manly said, they considered it "very certain that . . . the funds for the further prosecution of the work would be readily forthcoming, and that when these funds were obtained the many problems of control, rising and alighting, could be undertaken."

THE SECRETARY wanted to leave on his annual trip to Europe by the middle of June. Manly said it would be difficult to make a trial flight by then, and Langley agreed. "Considering the experience of ten years . . . nothing of this sort is ever ready when it is expected to be, and owing to the need of this journey for my health, I shall not delay it." He told Manly he would be gone at least until September, and to "expect . . . to be ready on my return. . . . In any case I shall wish to be present at the first trials."

With the engine now making forty horsepower, Manly thought he could make Langley's deadline. But the aerodrome's frame had been built to withstand the strain of only twenty-five horsepower. He worried that the greater power would shake the machine to pieces once the engine was installed in the frame. After "first one part and then another of the frame breaking," Manly got the vibration down to an acceptable minimum. Then the left propeller flew off, stripping a gear. "All of the work is progressing in a satisfactory manner," he wrote Langley, "except for continual set-backs due to various causes to which the work, as ever, has been particularly subject."

Langley read this news at the spa at Karlsbad, in Bohemia. He tried to reassure his long-suffering protégé. "The injury to the propeller and bevel gear is only in the normal course of accidents. Perhaps we ought to be glad rather than sorry that they have happened in the shop instead of a thousand feet above it! But as the time for flight draws very near I confess my anxieties, for the incalculable play of what is all but chance, increase. I admit that I feel almost a shrinking from the time of trial which I suppose is now so near."

AS THE MAN WHO was actually going to take the ride, Charles Manly may have worried more about the problem of balance than Langley did. The image of the skater skimming safely over thin ice at high speed still remained firmly fixed in Langley's mind. In fact, Langley was considering a scheme by which flying machines might do entirely without the imponderable complexities of wings and rudders and tails.

He had often thought about the clean trajectory of a stone hurled through the air, or of an arrow. Stones and arrows fall, of course, when they lose the momentum imparted by the thrower's arm or the archer's bow. But what if that momentum could be renewed indefinitely? Wouldn't they continue on their straight path? He had mentioned the idea to Manly one Saturday early that year in the upstairs shop of the South Shed, and made a note about it the next day: "The ultimate development of the flying machine is likely to be an affair of very small wings or no wings at all . . . ," he said. "It may depend for its velocity on what Mr. Bell calls 'its momentum' in the same way that an arrow or any other missile flies. It is known that the arrow derives its energy from the bow which projects it and that when this is spent the arrow will drop. We have, however, only to renew this energy and . . . the arrow may still be heading upward without limit. . . . It is a thing which deserves thinking over."

He was still thinking it over several months later. The idea was "so paradoxical," the secretary admitted, "that I hesitate to enunciate it even as a mere possibility. The very idea of the aerodrome as we have always conceived it, has been to obtain support from sustaining surfaces driven against the air. I seem to see my way to dispensing with the surfaces absolutely and altogether so long as the engine works. I do not mean that this is a hypothetical possibility, but something apparently practical and perhaps within our actual means, or very near it."

Langley was imagining an engine like the ones that would propel rockets out of the earth's atmosphere fifty years later. Perhaps he and Manly discussed it further, but he did not write about it again. It was an ingenious and visionary notion. But it would not help the great aerodrome fly.

OCTAVE CHANUTE'S WIFE died in April 1902. He himself was seventy, and slowing down. "I do not expect to experiment further," he confessed to Langley. Yet he clung to the hope that by sponsoring and coordinating the efforts of younger men, he might still become, if not the father of flight, at least its grandfather. And he still believed the solution lay in a wing that would adjust automatically to the vagaries of the wind, ensuring equilibrium without depending on the fallible instincts of the operator.

Chanute proposed to give two machines to the Wrights for testing at Kitty Hawk in 1902—one designed by Charles Lamson, a manufacturer and flight hobbyist who had experimented with manned kites, another designed by Chanute himself, with help from the experimenter Augustus Herring. The brothers were desperate to make the most of their precious time. They wanted nothing less than to be burdened with extra work. Their memory of the lazy, preachifying Edward Huffaker remained strong from the summer before, when "my brother and myself . . . could do more work in one week, than in two weeks after Mr. Huffaker's arrival." Augustus Herring sounded worse. They had heard that he was prickly, unpleasant, and "of a somewhat jealous disposition," always seizing more credit than he was due and picking fights with collaborators. Chanute chose not to tell Will that Herring had come to Chicago that summer with a plea to "let him rebuild gliding machines, 'to beat Mr. Wright.' " Had Will known that, his polite parries would have turned to frosty rejection.

As it was, the stolid Chanute either missed all Will's hints or simply chose to ignore them. He was determined to have a role in—or at least a look at—the Wrights' work at Kitty Hawk. The best Will could do was encourage him to delay his arrival at Kitty Hawk for several weeks, leaving the brothers time to work undistracted with their enormous new machine.

A YEAR IN THE WIND had so buckled the Wrights' wooden shed at Kitty Hawk that the roof sloped sharply at either end and the interior "strongly resembles the horror of an earthquake in its actual progress." Though both Will and Orv caught colds on the train, they went about the work of rebuilding the camp undismayed. The sun was shining and mosquitoes were scarce. They bolstered the shed's sagging floors and built a sixteen-by-sixteen-foot addition. With a device of their own invention they drilled the best well in Kitty Hawk, finding "splendid water" seventeen feet down. They upholstered their wooden chairs with excelsior and burlap and laid two layers of burlap and another of oilcloth over their dining table to make a soft top. More burlap made two comfortable bunks in the rafters.

In Elizabeth City they had picked up an oven and a barrel of gasoline. Orville had brought a rifle to shoot the small water fowl that skittered among the breakers, so they had occasional fresh meat, though the birds were so small that "after a bullet has gone through one of them there is just a little meat left around the edges." To ease the long round trip between the village and the camp, they had brought a bicycle, in parts, and fussed with the gears so they could ride it over sand. Their shelves were soon stocked with precise rows of canned goods.

"We fitted up our living arrangements much more comfortably than last year," Will told George Spratt. "There are . . . improvements too numerous to mention, and no Huffaker and no mosquitoes, so we are having a splendid time." He invited Spratt to join them.

In Kitty Hawk the Wrights were now treated as familiar and welcome guests. The locals had seen enough of them to respect them and finally to like them, though they had not been the easiest men to get to know. "They didn't put themselves out to get acquainted with anybody," said John Daniels, one of the regulars at the lifesaving station. "Just stuck to themselves, and we had to get acquainted with them. They were two of the workingest boys I ever saw, and when they worked *they worked.* I never saw men so wrapped up in their work in my life. They had their whole heart and soul in what they were doing, and when they were working we could come around and stand right over them and they wouldn't pay any more attention to us than if we weren't there at all. After their day's work was over they were different; then they were the nicest fellows you ever saw and treated us fine." The brothers shared good food cooked well and asked questions about the local economy, the land, the weather, and the families of the village. They were good with the children. That surely scored points, as did their "uniform courtesy to everyone."

The flying proposition remained dubious among the villagers. Two years earlier, they had regarded the Wrights as "a pair of crazy fools," Daniels said. "We laughed about 'em among ourselves." When the early gliders stayed in the air, they gained a shred of credibility. Yet their behavior, for all their hard work and obvious good character, remained decidedly odd. Sometimes the lifesavers would look over from the station and see the Ohioans standing near the beach, faces upturned, watching intently as gulls soared and banked overhead, even spreading their arms and twisting their wrists in imitation of the birds. An hour later the lifesavers would look again and there the brothers would be, still watching the birds.

IN FACT, the brothers spent less time watching the gulls than they did watching the eagles, ospreys, hawks, and buzzards that flew some distance inland from the crashing breakers, above the dunes where the brothers flew themselves. The gulls' gymnastics in the turbulent air of the shoreline were fascinating. But the big carrion-eaters and birds of prey spent more time in the aeronautical state of soaring—that is, hanging more or less motionless in the air, neither falling nor rising, their wings held steady. A soaring bird, as Will put it, is "gliding downwards through a rising current of air which has a rate of ascent equal to the bird's relative rate of descent." The soaring bird enjoys a perfect balance among the forces of lift, drift, and gravity. It was what the brothers aspired to.

On cold, cloudy days, the Wrights observed that the birds could soar only along the line where the forest met the dunes, taking advantage of the wind's upward turn as it collided with the trees. On clear days, the birds found what Will deduced to be columns of air rising from the sun-baked sand. They would flap their wings to reach a certain height, then "rise on motionless wings" in wide spirals. One day the brothers noticed a tiny flash of light far below a spiraling eagle; it was a falling feather reflecting the sun. They watched as the feather reversed directions and rapidly rose out of sight. "It apparently was drawn into the same rising current in which the eagles were soaring, and was carried up like the birds."

Will's favorites were the buzzards, which soared more often than the others. One day, atop the summit of the West Hill, he watched a buzzard at eye level only seventy-five feet away. It hung all but motionless over the steep slope. Will believed his own artificial wings were—or could be—as good as this bird's. He was less sure he could develop the buzzard's skill. "There is no question in my mind that men can build wings having as little or less relative resistance than that of the best soaring birds. The bird's wings are undoubtedly very well designed indeed, but it is not any extraordinary efficiency that strikes with astonishment but rather the marvelous skill with which they are used. . . . The soaring problem is apparently not so much one of better wings as of better operators." To develop that skill remained the brothers' chief desire, and they could attain it only with the sort of prolonged practice that long, safe glides could afford them.

If the wind tunnel tests had told the truth, and if the brothers had interpreted the tests correctly, then their new machine would carry them on long, flat glides, with the operator able to keep his balance in sudden gusts, and to turn left and right. Their ultimate aim remained the same as in 1900. If their sustaining surfaces were of the proper size and shape, and the wind just right and the operator canny, they might approach the feat of soaring. That is, the pilot might coast on the wind for moments at a time with only the most gradual loss in altitude. He could practice and eventually master the art of equilibrium that had eluded Otto Lilienthal. Indeed, he might conceivably hover above a single spot for a minute, two minutes, perhaps many more minutes, as the invisible forces of lift, drift, and gravity held the machine in perfect equilibrium. "When once it becomes possible to undertake continuous soaring advancement should be rapid."

IN THE SHED AT KITTY HAWK, the brothers took apart the old glider from the summer of 1901 to make space for the new machine. They opened their crates and began the assembly. Over eleven days, the machine took shape.

It was an extraordinary work of art, science, and craft. It was created to serve a beautiful function, so the form, following the function, took on its own odd and ungainly beauty. The leading corners of the wings were quarter-circles, the trailing corners shaped like scoops. In cross-section, the wings humped in front and trailed away in a graceful curve to the rear. The linen skin was taut under the hand, the wires tight. With 305 square feet of wing space, it was much bigger than Lilienthal's gliders and Chanute's double-decker, which it resembled only in the biplane configuration and the trusslike connections between the wings. Viewed directly from in front or from the side, there was hardly anything to see but a spare collection of lines—horizontal, vertical, diagonal, and curved. Only when viewed from above or below did the craft seem to turn solid and substantial, owing to the wings, thirty-two feet from tip to tip and five feet from front to back. Yet the glider weighed only 112 pounds. One man could stand at a wingtip, grasp a wooden upright with one hand, and easily lift his end several feet off the ground. Three men could pick the glider up and carry it around with little trouble. If they dropped it, it shuddered, but there was no damage. "It was built to withstand hard usage," Will said, and though it looked thin and spare, almost weak, it felt sturdy. When they faced it into a steady breeze, it no longer seemed ungainly. Suddenly they were no longer holding it up but holding it down.

On Wednesday, September 10, the brothers tested the completed upper wing as a kite. Two days later they tested the lower wing. Attaching a handheld scale to the ropes, they found that these curved surfaces, flown by themselves, exerted much less pull on the lines than their 1901 machine had. This meant the wind was guiding the wing into a flatter angle of attack, which promised flatter, longer glides.

Their first gliders, especially the one built in 1900, had flown as any child's kite flies, with the line at a slanting angle of about forty-five degrees. The closer a kite's line ascends to the vertical, the greater its efficiency. A kite whose cord runs on a vertical line down to the operator is, in effect, soaring. That is, its ability to generate lift counteracts its tendency to drift backward in the wind. It is aerodynamically perfect. If it could move forward under its own power, it would be flying.

Next, the brothers assembled the entire glider and carried it to a slope they measured at about seven degrees. In a steady wind, they let out their lines. The glider rose. The lines stood nearly straight up and stayed there.

IN DAYTON, Charlie Taylor was running the bicycle shop. The brothers had asked Kate to look in on him from time to time and offer help when he needed it. In their haste to get away from Dayton, the brothers had left their business affairs in a mess. This enflamed the strain that already existed between the taciturn mechanic and the opinionated schoolteacher. "It's a wonder I have any mind left," Kate told them.

> The business is about to go up the spout to hear Charles Taylor talk. Say—he makes me too weary for words. He is your judge, it seems. Everything that happens he remarks that it struck him that you left too much for the last minute. Today I got wrathy and told him that I was tired of hearing him discuss your business. But really, I don't see how we can keep things going. You didn't leave any checks or anything . . . I wish you would send me a check for $25. I don't enjoy going to the store

after money. Mr. Taylor knows too much to suit me. I ought to learn something about the store business. I *despise* to be at the mercy of the hired man.

As ever, the brothers were unmoved by the trouble they had caused "your lovin' swesterchen." "You will have to get used to some of Charles' peculiarities," Orville replied dryly. "They don't bother us."

More on their minds was the turn their father's ordeal had taken. From Kate and other correspondents, they learned that the elders of the Brethren's White River conference in eastern Indiana—one of Bishop Wright's own jurisdictions—had tried Milton in absentia and declared him guilty of libel against Millard Keiter, "insubordination to constituted authority," and "going to law" against a fellow Christian. They offered Milton sixty days to confess his errors or face expulsion from the Church.

At Kitty Hawk, the brothers' reactions differed sharply. On the Sunday following their receipt of the news from White River, after midday dinner with Bill Tate, both brothers wrote letters. Orville's went to Kate. In eight hundred words, he recounted the details of the boat trip to Kitty Hawk, the refitting of the camp, the laying in of supplies, and his own "merry chase," armed with a gun, after a mouse that was making nightly raids on their kitchen. About the bishop's problems Orv said nothing, though he was well aware that his sister was following every step in the drama and was deeply worried. Will wrote to Milton, who was traveling in the Midwest, visiting his conferences and performing his duties as bishop in defiance of his colleagues. Will analyzed the Keiterites' perfidies, offered moral support, and issued no fewer than five specific and detailed instructions on how Milton should respond. There could be no plainer illustration of the contrast between the two—Orville playing ingeniously at the trivial hunt for a mouse; Wilbur assuming responsibility for leadership in the family's crisis.

Kate wrote to Milton every few days. "We are not going to let this thing go—not by a long chalk. Just wait till the boys are back again, with Will feeling strong again!"

"We will bear in mind your caution to be careful," Will reassured the bishop. "We have no intention of being disabled while that gang of rascals is still attempting to injure you, and rob the church."

ON THE MORNING OF FRIDAY, September 19, Will made the first twenty-five test glides of the season, with Orv and Dan Tate running alongside with a hand on the wingtips. That day and the next, Will found that slight adjustments in the angle of the new front elevator offered him "abundant control" of the glider's fore-and-aft movements. And the tail seemed to prevent the perplexing problem of the previous summer, in which the glider occasionally pivoted around the tip of the higher wing.

But the new control device was tricky. To turn up, the operator had to push the elevator-control bar down—the reverse of the 1901 controls. With this movement not yet instinctive, Will found himself aloft in a "somewhat brisk" wind. Suddenly a cross-gust caught the left wingtip and pushed it skyward "in a decidedly alarming manner." Eager to return to earth, Will, in confusion, turned the elevator up instead of down and found the glider suddenly angling upward "as though bent on a mad attempt to pierce the heavens." He recovered from the stall and landed without damage. But he continued to have problems keeping the balky wingtips level in crosswinds.

For a long, rainy Sunday the brothers stewed and debated, "at a loss to know what the cause might be." What new forces had they summoned by lengthening the wings and adding a tail? The next day, they retrussed the wings so that the tips dipped slightly below the level of the center section. With this slight arch, the glider took on the droop-winged look of gulls, which fly well in high winds. Kite tests vindicated their intuition. Now crosswinds, if anything, seemed to improve their lateral balance. "The machine flew beautifully," Orv wrote that evening, "and at times, when the proper angle of incidence was attained, seemed to soar."

Even as a kite, this new machine inspired confidence. It simply looked more stable in the air than its bobbing and dipping ancestors of the previous summers. Perhaps this is why Orville now decided to take an equal part in piloting the craft.

He began the morning after the wings were retrussed, practicing assisted glides to get the feel of the controls. After their midday meal, he tried free flights, taking turns with Will, who stretched his glides to beyond two hundred feet. The tips were so responsive that in one flight he "caused the machine to sway from side to side, sidling one way and then the other a half dozen times in the distance of the glide." Orville managed one respectable flight of 160 feet at an admirably low angle of descent. Then, while concentrating on a tip that had risen too high, he lost track of the elevator controls and rushed upward to a height of twenty-five or thirty feet. Will and Dan Tate cried out. Orv stalled, then slid backward through the air and struck the ground wing first with a crackle of splintering spruce and ash. "The result was a heap of flying machine, cloth, and sticks in a heap, with me in the center without a bruise or a scratch." Only the seat of his pants suffered injury. This "slight catastrophe" meant days of repairs. But that evening the brothers were so pleased with the glider that "we are . . . in a hilarious mood." Orville told Kate: "The control will be almost perfect, we think, when we once learn to properly operate the rudders."

The control was not perfect. The winds of the Outer Banks blew in turbulent swirls altogether different from the pristine currents of the wind tunnel, and on the dunes there was no lift balance to hold the glider's wings safe and steady. In the next few days, the repaired machine made many more glides under good control. Will even managed to fly the glider in the path of an S, turning perpendicular to the wind and then back into it. But every so often, "without any apparent reason," one wingtip would rise and fail to respond to the pilot's insistent warping of the opposite wing. Tilting heavily to one side, the machine would go into a sickening slide sideways in the direction of the tilt, "just as a sledge slides downhill," Will said, "or a ball rolls down an inclined plane, the speed increasing in an accelerated ratio." One side of the glider rose and gathered speed, the other side dipped low and slowed, and the whole craft spun into a frightening, out-of-control circle. The low wingtip would crunch into the soft sand and continue to spin, gouging out a curved trough. The motion reminded the Wrights of the device they had used to sink their well, so they called it "well-digging." The problem was dangerous and bewildering, and they could not claim control of the glider until they had solved it.

To the brothers' delight, Lorin Wright walked into camp on the last day of September, and, equally welcome, George Spratt arrived the next afternoon. The barren expanse of sand increasingly took on the look of a sportman's camp. Spratt and Lorin snagged crabs for bait and caught an eel and some chubs. The three brothers competed in target shooting with Orville's rifle. The results were measured to the eighth of an inch. "W.W. beat me with a record of 5¾ inches total from center to my 5⅞," Orville recorded. "L.W. 7¾." To the rhythm of the nearby surf, they talked over the evening fire, the thoughtful farmer-physician lending his own assessments of the Wrights' glides and the well-digging puzzle. Spratt led the brothers on tramps through the nearby forest, pointing out birds and insects and plants by name.

Will climbed to his bunk early, often by 7:30. Orville stayed up later. He was working on his French and making headway in several books he had brought along. On the night of October 2, he drank more coffee than usual and lay awake for a long time. The glider's curious geometry floated through his mind—angles formed by the arrows of the wind and the flat cloth surfaces of the flying machine—and a perception dawned. In the well-digging episodes, he saw that as the glider went into its sideways slide, the fixed vertical tail in the rear not only failed to keep the glider straight, it also collided with stationary air, and thus pushed the machine into its dangerous spin.

Orville glimpsed a solution—to make the tail movable. If the pilot entering a turn could alter the tail's angle, then pressure would be relieved on the lower side of the glider and exerted on the higher side. The machine would turn under control, and neither slide sideways nor spin.

In the morning, Orville presented his idea. He expected his older brother to say he had thought of it already. Instead, after Orville spoke, Will was quiet for a moment. Then he said he saw the point—yes, the tail should be movable—and the movement should be brought about by the same hip-cradle action that operated the wing-warping. By shifting his hips, the pilot would twist the wings *and* alter the tail's angle at the same time. Suddenly it was clear to both of them. The two movements were intimately connected and ought to be performed simultaneously. Wing and tail and wind would act in concert.

Will's moment with the twisted box had been their first great epiphany. Orville's idea for a movable vertical rudder, and Wilbur's addendum, was the second, and it was striking that they conceived it together, as if their two minds really were halves of one whole.

They pulled off the two-surface rigid tail and replaced it with a single surface, five feet high and two feet across, that swung to either side on hinges. Wires connected it to the wing-warping cables. Now, when the man in the cradle swung his hips from side to side, the wings and tail shifted in a single, organic movement.

AUGUSTUS HERRING had wanted to fly since the early stories about Otto Lilienthal had appeared ten years earlier. Born in 1865 to the family of a well-to-do Georgia cotton broker, Herring attended the Stevens Institute of Technology, in Hoboken, New Jersey, then became the first American to build and fly a Lilienthal glider. He sought the centers of action in American flight research but never held a job for long. He worked for Samuel Langley and fell out with him. He worked for Chanute and fell out with him, too. Working alone, he made impressive glides in a machine of his own design. In 1898, Herring attached an engine powered by thirty seconds' worth of compressed air and claimed a motorized hop of fifty feet. That apparently taxed the machine's limits. Unable to proceed, Herring gave up and did other work for a time. But before long he succumbed again to his passion. In the summer of 1902, he went first to Hiram Maxim, then to Chanute. In Chicago he confessed he was out of work and offered to revise Chanute's old gliders in what he said would be an effort to overtake Wilbur Wright.

Herring's self-confidence was as powerful as the Wrights'. But without funds to support his own designs, he had no choice, if he wanted to fly, but to work for men of means and somehow make their work his own. It did not suit him. No matter how his superiors handled him, he grumbled and rebelled under direction. In their work in the late 1890s, Chanute told a friend, "Langley tried the cast iron way, and I tried the india rubber way, and we both failed. The trouble is that [Herring] tries sulkily those experiments which do not originate with himself, and is, as he admits himself, very obstinate."

On a dark, rain-soaked Sunday in early October, just as the Wrights were building their new tail, Herring and Chanute lumbered by horse cart into the camp at the Kill Devil Hills. They brought with them the parts for a three-wing glider, built by a Californian named Charles Lamson at Chanute's request, that combined ideas of both men. Mostly it was another effort to achieve Chanute's dream of wings that would shift by themselves to accommodate changes in wind pressure without reliance on the operator's instincts and dexterity. Herring assembled the machine the next day and took it out to the big hill. But on only his second try, the glider struggled through the air for no more than twenty feet, then fell, smashing its main cross-piece. The Wrights helped with repairs, but when Herring tried again he could do no more than hop into the air. Even when kited in a powerful wind blowing up a steep slope, the machine failed to lift Herring. He had no idea what was wrong with it. The Wrights blamed its flimsy wings. Orv was sure he had seen the surfaces flap and twist so violently that the wind must be striking the top on one side of the craft and the bottom on the other. "Mr. Chanute seems much disappointed."

Herring and Chanute, rooted to the ground, now watched as Wilbur Wright, in high winds, kicked off from the summit of a dune and sailed far down the slope with "no trouble in the control of the machine." Not only did he achieve a distance of nearly three hundred feet, but with the new apparatus for making turns, he banked the glider to both right and left in a long S shape. The two observers realized that Wright was gliding with the wind striking the machine not only from the front but from both sides, a feat of control that no other experimenter had ever even tried. With nothing of their own left to do, he and Chanute decamped nine days after they had come. In Norfolk they boarded a train for Washington.

Chanute wanted to see Langley, who had returned to the United States from Europe only a few days earlier. He called at the Castle without an appointment, only to find the secretary rushing off in a hansom cab. Apparently unable to resist a hint about the aerodrome, Langley paused only long enough to say, "You may hear some interesting news in a short time."

Chanute tracked down Charles Manly for a talk, then followed up quickly with photographs and letters to Langley. He made it clear that he suspected Herring of intellectual theft. Herring had "doubtless . . . got some new and valuable ideas by seeing Mr. Wright's machine," he said, "and he announced that he wanted to return to Washington. I suspected (although he did not say so) that this was to revive his former unsuccessful application for a patent, with perhaps modifications which had occurred to him."

Chanute said the Wright brothers had made remarkable glides by placing the operator prone in their machine and attaching a horizontal rudder at the front. He did not mention wing-warping, nor the movement of the fixed tail, which he seems not to have understood or even perceived. But he was keen to make sure that Langley understood the progress the Wrights were making.

> They are bicycle manufacturers and repairers who find their business so slack at this time of the year that they indulge themselves in a vacation which is fruitful in results. They have heretofore considered their experiments as a sport, likely to bring no money return, and have not intended to apply a motor, but they have acquired so much of the art of the birds that I think they should proceed further. They are very ingenious mechanics and men of high character and integrity.

Langley got the point. He shot a telegram to Kitty Hawk, asking if he might run down to the Outer Banks and see the experiments for himself. There was no immediate response by wire, but in a couple of days he received a letter from Wilbur Wright. He had not responded by telegram, Wright explained, because he and his brother planned to break camp within several days, and "as it would be exceedingly doubtful, owing to miserable transportation facilities, whether you could reach here before that time, and as there would be the additional risk of bad weather for the single day that you might possibly be with us, we did not think it advisable to telegraph you to come on. We shall if possible arrange to stop a day in Washington on our way home."

But the Wrights did not find it possible.

THE SKIES CLEARED and the wind blew steady and strong. Spratt had to go on October 20, leaving the brothers alone with only Dan Tate to help. Will and

Orv now looked to see what this glider could do. In five days they made hundreds of glides, stretching their distances to three hundred, four hundred, five hundred feet, and in buffeting winds up to thirty miles per hour. Tate looked on in bemused wonder: "All she needs is a coat of feathers to make her light and she will stay in the air indefinitely." On October 23 Will traveled 622 feet in a glide lasting nearly half a minute. Orville bubbled with excitement and pride. "We now hold all the records!" he wrote Kate on the night of October 23. "The largest machine ever handled . . . the longest time in the air, the smallest angle of descent, and the highest wind!!!"

SOME ONE HUNDRED YEARS LATER, Wright enthusiasts have spent many hours replicating the brothers' 1902 glides in precise recreations of their glider. The sensations of the pilot are as close as possible to Will and Orv's experiences. Leaving the ground is a coordinated rush of three men running and hauling. Then the operator feels one foot miss the ground. He knows the machine weighs more than a hundred pounds. Yet it defies gravity and rises up against his beltline, bearing his full weight and still rising.

*. . . one . . . two . . . three seconds . . .*

A beginning bicyclist often loses control by turning the handlebars too far to right or left. It is like that with the horizontal rudder in front of the glider. Too much of an angle makes the glider dip and bob.

*. . . seven . . . eight . . . nine . . .*

The wind feels like waves under a skiff, bucking the craft a bit, but yielding to it.

*. . . fourteen . . . fifteen . . . sixteen . . .*

A shift of the hip cradle brings a faint creaking and clacking as the wires pull at the wings and tail, and instantly the machine leans and banks like a bicycle.

*. . . nineteen . . . twenty . . . twenty-one . . .*

After a gentle descent the belly skids touch and slide, tossing skitters of sand. The flight is over, but the thrill is unforgettable. The pilot collects his dazzled thoughts and stands for another try.

The brothers' diaries show that Will made more glides than Orville, and that his glides were on average longer than Orville's. But this reflects only the 121 flights that the brothers noted out of some one thousand glides they made that fall. Orville said later: "Unfortunately, unlike many who use other people's money in their experiments, and therefore must keep records to make a showing of their work, we were working for ourselves, and just for the fun of it, so that some experiments were never recorded and the records of many others not as carefully preserved as they should have been." Usually they did not take turns, glide by glide. Rather, one brother would make several glides, then the other.

Only a tiny handful of humans had known the sensation—Lilienthal, Percy Pilcher, Chanute's assistants—but none for so many seconds or with such confidence of the ability to remain in balance and to land safely. The Wrights remembered their promises to Milton and Kate. So only occasionally would one or the other nose the craft above the height of a man. The thrill came less in flying high than in a dawning sense of oneness with the machine. With more practice, Will hoped "the management of a flying machine should become as instinctive as the balancing movements a man unconsciously employs with every step in walking."

Their long glides had grown out their particular aptitude for learning how to do a difficult thing. It was a simple method but rare. They broke a job into its parts and proceeded one part at a time. They practiced each small task until they mastered it, then moved on. It didn't sound like much, but it avoided discouragement and led to success. And it kept them uninjured and alive. The best example was their habit of staying very close to the ground in their glides, sometimes just inches off the sand. "While the high flights were more spectacular, the low ones were fully as valuable for training purposes," Will said. "Skill comes by the constant repetition of familiar feats rather than by a few overbold attempts at feats for which the performer is yet poorly prepared." They were conservative daredevils, cautious prophets. "A thousand glides is equivalent to about four hours of steady practice," Will said, "far too little to give anyone a complete mastery of the art of flying."

IF HE DID NOT PAY as much attention to matters of balance and control as the Wright brothers, Charles Manly at least never lost sight of the fact that he was building an engine that must perform in the unfamiliar environment of the sky. Late in the summer of 1902, he began to worry especially about the effect on the carburetor of a sudden acceleration through the air. To imitate the conditions of a launch, he turned a powerful fan on the engine during tests in the South Shed. Secretary Langley approved. With "goodwill and sympathy in your disappointments," he wrote from Europe: "I think you are very prudent to imitate the actual conditions under which the carburetor will work when the aerodrome is in flight, in its first few critical seconds." Langley, still in Europe, planned to arrive in Boston in the first week of October. He fully expected Manly to be ready for a trial over the Potomac that month, while the weather was still comfortable.

Then came a "totally unexpected" series of smash-ups to the propellers. Manly thought he had put them in workable condition a full year earlier. Now he faced a delay of four to six weeks, which "completely upset all of my calculations as to when I would be ready to start the houseboat with everything on board down the River, and until I overcome this trouble with the propellers, I can give no idea of when the aerodrome will be ready for trial." Then the propeller shafts, which had stood up to all the tests to date without a single problem, "twisted and buckled under the strain of driving the new propellers." Weeks passed.

In November Langley wrote again to Captain Lewis at the War Department. Lewis was no longer assigned to the Board of Ordnance and Fortification, but he and Langley had developed a friendly relationship, and Langley was keeping him informed. "I tell you privately that the engine for which we owe so much to Mr. Manly is finally showing itself perhaps the best anywhere. . . . Everything is ready, and were there still time this season, a flight would be made, but I do not expect one until after the ice has gone; then I, who have never known any great initiatory trial without some mischance, look for what I call our 'first smash.' When every human care has been taken, there remains the element of the unknown. There may well be something that forethought has not provided for, and I expect it, but excepting for the fact that a human life is now in question, I could make it without fear at all."

Langley's expression of confidence to Lewis was, indeed, private. The board's official reports were not, to the delight of observers in the press. "Experiments with the Langley aerodrome as a military flying machine reveal the possibility of the machine as a practical weight carrier," the *New York Daily Tribune* deadpanned that fall, "as it has successfully flown away with $50,000 of the allotment made by the Board of Ordnance and Fortifications. . . . The disproportion between the $214,000 remaining in the hands of the board and the $50,000 the flying machine has already cost leads to the conclusion that aerodromes come as well as go high."

STILL CURIOUS ABOUT THE Wright brothers, the secretary sent Manly to Chicago for a talk with Chanute. In a private memorandum, Langley recorded the crux of his aide's report—that "Mr. Chanute feels that the Wright Brothers have made a real advance even over his own ideas and experiments." Details of the Ohioans' machine were scrambled as they passed from Chanute, who did not understand them in the first place, through Manly to Langley. The secretary recorded that the Wrights controlled their balance by cords that allowed them to change the curve of the horizontal rudder at the front of the machine. He noted Chanute's belief that "though not automatic [this is] better than the Pénaud tail."

Langley's curiosity swelled. Clearly they had ideas about equilibrium that he should know about. Through Chanute, he again invited either brother to travel to Washington at his expense to speak with him, "especially of their means of control."

Chanute passed along the request, to which Wilbur Wright responded: "It is not at all probable that either Orville or myself will find opportunity to visit Prof. Langley in response to his suggestion. We have a number of matters demanding our attention just now. . . .

"It is our intention next year to build a machine much larger and about twice as heavy as our present machine. With it we will work out problems relating to starting and handling heavy weight machines, and if we find it under satisfactory control in flight, we will proceed to mount a motor."

LANGLEY AND MANLY had spent most of four years building an extraordinary engine to lift their heavy flying machine. The Wrights had spent most of four years building a flying machine so artfully designed that it could be propelled into the air by a fairly ordinary internal combustion engine. With the design of a powered aeroplane* now in its final form, the Wrights expended a minimum of thought and energy on their power plant. At first they hoped simply to buy an engine. But when they sent inquiries to manufacturers, specifying an engine of less than two hundred pounds that would make at least eight horsepower, only one said he had such an engine, and the brothers concluded he was overrating its power. So they sketched a design of their own and handed it to Charlie Taylor, who did most of the work in the back room of the bicycle shop while the brothers worked upstairs. He used a drill press, a metal lathe, and a few hand tools. The materials came from a Dayton foundry and the local hardware stores. At the end of six weeks, he had a simplified four-cylinder auto engine without a carburetor, spark plugs, or fuel pump. The spark came from dry batteries and a low-tension magneto manufactured by the Dayton Electric Company. The radiator was made from several pieces of metal speaking tube used in apartment buildings. Gasoline flowed into the engine by gravity from a small tank mounted on the wing strut overhead. The fuel valve was a petcock made for a gas lamp. In February 1903, the engine block cracked in a shop test. When a new block was delivered and the engine reassembled, it

* "Aeroplane" was the standard American term during the first decade of powered flight, so I have retained it throughout this narrative.

made twelve horsepower at 1,025 revolutions per minute. With four more horsepower than the brothers believed they needed, and twenty fewer pounds than their maximum, the engine was "a very pleasant surprise." If Stephen Balzer could have seen Charlie Taylor achieve twelve horsepower at 180 pounds in just a few weeks, it would have been a bitter sight.

THE BROTHERS HAD ASSUMED that propellers would cause them even less trouble than an engine, believing, as Langley had said ten years earlier, that "there is a very considerable analogy between the best form of aerial and of marine propellers." But they soon learned that ships' propellers were designed by trial and error, vessel by vessel. No one knew exactly how propellers worked, so no one had worked out a theory of propeller design, least of all for flying machines. The brothers had no reason to think either Langley or Hiram Maxim had solved the problem, since neither—so far, at least—had produced a machine capable of flying with a man aboard. With all the possible variations of the propeller blades' diameter, angle, and area, design by trial and error might take forever. So the brothers had no choice but to plumb the mystery of how propellers actually worked. If they could do that, they could make calculations, and the calculations would lead to a reliable design. "We have recently done a little experimenting with screws," Will told George Spratt, "and are trying to get a clear understanding of just how they work and why. It is a very perplexing problem indeed." They began to consider the problem seriously soon after their return from Kitty Hawk in 1902, and "it was not till several months had passed," Orville recalled, "and every phase of the problem had been thrashed over and over, that the various reactions began to untangle themselves."

Naval engineers had proposed that a marine propeller cuts through water as a screw cuts through wood. The brothers conceived a different image. To them, "it was apparent that a propeller was simply an aeroplane [that is, a plane surface in the curved shape of a wing, or airfoil] traveling in a spiral course. As we could calculate the effect of an aeroplane traveling in a straight course, why should we not be able to calculate the effect of one traveling in a spiral course?" Put that way, the problem sounded simple. But it "became more complex the longer we studied it. With the machine moving forward, the air flying backward, the propellers turning sidewise, and nothing standing still, it seemed impossible to find a starting-point from which to trace the various simultaneous reactions."

In the shop and at home, "they would get into terrific arguments," Charlie Taylor said later. "They'd shout at each other something terrible. I don't think they really got mad, but they sure got awfully hot." One evening at the shop Charlie heard the hottest of their arguments yet. The next morning, Orville arrived at the shop first and "said he guessed he'd been wrong and they ought to do it Will's way." Then Will came in and "said he'd been thinking it over and perhaps Orv was right." The argument reignited, each brother raising his voice on behalf of the position the other had taken the night before. "When they were through though, they knew where they were and could go ahead with the job."

They filled five pocket notebooks with formulas and calculations. By the spring, they thought they had propellers that would deliver the thrust they needed. In a letter to Spratt, Orville recorded jubilation tempered by the knowledge that only an actual flight could vindicate their judgment.

"We worked out a theory of our own on the subject, and soon discovered, as we usually do, that all the propellers built heretofore are *all wrong*, and then built a pair of propellers 8⅛ ft. in diameter, based on our theory, which are *all right!* (till we have a chance to test them down at Kitty Hawk and find out differently). Isn't it astonishing that all these secrets have been preserved for so many years just so that we could discover them!!"

ON THE DAY BEFORE CHRISTMAS 1902 a well-dressed stranger appeared at 7 Hawthorn in Dayton. His manner was engaging, his accent English. He handed the brothers a letter from Octave Chanute, which read: "Permit me to introduce Mr. Patrick Alexander, of England, one of the leaders in aeronautical investigations, who desires to meet you while in this country. I feel sure that you will be as much pleased to make his acquaintance as I have been myself."

Chanute, as usual, meant well and misled. Patrick Alexander did not simply happen to be in the United States, nor had he led any investigations. He was an independently wealthy aeronautical enthusiast and promoter, not an experimenter. Only three weeks and one day earlier, he had been in London listening to Major Baden Fletcher Smyth Baden-Powell, president of the Aeronautical Society of Great Britain, give an account of recent progress in aeronautics, including the lighter-than-air ships of Alberto Santos-Dumont and Ferdinand von Zeppelin, and "the wonderful progress with flying machines" made by "Mr. Wilbur Wright and his brother." It appeared, Baden-Powell said, that a practical aeroplane might be at hand, and "one can scarcely imagine any invention which could have a greater effect on the conduct of warfare. . . ." Patrick Alexander promptly left the meeting, canceled his schedule, and booked passage for America to meet the Wrights.

It was Christmas, a time the Wrights guarded for family and children. So the conversation was brief. But the brothers, if momentarily nonplussed by this sudden appearance, were charmed. "We both liked him very much," Will told Chanute.

Alexander's visit to Dayton was the clearest evidence yet that the Wrights were achieving a certain renown in the world's small community of flight experimenters, and all because of Octave Chanute's talents as a publicist. The elderly engineer was talking about the Ohioans to every experimenter he knew. In these bright young men, with their warmth and brains and rectitude, he could see the ripening of his fondest hopes for his own life. He still believed that a practical aeroplane, in its final form, must maintain its balance by adjusting automatically to changes in the wind—not, as the Wrights believed, through the constant control of the operator. But he was no fool. He had seen the brothers glide and turn, and he could see where they were headed. Still, he could not bring himself to regard the Wrights' gliders as entirely original. He did not understand precisely how they made their 1902 craft turn, but he believed their controls were like the ones devised by his French-African friend Louis Mouillard, who had tried (and failed) to turn a glider by applying more drag on one side than on the other. And certainly the basic biplane configuration of the Wright craft seemed very much to resemble Chanute's own double-decker glider of 1896. If the Wrights succeeded, Chanute could reason, surely their triumph would represent the culmination of his own long effort to help younger men overcome this most difficult of all scientific challenges. Surely, then, was he not merely the Wrights' friend but their teacher—if not the father of flight, then at least its grandfather?

On January 3, 1903, with his two daughters and enough baggage for a long stay abroad, Chanute set sail from Boston for Europe. He went as an ambassador of the coming Louisiana Purchase Exposition of 1904—better known as the St. Louis World's Fair—which would include a grand aeronautical congress. Chanute hoped to gather all the world's important men of aeronautics in one place, and simultaneously impress European flight enthusiasts with the progress being made in his adopted homeland.

Of these, the French were the most zealous in believing their own nation held a special claim on the sky. Alberto Santos-Dumont was only the latest Parisian to attract notice for flights; in fact, since the brothers Joseph and Etienne Montgolfier had developed hot-air balloons in the 1780s, the French had led the world in lighter-than-air aerial navigation. By 1900 ballooning had become a popular craze, especially among Paris's leisure class, and the pursuit of flight enjoyed far more credibility there than in the United States. Experiments, plans, and gossip about all things aeronautical radiated from the Aéro-Club de France on the exclusive Place de la Concorde. The club's leaders included Ernest Archdeacon, a wealthy attorney; Henri Deutsche de la Meurthe, an oil magnate; and Comte Henri de la Vaulx—all well-connected men of affairs who dreamed that France would lead the world in the conquest of the air.

It was with some excitement, then, that the Aéro-Club's members gathered on April 2, 1903, to hear an address by the Octave Chanute, whose name was well known in France by virtue of his authoritative work, *Progress in Flying Machines*. Most members were lighter-than-air men, by practice and belief, but they were well aware of the rival hopes of heavier-than-air men like Chanute. Even the name of Wright was not entirely new to them. They had heard it from one of their members, a soldier and captain of artillery named Ferdinand Ferber. In 1901 Ferber had flown a Lilienthal-type glider and become one of Chanute's many correspondents. When Chanute sent Ferber a copy of Wilbur Wright's "Some Aeronautical Experiments," with a description and photos of the 1901 glider, Ferber instantly wanted one of his own. He was no engineer. He paid a carpenter to work up an imitation out of bamboo and sheets, with no way to twist the wings. But on a slope in the maritime Alps near the Mediterranean coast, Ferber had made several glides in the machine. Since then, he had urged his fellow aeronauts to abandon lighter-than-air means of flight and pursue the banner of Lilienthal, which the American, Wright, had seized. Only the previous Saturday, *Le Monde Illustré* had published five photographs of the Wrights in flight and the tantalizing assertion that "the aviator turns with his wings." Still, Ferber's was very much a minority voice in the Aéro-Club. The spectacular dirigible flights of Santos-Dumont had a firm grip on the plans and dreams of most members.

Shaking hands all around, Chanute found himself "very kindly received." As patriots of the proudest and most comfortable kind, it would have pleased club members to regard this distinguished Franco-American as more or less one of their own. Yet as Chanute proceeded through his review of the latest developments in fixed-wing machines—and especially as he displayed his photographs of *les frères Wright* aboard their 1902 glider at Kitty Hawk—many among his listeners grew not more pleased but deeply unsettled.

No one in the audience recorded Chanute's remarks. But press accounts make it clear that he made a profound impression. He reviewed the work of "the great German scientist, Lilienthal," whose line of thought Chanute had picked up and extended after Lilienthal's death. Then he described the "extremely remarkable experiments" of Wilbur Wright, "the most eminent aviator in America," who, in a machine with twenty-eight square meters of supporting surface in a two-surface form, had achieved glides of 200 meters with an average angle of descent of only six degrees—and with no more damage than "a pair of torn trousers!"

The Wrights had told Chanute they were applying for a U.S. patent, but there is little doubt the engineer revealed patentable features of their glider to the French. It's clear, too, that he implied the brothers were his pupils and protégés—that their work was an extension of ideas and designs for which Chanute himself was chiefly responsible. In 1900, one paper reported, the Wrights had written to Chanute "to ask him for some details about his experiments," which they "desired to renew." Chanute, another said, "understood very well the necessity of helping one another in this tremendous task," and thus had taken "pains to train young, intelligent, and daring pupils, capable of carrying on his researches by multiplying his gliding experiment to infinity." The Wrights had experimented with "his modified apparatus." Ernest Archdeacon said he was struck by Chanute's humility. But it masked a subtle assertion of credit for what the Wrights were doing—an assertion that was utterly at odds with the truth.

Why did Chanute mislead so egregiously, and with such unfairness to his young friends in Dayton? And why, when he knew the danger of giving away secrets, did he blab so freely about the Wrights' control devices? Partly it was that Chanute simply did not understand the Wrights' work, as later events were to show. But surely the British flight historian Charles Gibbs-Smith was close to the truth when he said that "with all due charity, one can only conclude that this fine old man, finding himself in 1903 in the romantic country of his birth, and surrounded by the warmth, respect, and indeed adulation of his audience, succumbed to the temptation of acting the part of inspirer, teacher, and mentor of the Wrights." It was not the only time when Chanute yielded to this temptation, and in years to come, the Wrights would find themselves trying—first with extreme courtesy, then with increasing exasperation—to correct the deeply flawed version of events that began that evening in Paris.

For the moment the French airmen had no interest in fine distinctions over credit due to one American or another. That any American had seized so large an advantage over the true pioneers of flight—the French—unleashed a wave of patriotic indignation. "For most of the listeners, except Ferber and his friends, it was a disagreeable revelation," de la Vaulx recalled. "When we spoke in France rather vaguely about the flights of the Wright brothers, we did not doubt their remarkable progress; but Chanute was now perfectly explicit about them and showed us their real importance. The French aviators felt at last that . . . they had been resting on the laurels of predecessors too long, and that it was time to get seriously to work if they did not wish to be left behind."

Captain Ferber was away from Paris when Chanute gave his talk, but he learned of it. He wrote immediately to Ernest Archdeacon, whom he knew to be interested in fixed-wing flight, and urged that he use his influence with the Aéro-Club to establish a hefty prize for distance in fixed-wing gliding. "With our experience of the automobile," Ferber told Archdeacon, "we know that it is racing which leads to the machines being improved; and the aeroplane must not be allowed to reach successful achievement in America."

Archdeacon needed no special urging. He was near a state of patriotic apoplexy. In the next issue of *La Locomotion*, the lawyer issued a desperate, heavily italicized call to his countymen—which, in turn, was reprinted in the next issue of the more influential *L'Aerophile*. Not only Wilbur Wright but Samuel Langley, he declared, was on the verge of flight. Archdeacon had it on Chanute's own authority that Langley already had "entirely *solved the very difficult question of equilibrium*" and was "counting on showing the universe, in the course of 1903, *the first self-propelled, man-carrying aeroplane!!!*" And Langley had government money behind him—"*an indispensable condition of success*, for definitive flying machines assuredly cannot be built without very long and very costly trials. . . ."

> France, this great homeland of inventors, assuredly does not hold the lead in *the special science of* AVIATION, even when the majority of good minds are today convinced that that alone is the true way.
>
> Will the homeland of the Montgolfiers have the shame of allowing that ultimate discovery of aerial science, which is assuredly imminent, and which will constitute the greatest scientific revolution that has been seen since the beginning of the world, to be realized abroad?
>
> Gentlemen scholars, to your compasses! You, you Maecenases, and you too, gentlemen of the government, your hands in your pocket . . . or else we are beaten!

WHEN CHANUTE RETURNED to Chicago soon after his talk in Paris, he invited Will back to Chicago for a second talk to the Western Society of Engineers. On the evening of June 24, 1903, Will described the breakthrough experiments at Kitty Hawk the previous fall. When Chanute invited questions, one engineer asked Will what he thought of Alexander Graham Bell's efforts to lift a man in a gigantic kite.

Will executed a courteous dodge.

"It is very bad policy to ask one flying machine man about the experiments of another," he replied, "because every flying machine man thinks that his method is the only correct one." Bell, he said, seemed to be interested principally in "the method of construction—to get something strong."

Another man rose to ask whether a glider's wings wouldn't be stabler if attached at a dihedral angle—a V-shape, like many birds of prey.

That was Maxim's approach, Will said, and Langley's. Maxim's craft had flipped over in "a side gust of only moderate force," and "the Langley machine was tested only in dead calms when there were no side gusts to contend with." He and his brother had tried the dihedral pattern on one of their gliders, he said. "But when we found that every little side wind threatened to capsize it, we drew the tips down like the wings of a gull." When the wind blows hard, buzzards stay in their trees, he pointed out, while gulls navigate easily. "We found the gull position much the best. The dihedral angle is the proper solution of the problem for flight in still air, but it makes matters worse instead of better when the wind blows.

"Unfortunately, the wind usually blows. So we have found it best to abandon this method and employ other means of securing lateral equilibrium."

Wilbur Wright could pack a punch of enormous rhetorical power in the space of a few words—so few, in fact, that some such punches flew so fast that bystanders failed to notice them. So it was with these five words—"*Unfortunately, the wind usually blows*." No one who heard them that evening in Chicago recorded whether the audience understood that Wright had just gutted the experimental strategy Samuel Langley had been following for more than ten years.

The Smithsonian secretary was trying to invent a flying machine that never would fly in a wind.

IF SAMUEL LANGLEY had ever ridden "a fractious horse," he had not made a habit of it, and it is doubtful that he had tried, let alone mastered, a bicycle. He had neither flown in a glider nor seen one fly. He had chosen the alternative method of learning to ride a fractious horse—"to sit on a fence and watch the beast a while, and then retire to the house and at leisure figure out the best way of overcoming his jumps and kicks." The figuring had now consumed ten years and filled many bound volumes of memoranda, letters, calculations, tables, diary entries, sketches, and plans.

On the basis of those ideas and findings—especially the performance of the 1896 models—Langley had designed the great aerodrome. He had delegated the work of actually building the machine to a highly competent and dedicated young engineer, educated at a fine university, who in turn had delegated most of the hands-on jobs of machining and carpentry to skilled tradesmen, working for wages. These men came and went. While they stayed, they did not participate in the secretary's meetings and correspondence with Charles Manly. Nor did any leave a record of his private opinion of the enterprise, so one can only imagine what these men and their families thought of Secretary Langley and his flying machine. The records do show that not a few of Manly's setbacks were the result of workmen's errors. Perhaps they slipped up occasionally because they worked without much understanding of the overall plan, or because they felt little personal stake in the result. Perhaps, as Manly believed, they simply needed very close supervision.

The work went on six full days a week, with Sundays off. Like any good student of the nascent science of industrial engineering, Manly broke the process into parts, arranged the parts in their proper sequence, and assigned a man to each. His detailed plan of work for early December 1902 is a good example. R. L. Reed, the long-serving carpenter and shop foreman, was to concentrate on the top priority of new flywheels for the engine. C. H. Darcey would assist Reed, except when Reed didn't need him; then Darcey would complete a new propeller for the quarter-scale aerodrome. G. D. McDonald, after finishing a little side job for the secretary, was to complete new porcelain castings, make new exhaust springs, and "hob the blank for the extra worm gear for the starting mechanism of the large engine." H. O. Webb was to bore and counterbore blanks for extra flywheel hubs, while Richard Newham was to make forgings needed for the derrick on the houseboat—though when Webb had finished the first flywheel blank, Newham should slot it to fit the engine shaft, and "if a new goose-neck tool is required for this job, he should make the tool *first thing*." Mr. Endriss, the new machinist, was to finish a new case for the tachometer, test the instrument to make sure its readings were correct, then complete the combined main piston rod and so on.

And these details were the work of only a few days. There were "innumerable other details," Langley remarked later, "for the whole question is one of details. . . . It is impossible for anyone who has not had experience with such matters to appreciate the great amount of delay which experience has shown is to be expected in such experiments."

Yet by the spring of 1903, the delays at last seemed to be over. Manly had brought the engine to a state of perfection. Anyone who looked at it could see that it was the result of superb workmanship. It looked like a five-pronged star inside a wheel. It was built almost entirely of steel, the strongest metal for its

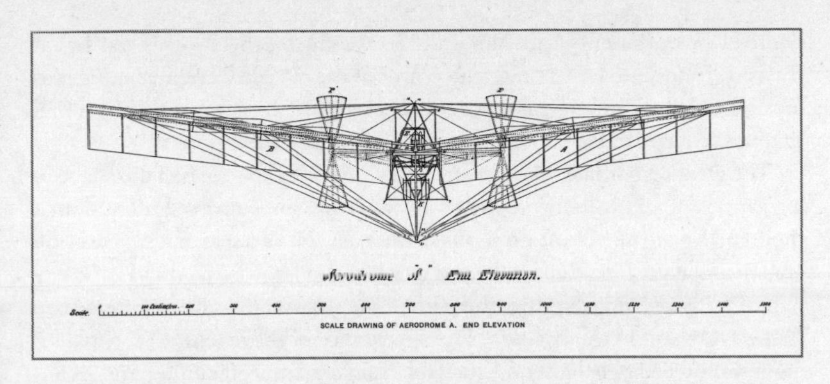


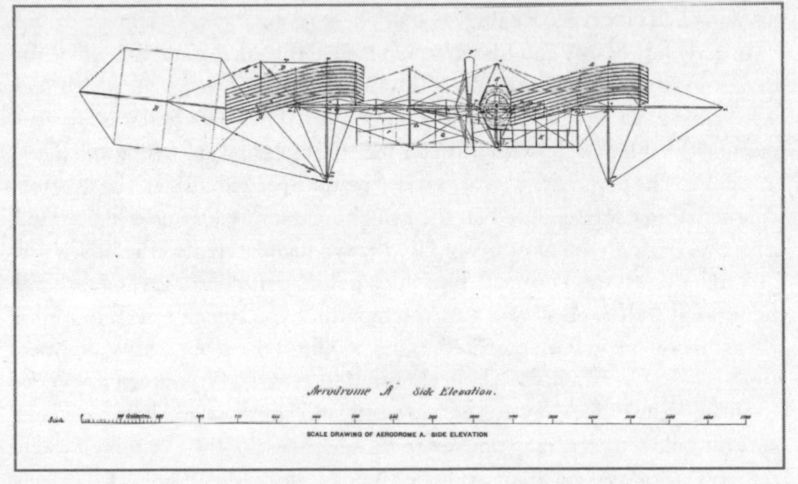


"THE DELAYS AT LAST SEEMED TO BE OVER."

Two diagrams of Langley's great aerodrome: front-view sketch (top); side-view sketch (bottom) with Pénaud tail at far left

weight, with brass bushings, cast-iron pistons, and cast-iron liners for its five cylinders. Every fitting gleamed.

In March Manly ran the engine nonstop in the South Shed for three flawless tests of ten hours each. The dynamometers recorded its best performance at 53.5 horsepower—nearly five times the power that Langley had originally demanded of Stephen Balzer. With cooling water, flywheels, batteries, and accessories, it weighed 207.5 pounds, or 3.96 pounds per horsepower. Of all the engines built so far to power a flying machine—Clement Ader's, Hiram Maxim's, the Clement Autocyclette engine just completed for Alberto

Santos-Dumont's newest airship, and the Wright brothers'—this was by far the best. It possessed four times the power of the Wrights' engine, and it was four times lighter in comparison to its horsepower. If pure strength were all that was required to fly, Langley was sure to win the race.

The great aerodrome, likes its ancestors, No. 5 and No. 6, had the shape of a gigantic dragonfly. Its two pairs of bowed wings encompassed more than a thousand square feet of lifting surface. Between the two sets of wings was the engine and its two propellers. In the rear was the large Pénaud tail. In every part of its frame and covering, the aerodrome showed the same supreme care that was obvious in the engine. "The appearance of the machine prepared for flight was exceedingly light and graceful," said an Army officer, "giving an impression to all observers of being capable of successful flight."

In early July Manly and his workmen loaded the engine and the two aerodromes—great and quarter-scale—into the enormous houseboat, which had been waiting at a slip on the Potomac since 1901. The boat's hull was fifty feet long and thirty feet wide, with a fifteen-ton turntable and launching apparatus on its roof. The parts of the great aerodrome were packed inside. The quarter-scale aerodrome was mounted on the launching track and covered in a shroud of canvas. Early on the morning of July 14, two tugboats pulled the houseboat forty miles down the Potomac, past Alexandria and Mount Vernon, around the bend at Indian Head and past the mouth of Occaquam Creek to a spot where moorings had been prepared in the middle of the river near Widewater, Virginia. The river there was about three miles across and seventeen feet deep. A soldier of the U.S. Army was posted as a guard. For room and board, a steam launch would ferry the men upriver to the clubhouse of the exclusive Mount Vernon Ducking Association on little Chopawamsic Island, just off the arsenal at Quantico. (Langley, though not a duck hunter, had joined the club.)

Manly was to keep the secretary constantly informed of events, and to notify him when a trial was impending, so he could make the seventy-minute trip by train from Washington to the Quantico station.

Langley insisted on caution. He instructed Manly to hold the length of the first flight to no more than ten or twelve minutes.

*Chapter Seven*

# "Our Turn to Throw"

"THE MACHINE GIVES AN IMPRESSION TO ALL OBSERVERS OF BEING CAPABLE OF SUCCESSFUL FLIGHT."

Langley's great aerodrome awaiting its first test

THE HOUSEBOAT WAS no sooner moored than Secretary Langley began to complain of the rising humidity. Then he began to feel ill. He was about to turn sixty-nine years old. He was determined to witness the impending test, but he dreaded the malarial vapors of summer on the Potomac. On doctor's orders, Langley went up to Boston with Cyrus Adler, and Richard Rathbun prayed he would stay there—or, better, that Langley would move farther north for a full-fledged vacation on the New England coast, then go on to Europe. Mosquitoes and reporters were giving the long-suffering assistant secretary all he could handle at the Castle. The only thing worse would be to fight them off with the secretary on the premises.

Rathbun pleaded with Adler to do his best to keep Langley off the Washington scene for as long as possible. "The weather is warm, moisty and heavy," he warned, "and the Secretary is bound to succumb. There are not less than twenty-five newspaper men lurking in small boats around the houseboat, seeking admission, making constant inquiries for Mr. Langley, and causing a great deal of annoyance, in fact, to the extent of playing practical jokes upon the people on the boat. Here at the office he will be disturbed with newspaper men in the same way. Several come here every day. This flying machine business has become the most prominent affair in the world, and each paper is bound to come out ahead. You and I could stand such matters, but you know what an irritating effect they have upon the Secretary. I may not be a proper judge in the matter, but I feel that Mr. Manly at this point will do the work better alone than with the Secretary."

Manly did his best to pretend the press did not exist. No schedule of trials was released or promised. The reporter of the *New York Daily Tribune* said no one outside Langley's circle really knew how the aerodromes worked. "Although plausible attempts have been made in public to describe the mechanism, it is said authoritatively that they have been largely poor guess work." Some of his colleagues might not let secrecy stand in their way, he warned. Thus "the public may regard with suspicion astonishing tales that are likely to be printed within a few days."

The *Tribune* reporter undoubtedly was recalling the spectacular narrative lie that had appeared in the rival *New York Sun* some months earlier. The *Sun*'s correspondent, overcome by competitive fever or duped by a persuasive hoax, reported that a gale sweeping across Washington had snatched the great aerodrome from the roof of the houseboat and sent it soaring into the air—with the houseboat trailing behind by its ropes. "Rivermen say the flying machine test that followed was remarkable," the writer attested. "The aeroplane dragged the houseboat around the Potomac for a while, but, hampered as it was, its flight was erratic. Finally after a number of peculiar manoeuvres the flying machine and the boat ran into the steamer *Harry Randall*, lying at her pier, the boat smashing twenty feet of the steamer's guard rail and the aeroplane lighting on the flagpole." Chasing after his bird, Secretary Langley had "arrived in time to view the wreck and receive the congratulations of those who saw his flying machine in successful operation."

The whole thing was a confabulation. Perhaps it was understandable that Langley, in Boston, was trying not even to look at the newspapers—except, coincidentally, the Sunday edition of the *New York Sun*, whose book reviews he admired.

OUT ON THE RIVER, Manly was joined by his brother, John, a professor of English at the University of Chicago, who had come to see a Manly make history. Instead, he witnessed the next battle in the Smithsonian's war with the press.

Even in fine weather, "considerable nervous excitement and tension" would have prevailed. But wind and rain were making things even more tense. No doubt partly to get dry, and perhaps partly to quench their thirst, the reporters were becoming regular visitors to the clubhouse of the Mount Vernon Ducking Association. When Langley heard about this in Boston, he issued orders, through Manly, that the clubhouse was off-limits to the press.

As the reporters pondered this turn, one fact stood out in sharp relief. Their assignment was to report on an official U.S. government project for which the public—their readers—had paid fifty thousand dollars. Mindful of this injustice, or simply fed up with Langley and Manly, one or more of them had a private chat with Mr. Truxton Beale, former U.S. ambassador to Persia and a charter member of the Ducking Association, who had just arrived on the island. Beale must have had disagreements with Langley already. For while Manly and his team worked offshore, Beale summoned the reporters to lunch as his personal guests; invited them to return as often as they wanted; and made critical remarks about "the air of proprietorship exhibited by the Smithsonian employees." When the aerodrome men returned to the island, Beale informed them that "an indefinite stay of the workmen was beyond membership privileges." H. C. Yarrow, the sitting president of the Ducking Association, affirmed Beale's decision in a telegram from his summer home in Quebec, and Manly, steaming, had no choice but to move men and equipment into hastily rented tourist rooms at Clifton Beach, Maryland, five miles downriver.

Reporting this development with suppressed glee, one New York correspondent said the Smithsonian's "secrecy . . . and the lack of consideration for the press have occasioned severe criticism of the entire undertaking in Washington, where little faith is placed in the utility of the enterprise."

As a *Tribune* editorialist noted, with some sympathy for Langley, close public scrutiny of any long-term scientific investigation carried "the danger of misinterpreting a single incident." In other words, a single flubbed experiment could appear to be a failure of the entire enterprise. This would be true even if reporters were watching the work as disinterested observers. Now, with the controversy over access to the island, Langley and Manly had turned the newsmen into cold-eyed adversaries.

In a rare public statement, Langley tried to clarify matters. His funding was partly public and partly private, he said. His experiments were "believed to be the first in the history of invention where bodies far heavier than the air itself have been sustained in the air for more than a few seconds by purely mechanical means." All his experimental successes of the past had been built upon instructive failures, he explained, and "I know no reason why prospective trials [of the great aerodrome] should be an exception. . . .

"It is to be regretted that the enforced publicity which has been given to these initial experiments, which are essentially experiments, and nothing else, may lead to quite unfounded expectations. . . . It is the practice of all scientific men, indeed, of all prudent men, not to make public the results of their work till these are certain."

But the statement came too late to head off yet another misstep in public relations.

On August 8, still seething over the island affair, Manly "caught all the reporters napping" by preparing to launch the quarter-scale aerodrome at 9:30 A.M.—an early hour by newsmen's standards. The excitement among the crew was palpable. In every detail the model was a scaled-down twin of the great aerodrome. They knew, of course, as John Manly said, that "this was . . . not the great test, the final test, the test of the man-carrying aerodrome, but it was felt by all to be of almost equal importance, for if the balancing of the small aerodrome was correct, the large one would maintain its equilibrium, and the problem of human flight would be solved practically as well as theoretically."

Manly acted quickly because the air that morning was very still. First, the crew mounted the model on its launch car. John Manly, stationed on the tugboat with a camera several hundred yards away, saw a signal rocket shoot skyward, and "instantly there rushed towards us, moving smoothly, without a quiver of its wings, with no visible means of motion and no apparent effort, but with tremendous speed, the strange new inhabitant of the air." It came on with "wonderful speed" and "strange, uncanny beauty. It seemed visibly and gloriously alive." For a moment Manly felt "a wild fear" that the aerodrome would crash into the tug, but then it swerved gracefully to one side. For ten more seconds it flew straight; then it curved again and skimmed into the water.

This "entirely successful" flight, Manly told Rathbun, "completely reassured me as to the equilibrium, power and supporting surface of the large machine."

But with the lovely spectacle over, an ugly scene ensued. As the tugboat hauled the machine out of the water, its grappling hooks splintered some of the woodwork. So the reporters, rushing to the scene in boats, arrived to see what looked like the aftermath of a wreck. One of the Smithsonian men only fueled their suspicions by trying to throw a sheet over the machine. "As the reporters were swarming about us only a few feet away," Manly told Rathbun, "the machine appeared to them to have been seriously injured and caused them to think that an accident had occurred and the flight had been a failure."

At first Manly refused to answer the shouted questions. Then, seeing that the reporters were about to write up the experiment as a bust, he "thought it wise to make a brief statement to them correcting this impression, in order that the Secretary may not be disturbed by another lot of bitter and unfair criticism." So he called the reporters back and assured them that "all the data which this machine was designed to furnish were obtained," that "the equilibrium was perfect, the power adequate and the supporting surface ample," and that no accident had occurred. "I can give you no further information at this point."

The headlines were grudging: "Langley's Ship Flies But Falls Into River."

DESPITE HIS IRRITATION with the reporters, Manly had been transfixed by the flight of the quarter-scale model. He said photographs gave "no adequate idea of the wonder and beauty of the machine when actually in flight."

> For while the graceful lines of the machine make it very attractive to the eye even when stationary, yet when it is actually in flight it seems veritably endowed with life and intelligence, and the spectacle holds the observer awed and breathless until the flight is ended. It seems hardly probable that anyone, no matter how skeptical beforehand, could witness a flight of one of the models and note the almost-bird-like intelligence with which the automatic adjustments respond to varying conditions of the air without feeling that, in order to traverse at will the great aerial highway, man no longer needs to wrest from nature some strange, mysterious secret, but only, by diligent practice with machines of this very type, to acquire an expertness in the management of the aerodrome not different in kind from that acquired by every expert bicyclist in the control of his bicycle.

That night Manly sent the houseboat back to Washington for final preparations on the great aerodrome. "The situation at the club made me very angry," he told Rathbun, "but fortunately I had no time to think of my anger." His mind was already focused. He stayed around Quantico for another day, going up and down the river and making notes. "I must . . . become thoroughly familiar with this portion of the river, so that I can readily recognize all the landmarks while flying through the air."

• • •

IN DAYTON, the Wrights were following all this through the newspapers, and not without sympathy. "Prof. Langley seems to be having rather more than his fair share of trouble just now with pestiferous reporters," Will wrote Chanute. "But as the mosquitoes are reported to be very bad along the banks where the reporters are encamped he has some consolation."

IN OCTOBER 1902, George Spratt had left the camp at the Kill Devil Hills brimming with excitement over a "bright idea." He became so excited on his way home to Pennsylvania that he stopped in Norfolk, bought a set of drawing instruments he could ill afford, and found a room at the YMCA where he could figure and sketch. Before long, "confusion set in deep." But at home he kept after it—the beginning of a conviction that wings with a circular curve would be more stable than the flat parabolas of the Wright machines. He began to regard the idea as a "nest egg" of great value, which would lead him one day to the breakthrough he wanted so badly. Not least of all, he might finally win an argument with his accomplished friends in Dayton. "I am anxious to have good grounds for saying 'I told you so.' "

Spratt's letters never were long on clarity, and the Wrights sometimes struggled to understand just what he was up to. When they did understand, they were dubious. Still, Will had come to believe that even a smashed hypothesis often pointed in a more promising direction, and he knew how much the work itself meant to "dear friend Spratt." He encouraged Spratt to keep at it "till you get everything clear," as "there is evidently something there worth knowing, though it may turn out different from what you expected." When he had to press Spratt to define his terms, or support a contention with proof, or hint that he had misread his own experiments, he was gentle: "I know that an ounce of fact outweighs a pound of theory. We must be sure though . . . that we do not misapprehend what the facts really are."

Will also tried to coax Spratt out his periodic fits of "the blues." But in the spring of 1903 they were deepening, for Spratt's sense of his destiny and the demands of his daily life were colliding. He was now certain—or as certain as his perpetual self-doubt allowed—that he had "solved the mystery of the lift of curved surfaces and proven it experimentally." It was "the keys to the heavens . . . the track in the air." Yet he was newly married and building a house, and still responsible for the family farm, where the work left him feeling exhausted and ill. He started to build a glider to test his theory, but made only a little progress, for "I cannot get the time to work at it, and even if I could I haven't the means to carry on such work. . . . You see I have more than I can handle."

In this state, the letters he received from Dayton—with news of an engine and propellers and a fine new flying machine, and extensive tables of wind tunnel data—became a little hard to bear. He cherished his own ideas yet seemed to get nowhere, while the Wrights, with theories he doubted, moved forward so confidently and decisively.

With all this on his mind, he wrote a long letter to Wilbur on April 15, 1903. "How I envy your ability to act quick and to the point. I wish I had more of that trait, but alas by general make and training I was moulded differently. I am never certain even when I am sure! And experiments that I make, and note the result with all accuracy and positiveness, in a few days I begin to wonder if it was really so, and fear a mistake might have been made and I do it again."

He could see only one sure route out of his quandary. Perhaps, he proposed, there could be a partnership. He could supply the theory; the Wrights, with all their industry and mastery of mechanics, could apply it. He had thought of asking Chanute for his opinion, but since he had unburdened himself to Will, he plunged ahead and asked. "The application of the principal [sic] is I think patentable in such a way as to cover all possible flying machine construction," he said, "and if so could it be possible, and would it be advisable for you, Orville and myself to hook up together and develop the thing. I am willing to share and do my fair part in such an arrangement, and would be even tho I could develop the thing myself for I have no burning lust for money, and prefer to be satisfied with making progress."

He asked Will for advice. Should he give up his plan to build a machine embodying his ideas, and instead publish the ideas in a scientific article? If he did, he would "put others on a par with myself." Or should he patent the idea and make common cause with the Wrights?

This presented Will with a delicate task. His brother and he had no wish to hitch their fate to anyone else's notions, least of all Spratt's ideas about wings. Yet he wished to do no damage to his friend's fragile hold on self-confidence and hope. His answer was a diplomatic tour de force.

As for advice, Will said:

> I must confess I am at a loss just what to say. . . . Regarding a matter which might affect the whole course of a man's life, I almost fear to give any, lest injury might result from it, instead of good as intended. I can suffer the consequences of my own mistakes with some composure, but I would hate awfully to see some other person suffering from an error of judgment of mine. Nevertheless I have a great desire to see you succeed and if you feel free to communicate the matters you have

> in mind, I will promise to do what I can to help you either with advice or with assistance in obtaining help to carry forward the work, provided of course that the matters communicated are in my judgment meritorious.

He silently side-stepped the invitation to become partners. But he gently addressed Spratt's own sense of inadequacy in the face of what the Wrights were doing. "You make a great mistake in envying me any of my qualities. Very often what you take for some special quality of mind is merely facility arising from constant practice, and you could do as well or better with like practice."

He said Spratt must not think too much of the Wrights for their apparent skills in debate; they came with a price. "It is a characteristic of all our family to be able to see the weak points of anything but this is not always a desirable quality as it makes us too conservative for successful business men, and limits our friendships to a very limited circle. You envy me, but I envy you the possession of some qualities that I would give a great deal to possess in equal degree."

If Will had meant to send a hint about the partnership idea, Spratt missed it. He sent back a draft of a scientific article and renewed his invitation. He hoped the article would be published by the prestigious Franklin Institute, in Philadelphia, but he had held back that "good nest egg—which is really the application of the principles set forth by the experiments, which enters . . . into construction of machines. . . .

"If one of you could be spared from Dayton and go to Kitty Hawk with me, my wife would go to be our house keeper. I would move my machines and tools . . . and we could start a series of machines, and develop a machine on the quiet which I think would reward us in full for all money spent." The Wrights' plan for a motor was premature and dangerous. Their machine could not bear it safely. For in the building of a ship, "the hull of a boat . . . must first be obtained that will float, the oars or pole next, the sails next, the motor next." Before a powered flyer, there must be a true soaring machine, with automatic stability. That—or an idea for it—was what Spratt was offering them.

Will now had to be frank.

> Orville and I expect to go to Kitty Hawk again this Summer or Fall. We have designed a 500 ft. machine and propelling mechanism and have a large part of both already constructed, but much yet remains to be done before we are ready for our trip. We would be very loath to take up any thing new that would prevent us from carrying through the experiment

we are already started upon. We would be very glad to give you such assistance as we are able to give, provided it did not take too much of our time, or interrupt our own work. . . . At all events we will expect to see you in Kitty Hawk when we go into camp this year.

The Franklin Institute rejected Spratt's article for publication, and his ardor faded. "Hay and wheat have taken all my time and ambition," he told the brothers that summer. "My machine I nearly completed but now it is thrown back in the hay mow. . . . It looks very very tired and uncomfortable and I wonder, when I pass that way, if it will ever be able to fly—or in my imagination even, to soar as it did once."

FOR A YEAR Bishop Wright had ignored his official suspension from leadership in the Church of the United Brethren. He said the suspending body—the bishops of the church's White River Conference—had been constituted illegally, thus its action was void. Milton simply had proceeded with his duties, defying his fellow bishops. Many laymen and clergy supported him.

The showdown came at the annual meeting of the White River Conference, in the small town of Messick, Indiana, in the first week of August 1903. In a chaotic scene, Bishop Wright and the presiding bishop attempted simultaneously to gavel the convention to order. Will, who had taken a separate train, arrived somewhere in the midst of this scene, apparently in time to see the local sheriff serve an injunction on his father, ordering him to cease and desist. Three days later, the conference's elders voted twenty-two to two to expel Milton from the United Brethren.

THE BROTHERS ARRIVED at the Kill Devil Hills on September 25, 1903. In past years, in torrents of rain and clouds of mosquitoes, "We had supposed that nature had reached her limit," Orville said, "but far from it!" Their long shed was several feet closer to the ocean than where they had left it—a wind of ninety miles per hour had shoved it there—and great pools of seawater stretched toward the horizon in all directions. They built a shed for the new machine, calling it their " 'hand-car,' a corruption of the French 'hangar' used by foreign airship men." The old shed became "the summer house," which they used themselves.

In mid-October a gale rose to forty, sixty, then seventy-five miles per hour. It whipped the sand into stinging furies, tore at the tarpaper on the roof of the shed and brought a tide of water over the floor. Orville scurried up the ladder to nail the paper down tighter. But the wind seized his coat, "blew it up around my head and bound my arms till I was perfectly helpless." Will had to climb up and literally hold his brother's coat while Orv furiously pounded nails and fingers. The wind's effects on the sand hills were more welcome. They were in "the best shape for gliding they have ever been. . . . Every year adds to our comprehension of the wonders of this place."

Earthbound for a year, the brothers found the strong winds tempting. They had meant to get the new shed done before trying anything, but one day, "the wind was too good for gliding, so we got out the old [1902] machine and took to the hills."

They made scores of glides that day, hanging all but motionless for twenty and twenty-five seconds at a time. One glide lasted more than thirty seconds, the longest on record. In a glide of twenty-six seconds one of them landed only fifty feet from where he had left the ground—a span of time in which they had covered five hundred feet the year before. In another flight of ten seconds, the glider landed *behind* the spot where it had started.

These glides showed that the machine was approaching the aerodynamic efficiency of the gulls and buzzards around them. The brothers were grasping the achievement that Will had imagined in the summer of 1899. It was "the nearest [to] soaring that has ever been done, probably . . . ," Orv exulted to Charlie Taylor that evening—and if they could soar, they could fly. "A few more days like today," Orv said, "and we will be able to stay up whole minutes at a time without descending the hill at all. Every thing now seems on the up grade to success."

Five days later, after they "arranged to get more twist to the wings with less motion of the body," Will remained in the air, "practically stationary," for forty-three seconds—"about three times the best of anyone else." The wind had slackened a bit, yet both brothers made flights that were "higher and more spectacular than any heretofore." After another severe storm they took the glider out again. "But we soon found that it was only too anxious to soar and we had great difficulty in keeping it from going up too high." The machine would rise "without making any descent of the hill at all, and so rapidly that it would fairly take our breath." One day they floated at altitudes of forty and sixty feet.

After a delayed shipment of parts arrived, they spent less time gliding and more time inside the thin, wind-buffeted walls, attending to the main task of the season. "I think Nature was just storing up all her energies for this terrible blow she has just dealt us," Orv told Kate. "We are now working on the new machine. . . . It's a 'Wopper.' "

As the weather had put the Wrights in a mood to personify Nature, they may have taken hope from the visitor she now sent them. The year before, an earthbound mouse had taken up residence in the shed. This year their guest was a bird—a sparrow-sized beauty, yellow and black, with a thin, sharp bill. They couldn't identify it. The little creature hopped about the kitchen, "perched on the cups, walked over the plates and made itself at home generally." It was so at ease with the brothers that Orville could reach out and touch it.

IN HIS BELATED REMARKS to the press about the quarter-scale aerodrome, Charles Manly had withheld at least one finding—that he would need more horsepower to drive the great aerodrome than the twenty-four he had estimated earlier. This was all right, he assured Langley, since the engine had been performing consistently at over fifty horsepower. But the secretary, still in Boston, was "deeply troubled" by the "painfully small margin" of power. "Perhaps you can relieve my anxiety, but I am almost ready to ask if it is desirable to try the momentous experiment at all, *under such conditions*, or might the flight be tried without the weight of an aeronaut?"

Manly managed to reassure his chief that all would be well, and he went about his final preparations. Considerable planning and coordination was required, much of it having nothing to do with actually flying the craft. Manly was well aware that Langley insisted on being present for the climactic trial, though he also wanted it accomplished soon enough for him to make a quick European trip before winter. Members of the Board of Ordnance and Fortification and of the U.S. Geological Survey were to be summoned as official witnesses, along with Richard Rathbun, two members of Congress, a surgeon, the Smithsonian photographer, and the photographer's assistant. Yet because the test must be held on a day of perfect weather—meaning virtually no wind at all, a rarity on the Potomac—Manly would have to be a shrewd forecaster indeed to tell everyone when to come down from Washington. Langley was pressuring Manly to fly on the earliest possible date, yet he did not want to come unnecessarily. And Manly, one can be sure, wanted the secretary underfoot as little as possible.

Even more important, Manly had to take steps to prevent disaster from befalling the aerodrome before it ever left its traces. From long experience he knew that once the machine was assembled, a rogue gust of wind could wrench it from its handlers' grip and send it tumbling. Even in favorable weather, no fewer than ten men were required to handle it safely, and Manly insisted they be thoroughly drilled in the elaborate assembly procedure. One by one, each of the four huge wings had to be carefully removed from inside the houseboat and placed on a raft alongside. From there, a derrick designed just for this purpose lifted the wing up to the houseboat's roof, where it was placed inside a big bin to protect it from the wind. Then the heavy frame-and-engine assembly, too, was hoisted over to the raft, then to the roof, where it was placed on its launching car. Next the men lifted the wings out of the protective bins and attached them to the frame—first the rear pair, then the front pair. Finally the guy wires were attached, to keep the wings anchored. Manly discovered this process alone could consume two hours for each pair of wings. He supervised drills with his heart in his throat. The men were already complaining of being overworked, and "the slightest carelessness on their part in handling the wings etc. would seriously injure them so that many days would be lost in repairing."

Even so, Manly was less worried about mechanical problems than the effect on the secretary's ragged nerves. "I sincerely trust that you will not let the work cause you any concern that you can possibly avoid," he wrote Langley, "for the prospects for success are really much brighter than I fear you have allowed yourself to imagine."

Manly knew his boss well. Langley was indeed feeling pessimistic, and he confirmed his opinion on a visit to the houseboat after the flight of the quarter-scale. "Seeing things on the spot here, they look worse than from Washington. The summer's experience indicates that the chances are about 5 to 1 against a launch being possible on any given day even if everything were ready." He had hoped to stay downriver until a test was accomplished, but now he thought there might be no trial until as late as October, and "I *cannot* remain here for any such time."

In a letter to Mabel Bell, the inventor's wife, he said the experiments were going forward "in spite of every impediment that ill fortune can visit us with. . . . I need hardly say to you and to Mr. Bell, in confidence, that there is no certainty what the result of this may be, only that I have always expected from the experience of past years that the first attempt would be a failure. I admit here to an anxious mind, but only to this, so long as I escape from malaria on the field of action."

On September 3 Manly thought everything was ready. The aerodrome was assembled. The photographers took their stations. Two tugboats steamed to points where they would stand ready to rescue Manly in case he could not circle back to the houseboat through the air.

But the engine would not start. During the long weeks of delay, fog had ruined the dry cells.

Several days of diagnosis and repair followed. Manly put the engine through several tests. He fixed a cylinder, made minor adjustments, and started it again. Then, "without any warning one blade of the starboard propeller broke at the hub, smashing into the cross frame and the guy wires." Manly leaped to cut the power, but too late to save a crucial part on the shaft of the off-balance propeller.

"The Usual Accident," one paper reported.

When Langley and Manly constructed the hollow ribs of the aerodrome's wings, they had searched long and hard for a waterproof varnish that would protect the meticulously bent wood against moisture, which could soften the glue and cause the ribs to relax. To test the varnish, they submerged ribs in water for twenty-four hours. The varnish worked; the ribs kept their curve. But the tidewater fogs proved more destructive. Waiting for a turn in the weather, Manly discovered the glue in the ribs had softened after all. The wings had flattened into useless flat planes.

They opened up each rib, scraped out the old glue, applied fresh glue, and wrapped the ribs in surgeon's tape. The weather got worse. "Every storm which came anywhere in the vicinity immediately selected the river as its route of travel," Langley said later, "and although a 10-mile wind on the land would not be an insurmountable obstacle during an experiment, yet the same wind on the river rendered it impossible to maintain the large house boat on an even keel and free from pitching and tossing long enough to make a test." During two or three brief periods of calm, the men rushed to get the machine ready to fly. But each time, heavy winds and rain swept over the river. Once they had to leave the machine on top of the houseboat all night because the waves were too high to permit the safe use of the raft.

AT 10:00 A.M. ON OCTOBER 7, the wind was blowing at twelve miles per hour—much higher than ideal, but Manly had become desperate. There was no time to call Langley or anyone else. Only the reporters were there to watch as the Smithsonian crew hustled to get the machine ready before the wind worsened. The engine was started. At twenty minutes after noon, Manly stepped into the little aviator's compartment and checked his control devices. He signaled an assistant, who fired off two skyrockets in an "all ready" signal. One of the tugs answered with two toots of its whistle. Reed leaned down and slipped off the loop that held the catapults. There was "a roaring, grinding noise." In three seconds the machine reached the end of its sixty-foot rail.

Manly felt "a sudden shock," then, for just a moment, "an indescribable sensation of being free in the air."

Then his mind grasped "the important fact . . . that the machine was plunging downward at a very sharp angle." To bring the tail down and the nose

"THE MACHINE WAS PLUNGING DOWNWARD."
The test of October 7, 1903

up, he seized the wheel that controlled the Pénaud tail and shoved it as far as it would go.

Nothing happened.

The front wings struck the water and disintegrated. Manly found himself entirely underwater. He grabbed the guy wires over his head, pulled himself out of his seat, and kicked up to the surface. The first thing he saw was a reporter, "his boatman expending the utmost limit of his power in pushing his boat ahead to be the first one to arrive."

GEORGE FEIGHT, THE WRIGHTS' neighbor on Hawthorn Street, sent them a newspaper account of events on the Potomac.

"I see that Langley has had his fling, and failed," Will wrote to Chanute. "It seems to be our turn to throw now, and I wonder what our luck will be."

AS THE BROTHERS ASSEMBLED their new machine, Patrick Alexander, the British flight enthusiast, was en route to New York for the second time in a year, this time to prepare for the aeronautical congress at the St. Louis World's Fair. Chanute secured an invitation for Alexander to visit the Wrights' camp, then cabled Alexander in New York, giving instructions to meet him in Washington so they might travel to North Carolina together. But either the telegram failed to reach Alexander or he decided other tasks were more pressing. Instead of heading for North Carolina, he boarded a train for meetings in Boston. After the Englishman learned what happened at Kitty Hawk that fall, and recalled the bungled connection, he blamed Chanute and never forgave him.

GEORGE SPRATT APPEARED in camp on October 23. The extra hands were welcome, for Dan Tate had left the Wrights' employ in a wage dispute. Even on his good days, Tate failed to meet the Wrights' standards for industry, and "whenever we set him at any work about the building, he would do so much damage with his awkwardness that we found it more profitable to let him sit around." Even so, the brothers had granted Tate's demand of a guaranteed weekly wage of seven dollars, nearly twice the going rate in Kitty Hawk. But when Will asked him to fetch oak stumps for a fire, he refused the assignment as unreasonable, "took his hat and left for home."

So the brothers hauled their own wood—more and more of it as the calendar advanced. They never had stayed so late in the year, and some days it was "so cold that we could scarcely work on the machine." They improvised a stove from a carbide can, and Will began to take to his bed with five blankets, two comforters, and a hot water jug.

They assembled the upper wing—"the prettiest we have ever made," Orville thought—then the lower, then the forward rudder, the tail, and a pair of skids. On November 1, Orv predicted, "The new machine will be ready for trial some time this week." They decided to go straight to a powered trial, without testing the machine as a glider.

The next day they began to attach the engine to the airframe and to attach uprights to hold the propellers. With one brother aboard, the machine would weigh a little more than seven hundred pounds.

Until now the construction had proceeded smoothly. But in a test on November 5, a misfiring engine conspired with loose propellers and loose sprockets to make a disaster. The propeller shafts tore loose from their mountings and twisted. For metal-working, the Wrights had no choice but to send the shafts back to Charlie Taylor. George Spratt, seeing a flight delay of many days and unable to stay away from his wife and farm any longer, departed the same day, carrying the damaged shafts with him for shipment from Norfolk to Dayton.

Spratt's route of departure took him through the village of Manteo on Roanoke Island. Here he ran into Chanute, just arriving. They had a frank talk. Spratt believed it was dangerous to try the machine with an engine before testing it as a glider, and he expressed "apprehensions of disaster." In a follow-up letter, Spratt said: "I do not believe their machine will 'pick up' from the ground and if launched free I am very fearful for the operator."

Chanute was worried about the Wrights. Without the propeller shafts, there would be no powered flight for many days, and Chanute said he couldn't stay that long. For his benefit the brothers labored up the slopes to make a few more glides in the 1902 machine. But the wood had grown dry and rickety in the heat of the shed, and they decided the glider was no longer safe. For most of their friend's stay the weather remained so bad the three men did little but sit close to the stove and talk.

Chanute questioned the brothers closely about the mathematical calculations they had used in building their engine, and he didn't care for what they told him. Engineers usually allowed for a 20 percent loss of an engine's power in the friction of transmission chains and sprockets, he said. Yet the Wrights had built in a safety margin of only 5 percent. This worried the brothers. Unable to work because of the missing shafts, "We had lots of time for thinking, and the more we thought, the harder our machine got to running and the less the power of the engine became . . ." "We are now quite in doubt as to whether the engine will be able to pull [the flyer] at all with the present gears," Orv told Milton and Kate.

They were less impressed when Chanute tried to buck them up, saying, not for the first time, that the brothers' expertise in operating their machines counted more heavily in their favor than the machines themselves. "We are of just the reverse opinion," Orv remarked. They estimated their odds of success at no more than even.

It grew colder. The sky turned winter white. Chanute left the camp on November 12.

Concerned about Chanute's warning, the Wrights devised a new mechanical test. The results confirmed their own earlier predictions of the engine's efficiency, and they breathed easier. Orv told Milton and Kate of Chanute's worries, but that "he nevertheless had more hope of our machine going than any of the others. He seems to think we are pursued by a blind fate from which we are unable to escape."

AFTER THE GREAT AERODROME fell nose first into the Potomac on October 7, Manly told the reporters—and Langley, by telegram—that it had failed to fly because it was "too heavy in front." He quickly changed his mind, but it was too late. The press had been waiting endlessly for a trial with little faith in its success, and the Smithsonian men from the secretary down had treated them like uninvited guests at an exclusive party. It was a deadly combination, for the reporters and their editors now unleashed a fusillade, saying the disaster discredited Langley's entire project and raised questions about his worthiness to hold his high post, if not his sanity. "Dismal if not altogether unexpected failure is the outcome of Professor Langley's elaborate and expensive experiment in aerial navigation," one of them reported. The "total wreck" of the secretary's aerodrome "demonstrated not only its complete inability to fly, but the impossibility of alighting without self-destruction, even could it be so perfected as to make short flights." *The Washington Post*'s man called it "a crushing blow to his theory . . . The aeroplane . . . was too frail for the great strain put upon it, its wings were too feeble to sustain the weight put upon them, and its motor and propellers were incapable of doing what the inventor hoped they would." The *Post*'s editorial page said "any stout boy of fifteen toughening winters could have skimmed an oyster shell much farther, and that without months of expensive preparation or . . . government fleets, appliances and retinues."

The *Chicago Tribune* was a little kinder: "Notwithstanding the outcome of the experiment, which, of course, everyone anticipated, it is impossible not to admire the patience of Prof. Langley and the pluck of Prof. Manley [sic]. . . . The accomplishment is still a long ways off, but apparently just as marvelous and difficult problems have been settled. We may all sail through the ether yet."

Langley asked his own questions and drew quite different conclusions. After closer examination of the wreckage, Manly had concluded—and several trusted eyewitnesses agreed—"that the front portion of the machine had caught on the launching car." That meant, Langley said, that "the machine . . . had never been free in the air," and thus no flight test had occurred at all, but only an experiment that ended before it had begun.

He issued a statement, saying his faith in ultimate success "is in no way affected by this accident, which is one of the large chapter of accidents that beset the initial stages of experiments so novel as the present ones. . . .

"Whether the experiments will be continued this year or not has not yet been determined."

The money was all but exhausted. The cost of keeping a tugboat constantly at the ready for many weeks had been a terrible drain on Langley's remaining monies. And on the night after the crash, a major storm wrecked the steam launch, raft, and rowboats that had hovered around the houseboat. Replacements became yet another line in Langley's starved budget.

Yet there had to be another test, a true test. Contrary to the reporters, who had seen "an unrecognizable mass" lifted from the water, only the front wings, rudder, and tail had been damaged, and Manly had an extra rudder, tail, and set of wings ready to go at the South Shed. The engine was fine. Manly estimated it would take him only a few weeks to get ready. The weather was dangerous, but there was one advantage to working so late in the year: There would be fewer boats on the Potomac. That meant they could run a test much closer to home. They selected a spot just below the Washington Navy Yard, off Arsenal Point, where the Potomac and the Anacostia meet. It was twenty blocks south of the Castle.

Manly made only one change to ensure that the mishap of October 7 would not be repeated. He removed one small lug from a metal rod that projected out from a guy post.

ON THE MORNING OF NOVEMBER 19, the Wright brothers emerged to find ice on the ponds around the camp. They huddled inside all day, trying to keep warm and stewing over the long wait for the new propeller shafts. At noon the next day a friend from Kitty Hawk arrived in camp with groceries and the new shafts. But to their mounting disbelief, they could not get the chain sprockets locked securely on the shafts, nor would the magneto provide enough spark. They gave up and went to bed. "Day closes in deep gloom."

In the morning one of them had a new idea and reached for the container of Arnstein's hard cement. At home, they used this not only to fasten bicycle tires to rims but to "fix anything from a stop watch to a thrashing machine." "We heated the shafts and sprockets, melted cement into the threads, and screwed them together again. The sprockets stayed fast."

The weather grew still worse, too stormy and windy to take the machine out. They fiddled indoors, making a gadget that would gather data on the duration, air speed, and number of propeller revolutions for each flight.

Then, during an indoor test of the engine on November 28, one of the new propeller shafts cracked.

They may have seen or heard about press reports that Langley was going to try again before Christmas. Certainly the prospect of returning to Dayton without a single test of their powered machine would have seemed worse than a couple more weeks in the cold. So they decided that Orville would travel back to Dayton, where he and Charlie Taylor would supervise the manufacture of new shafts of solid steel, not of tubes. Will waited alone.

IN WASHINGTON, the great aerodrome was repaired and ready. Every day Langley and Manly hoped for the wind to die. On the evening of December 6 they thought the next day might be all right. But overnight it grew colder, and in the morning there was ice along the riverbanks.

On the morning of Tuesday, December 8, Langley participated in the annual meeting of the Smithsonian regents. He informed them that a contract had been signed with the architectural firm of Hornblower and Marshall for the design of a grand new building for the National Museum. He discussed concerns—raised by Alexander Graham Bell and other regents—that he had overstepped his authority in appointing a new head of the Bureau of Ethnology. And he listened as Bell renewed his call to move the remains of James Smithson, the original English benefactor of the Institution, from the British cemetery in Genoa, Italy, to a tomb at the Smithsonian Castle.

At lunchtime, Langley was notified that "a dead calm" had descended on the Potomac. He directed Richard Rathbun to notify the War Department that a flight might be tried. Manly had to rush around to line up a tugboat. Langley came down to the point at two-thirty with Richard Rathbun and Cyrus Adler. Hundreds of spectators already had gathered. Langley waited on a wharf with his colleagues and the official representatives of the War Department. Several more cold hours passed as Manly and the crew went through the laborious job of attaching the wings and rudder to the frame. By the time everything was ready, the light was fading and the wind had become "exceedingly gusty."

Langley and Manly looked out over a choppy gray river dotted with chunks of floating ice. The secretary apparently left the decision to his aviator. "It seemed almost disastrous to attempt an experiment," Manly said. Yet, "it was practically a case of 'now or never.' " Their money was gone. There was no way to keep the operation going until the spring. So he "decided to make the test immediately so that the long-hoped-for-success, which seemed so certain, could be finally achieved." In an ambulance boat, a crew had blankets at the ready, and "a black bottle without a label." Langley moved to the tugboat to watch.

The engine was running very well—the best ever, according to McDonald, the chief machinist. Manly, wearing a canvas jacket lined with buoyant cork, climbed into his seat, the propeller blades whirling a few inches from his head. Manly signaled to Reed, who tripped the catapult. The aerodrome leaped forward and skimmed down the sixty-foot track.

As the machine sped along the launchway, Reed thought the tail dropped and dragged.

Just before leaving the track, Manly felt "an extreme swaying motion im-

"THE WHOLE REAR OF THE WINGS AND RUDDER WERE COMPLETELY DESTROYED."

The aerodrome's final test, December 8, 1903

mediately followed by a tremendous jerk which caused the machine to quiver all over."

Witnesses saw "the whole rear of the wings and rudder being completely destroyed as the machine shot upward at a rapidly increasing angle." To some, the aerodrome even appeared to break in two before it left the track. The enormous wings apparently could not withstand their sudden introduction to the forces of flight. They crumpled as soon as they were asked to fly.

Langley, on the tugboat, "was not far enough forward to see certainly what happened."

To bring the nose down, Manly swung the wheel that operated the Pénaud tail. But "this had absolutely no effect."

An instant later the machine was vertical, the nose pointing straight up. For a moment it hung in the air, its 730 pounds counterbalanced by the propellers' upward thrust. A photographer from *The Washington Star* caught the machine from a distance at just this instant. The image, enlarged, shows the rear wings already crumpled, the front wings beginning to contort. The frame, contrary to witnesses, does not appear to be broken, but only the wings. At dead center, unaffected by the wreckage around it, still driving the propellers, is the circular form of the Manly-Balzer engine.

Then the wind struck the exposed underside of the wings with full force, driving the aerodrome backward toward the houseboat and down to the water sixty feet below, with Charles Manly pinned underneath.

After several "most intense moments," Manly disentangled himself from the machine's framework and swam out from under the aerodrome, the houseboat, and a sheet of floating ice.

Manly was bundled in blankets and given a dose from the black bottle. He went inside the houseboat to change his clothes. When he came out, he learned that the tugboat crew had gotten a line around the nose of the upside-down aerodrome. When they hauled it forward, the angle of the wings caused the machine to descend all the way to bottom of the river, where it lodged in the soft mud.

ORVILLE WRIGHT, carrying the new propeller shafts, read the news about Langley on the train from Dayton to North Carolina.

It took less than a day to install the new shafts. With Dan Tate still observing his boycott, the brothers made an arrangement with the crew at the lifesaving station. They would raise a flag when they needed a hand with a trial of the machine. But on the first day the machine was ready, no flag appeared. The wind was too slack for a start from level ground—a requirement, they felt, for a true powered flight. They practiced running the machine along the track.

The next day was Sunday. Their reading was interrupted only by the visit of A. D. Etheridge, a member of the lifesaving crew, who brought his wife and children to look at the machine.

On Monday, December 14, the Wrights hoisted the signal flag. The breeze blew at a listless five miles per hour, but they were impatient for action and decided to run the machine down a slope. They laid a sixty-foot wooden launch rail. The aeroplane's skids would rest atop a small, one-wheeled truck that would roll down the rail at the urging of the engine and propellers. A man at either wingtip would keep the machine balanced as it rolled. If all went as planned, it would lift off the truck and fly.

Five men trudged over from the lifesaving station. A couple of small boys came, too, but they ran for home when the engine roared to life.

Together the men trundled the machine up the sand hill on its creaky truck and maneuvered it into position on the sixty-foot rail. One of the brothers tossed a coin. Will won the toss. He fit himself into the hip cradle, ducking under the chain that led from the engine, on the operator's right side, to the propeller shaft on his left. The machine began to roll before Orville, at the right wingtip, was ready to steady it properly. It raced downhill for thirty-five or forty feet and lifted away from the rail, but the elevator was cocked at too sharp an angle, and the machine rose abruptly to fifteen feet, stalled, and thunked into the sand after only three seconds in the air, breaking a few parts. But Will was encouraged. "The power is ample, and but for a trifling error due to lack of experience with this machine and this method of starting the machine would undoubtedly have flown beautifully. There is now no question of final success."

IN DAYTON, Bishop Wright was preparing copies of a description of the machine for distribution to the press. On the fifteenth he received a telegram from Wilbur which read in part, "SUCCESS ASSURED KEEP QUIET."

REPAIRS TOOK A DAY and a half. Late on the afternoon of the sixteenth, with the machine finally ready for another try, the brothers felt the wind fade. As they waited on the beach, tinkering and still hopeful, a man they had not met before approached the camp. He introduced himself as W. C. Brinkley, of Manteo, a salvager. He looked the machine over for a moment or two, then asked what it was. A flying machine, he was told. He asked if they intended to fly. Indeed they did, the brothers said, given "a suitable wind." Brinkley looked for another moment or two at the thing, then remarked, with an obvious desire to be courteous, that it certainly looked as if it *would* fly—"with a suitable wind," meaning, apparently, one of hurricane strength. But if Brinkley was skeptical, he was also curious, and he asked if he might stay the night to see a trial the next day. The Wrights said he was welcome.

Overnight a northerly wind put a new skim of ice on the puddles and ponds. In the morning the brothers bided their time for a couple of hours. Then, convinced the wind would stay strong for a bit, they raised the flag to signal the lifesavers and went to work. It was so cold they had to run in and out of the shed to warm their hands. Brinkley was with them, and from the station came A. D. Etheridge, W. S. Dough, John Daniels, and a kid from Nags Head named Johnny Moore. One or two lifesavers who had stayed at the station kept an eye on the activities from a distance, through telescopes.

The wind was blowing at about twenty-five miles per hour, strong enough for a launch on level ground. The sixty-foot launching track was relaid to face north-northeast, directly into the wind. The machine was hauled into its starting position. To the south, the hump of the big hill loomed over their shoulders. Ahead, the machine faced a blank, barren plain.

By the coin toss of two days before, it was now Orville's turn. The brothers padded through the sand around and around the machine, checking things. They cranked the engine and let it run for a few minutes. The camera was put in position, and the brothers asked John Daniels to pull the cord to the shutter if the machine got into the air. Daniels remembered that the brothers "walked off from us and stood close together . . . , talking low to each other for some time," then shook hands. Will called to the men not to look so downcast, but to give Orville a little applause. They tried, Daniels said, but there was "no heart in it."

At 10:35 Orville inched into the cradle. He released the rope. With Will jogging alongside, his left hand on the right wingtip, the craft lumbered forward, reaching a speed of seven or eight miles per hour.

Between the two spruce skids and the little one-wheeled truck running along the rail, a space appeared. An inch became a foot, two feet, three feet. A long shadow ran across the sand.

John Daniels squeezed a rubber bulb to open the shutter of the camera.

Will, still jogging, saw the machine rise abruptly to a height of about ten feet, then dip just as suddenly, then rise again.

Spread-eagled on the wing, Orville struggled to keep the elevator controls level. The craft dipped a second time, a wing tilted, and he was back on the ground, 120 feet from where he had left the launch rail.

A couple of parts were cracked, so an hour passed before Will could take the next turn. He bettered Orville's distance by about fifty feet. Orville, on his second try, went a little farther still, and kept the machine steadier than in his first try. A gust came at him from the side, lifting the tip. When he twisted the wings to bring the tip back to level, he found the lateral controls strikingly responsive, much better than on the glider. But the forward rudder was too sensitive. The machine bobbed and dipped in an "exceedingly erratic" path.

At noon Will tried again, and again came the bobbing and dipping. But somehow he found the proper angle for the forward rudder, and the men at the launch rail realized he was not going to come back to the ground right away. The machine was leaving them far behind—two hundred, four hundred, six hundred feet, the noise of the engine fading, the wings on an even keel.

He was flying.

The machine approached a hummock in the plain. Will moved to adjust the forward rudder "and suddenly darted into the ground." He had gone 852 feet, a sixth of a mile, in fifty-nine seconds. The rudder frame was cracked but otherwise the machine was fine, as was the operator.

This fourth flight had been by far the most impressive, the fulfillment of their hope for sustained, powered flight. But they also realized that Orville's brief first try could be characterized in words that applied to no previous effort by any experimenter. Orville himself, who took excruciating care in later years to express their history in precise terms, fashioned a description of what the first trial of the day had achieved. It was "a flight very modest compared with that of birds," he said, "but it was nevertheless the first in the history of the world in which a machine carrying a man had raised itself by its own power into the air in full flight, had sailed forward without reduction of speed, and had finally landed at a point as high as that from which it started."

That wasn't an exciting or inspiring way of saying that two human beings had learned how to fly. But it was the way the Wrights thought about things. Hyperbole about events of this day would come from others—but not for years. The magnitude of what they had done could be appreciated only by those who fully understood the steps they had taken and the problems they had solved through four years of work. That included the two of them and no one else in the world. They had flown, barely. But they were utterly alone in their comprehension of what that really meant.

THE OFFICIAL COST of Langley's enterprise now approached $70,000. The Wrights figured up the total cost of their experiments of 1900 through 1903, including train and boat fare to and from the Outer Banks, at a little under one thousand dollars.

# Interlude

"What they wanted was not so much secrecy as peace."
The brothers at Huffman Prairie, Ohio, April 1904

On the afternoon of December 17, 1903, the wind snatched up the flyer and hurled it into a splintering somersault. John Daniels, trying to save it, got tangled in the uprights and wires and was shaken up and bruised for his trouble. The machine had made its final flight.

When the brothers were sure Daniels was all right, they stowed the wreckage, then walked the four miles to the weather station, the closest place where they could send a telegram. By coincidence, the man on duty was John Dosher, who had answered Will's first letter to the Outer Banks. Orville dictated the message. Thanks to Dosher's spelling, and someone's error about the duration of the fourth flight (it was fifty-nine seconds, not fifty-seven), the message went out this way:

SUCCESS FOUR FLIGHTS THURSDAY MORNING ALL AGAINST
TWENTY ONE MILE WIND STARTED FROM LEVEL WITH ENGINE

POWER ALONE AVERAGE SPEED THROUGH AIR THIRTY ONE MILES LONGEST 57 SECONDS INFORM PRESS HOME CHRISTMAS. OREVELLE WRIGHT

At the weather station in Norfolk forty miles to the north, James Gray translated the dots and dashes into words, then tapped a return message: Would it be all right if he shared the news with a friend at the Norfolk newspaper?

"POSITIVELY NO," the brothers replied.

But Gray chose to ignore the answer. By dinnertime, a reporter and an editor at the *Norfolk Virginian-Pilot* were rushing to put together a story for the next morning's edition. Unable to get further information from the Wrights themselves, they borrowed some facts from earlier accounts of the gliders, acquired a few new ones about the day's flights—apparently through a telegraphed exchange with one or two of the Kitty Hawk witnessess—and filled in the gaps with audacious guesswork. Keville Glennan, the city editor, ordered a banner headline spread across the front page: "FLYING MACHINE SOARS 3 MILES IN TEETH OF HIGH WIND OVER SAND HILLS AND WAVES AT KITTY HAWK ON CAROLINA COAST."

AT 5:30 THAT AFTERNOON, at 7 Hawthorn, Carrie Kayler answered the Western Union man's ring. She carried the telegram upstairs to the bishop's study and returned to the kitchen. He came down smiling. "Well, they've made a flight," he announced. Kate happened to come through the door at just that moment. They rejoiced to learn that the boys would be "HOME CHRISTMAS." And they obeyed the instruction to "INFORM PRESS." Kate ran the telegram over to Lorin, who—after his supper—crossed the Miami to the newspaper offices downtown.

At the *Dayton Journal*, Lorin spoke to city editor Frank Tunison, who doubled as the Associated Press's representative in Dayton. Tunison was tired of flying machines. Claims like this one found their way into newspaper offices month after month, and each was much like the last. He looked down at the black teletype.

"FIFTY-SEVEN SECONDS," he repeated. "If it were fifty-seven minutes it might be worth mentioning."

So the *Journal* carried no item on December 18. Editors at the *Dayton Daily News* were either more credulous than Tunison or more in need of local copy that night. They carried an accurate six-inch item on page 8, based on the telegram to Bishop Wright, though it ran under the misleading headline "DAYTON BOYS EMULATE GREAT SANTOS-DUMONT."

Kate had sent her own telegram to Octave Chanute on the evening of the seventeenth: "BOYS REPORT FOUR SUCCESSFUL FLIGHTS TODAY . . . LONGEST FLIGHT FIFTY-SEVEN SECONDS." One can imagine the engineer's surprise when he opened the *Chicago Tribune* of Sunday, December 20, to see a detailed drawing of the Wright machine featuring a horizontal, six-bladed propeller like a helicopter's below the lower wing, presumably to push the craft off the ground. The *Tribune*'s version was echoed in *The New York Times*, which said the Wright machine carried "a propeller working on a perpendicular shaft to raise or lower the craft, and another working on a horizontal shaft to send it forward."

The phantom propeller made several appearances in the handful of U.S. newspapers that carried items about the "The Latest Flying Machine," as a Boston paper put it, over the next several days. The articles were rewrites of the Norfolk men's quick work. Readers learned that a "canvas fan" steered the craft; that it was launched from a track that ran down a slope; that it flew for three miles with a peak altitude of sixty feet; that Orville had made the longest flight; that Wilbur had made the longest flight; that "a small crowd of fisher folk and coast guards . . . followed beneath it, with exclamations of wonder"; and that Orville, upon landing, shouted, "Eureka!"

Back in Dayton, the brothers read the newspapers and were appalled. They tried to set matters right by issuing a new statement to the Associated Press. Will took a scolding tone that was hardly likely to win newsmen's sympathies. Their private telegram had been "dishonestly communicated" to newsmen, he said, leading to the dissemination of "a fictitious story incorrect in almost every detail," then of "fakes pure and simple." Beyond that, he offered only the precise facts of the four flights—speeds, heights, distances, durations, and wind velocities, with an emphasis on the final flight of fifty-nine—not fifty-seven—seconds.

He allowed himself one flourish. He said he and his brother had wanted to know if the machine had enough power to fly, enough strength to land safely, "and sufficient capacity of control to make flight safe in boisterous winds, as well as in calm air. When these points had been definitely established, we at once packed our goods and returned home, knowing that the age of the flying machine had come at last."

Few papers had printed the original, error-ridden reports. Fewer still now printed the truth. Like Frank Tunison, editors undoubtedly thought this sounded much like many other flying-machine claims, all of which had turned out to be unfounded. The only American widely credited with a real chance of flying—Langley—had just failed spectacularly. And how could this pair of unknowns claim to have inaugurated "the age of the flying machine" when Alberto Santos-Dumont already had done so with his gas-bags?

A handful of people knew the difference. Just after Christmas, the Wrights received a letter from Augustus Herring, their old, unlamented campmate from 1902. It was a galling attempt at blackmail. Herring said he had built his own biplane machine with a gasoline engine. It was so similar to the Wrights' that "it seems more than probable that our work is going to result in interference suits in the patent office, and a loss in value of the work owing to there being competition." He had just been offered cash for "all rights I might have to interference suits" against the Wrights. He had turned the offer down, he said, since litigation would only hurt both parties, and "there will be enough money to be made . . . to satisfy us all"—if the Wrights would agree to "joining forces and acting as one party in order to get best terms, broadest patent claims, and to avoid future litigation." He already was talking to two foreign governments, he said, and he would settle for only a one-third interest.

Will told Chanute about Herring's letter. Even at Kitty Hawk in 1902, he said, the brothers had "felt certain that he [Herring] was making a frenzied attempt to mount a motor on a copy of our 1902 glider and thus anticipate us. . . . But that he would have the effrontery to write us such a letter, after his other schemes of rascality had failed, was really a little more than we expected. We shall make no answer at all."

Will found a good patent attorney. In the middle of January 1904, he took the interurban trolley twenty miles northeast of Dayton, to Springfield, Ohio. There he had a long talk with Henry Toulmin, who had been recommended by friends. Toulmin advised Will not to worry about Herring's threat. He said he would prepare a new application for a patent—on the 1902 machine, with its perfected apparatus of control, not the 1903 machine, which the Wrights now began to call the "flyer," to distinguish it from its glider ancestors. And he apparently confirmed Will's inclination to say nothing more about the machine to anyone until a convincing public demonstration could be made.

The brothers made a critical decision. They would push their bicycle business to the side, at least for the time being, and concentrate all their energies on flight. They began to build a new flyer. Frank Tunison had said they would deserve to be noticed if they could fly for fifty-seven minutes, not fifty-seven seconds. They accepted the challenge.

IT HAD BEEN NEARLY DARK when Secretary Langley's aerodrome took its second plunge into the Potomac on December 8, and the secretary left the scene in time to dine at home. He invited his friend Cyrus Adler, the Smithsonian librarian, to join him. They drank excellent wine, smoked Langley's fine but very mild cigars, and referred not once to the failed experiment. Instead, Adler recalled, they discussed fairy tales and other matters of mutual intellectual interest. The secretary "seemed in perfectly good humor and quite philosophical."

When Langley and Manly saw the photograph of the collapsing aerodrome in *The Washington Star,* they embraced the notion that this second accident, like the first in October, could be blamed on the launching mechanism, not the aerodrome itself. It became an article of faith with them that the aerodrome's failure was attributable solely to a bracket attached to a steel pin one inch long and one-sixteenth of an inch in diameter. If this tiny implement had remained true, Manly said, then "certain it is that . . . success would have crowned the efforts of Mr. Langley, who above all men deserved success in this field of work, which his labors had so greatly enriched." Yet their funds were exhausted, the river nearly frozen. No more trials were possible that winter. But if they could secure more money, they might look forward to a new flying season and ultimate vindication.

When Manly saw reports of the flights of December 17 at Kitty Hawk, he was sure someone was exaggerating. He checked with Chanute, who reassured him that "the press accounts I have seen are all inaccurate." Chanute relayed the contents of the telegram he had received from Katharine Wright—four powered flights against a wind of twenty-one miles per hour, the longest just under a minute.

On Christmas Day, at his family's home in Lexington, Virginia, Manly jotted a soothing note to Langley. He told his chief what Chanute had said about the Wrights. Of course, Manly conceded, even the actuality "marks quite an advance. . . . Yet until the Wrights do much better than this, I think we may safely contend that the upward trend of the wind was a large factor in enabling them to accomplish the short flight." Manly missed the implication of Chanute's plain statement that the flights of December 17 had been "from level," not on the slope of a dune, the only place at Kitty Hawk where the wind had an "upward trend."

Then he said something that revealed the terrible gap between the Wrights' conception of the flight problem and the one to which he and Langley still clung. "To advance 10 miles per hour against a 21 mile wind," the young engineer said, "was quite a different thing" than "to fly in a calm at the rate of 31 miles per hour"—a reference to the aerodrome's capacity for speed. After all, he pointed out, "the buzzard soars indefinitely in a 21 mile wind but can't do much when there is no wind."

Manly still was thinking of that superb engine. It possessed enough power to propel the aerodrome at a great speed. Therefore it must be the magical force that Langley had imagined all these years. The two were trapped inside an utter misconception of the problem they were trying to solve. It still seemed better to them to be able to fly in a calm than to be able to fly in the wind—even though "the wind usually blows." In fact, the aerodrome was structually incapable of sustaining speed through the air, calm or not. Even if it could have flown, it could not be balanced in a wind. It probably could not be steered in either a wind or a calm. And it had no means of returning to Earth without endangering the life of its pilot.

But Langley and Manly sensed no trap. To their friends and colleagues, they spoke of a tragic mishap—an experiment gone awry, an endeavor interrupted. The aerodrome had not been tried and found defective; it simply had not been properly tried. It did not fly, but it was capable of flight. Among Langley loyalists, this argument seemed plausible, and they kept it alive. The man who became, for a time, its most prominent spokesman was perhaps the one man in the world best suited to guarantee it a measure of credibility—Langley's best friend, Alexander Graham Bell.

Bell for the rest of his life would contend not only that Langley's 1896 models had "demonstrated to the world the practicability of mechanical flight," but that the great aerodrome itself "was a perfectly good flying machine, and . . . the first flying machine ever constructed capable of carrying a man . . . There was nothing the matter with it. It stuck in the launching ways, and the public were no more justified in supposing that it could not fly, than they would be were they to suppose that because a ship stuck in the launching ways it would not float."

In the meantime, Bell had flying plans of his own. He was working on a new invention. It was based on a complete departure from the ideas of Langley and the Wrights alike. Perhaps the Wrights had flown; perhaps they had not—for a time Bell found it hard to believe that such a feat could escape broad coverage in the press. But whatever they had done, it was no match for the image in his mind—the image of a man-made butterfly, hovering safely and stably in midair, oblivious to the wind. If he could turn the image into reality, the Wright brothers would sink back into obscurity.

COMMENTATORS COULD NOT resist the spectacle of the high-minded "professor" coming to smash. Langley's debacle brought to life ancient folk images of the brilliant but foolish intellectual (this one with his head all but literally in the clouds)—how he seeks to rise above the dictates of common sense and receives his just comeuppance. The *Chicago Tribune*'s editorialist was among the most scathing:

> It seems as if it had been sufficiently demonstrated that Prof. Langley's aerodrome will do nearly everything except what the professor intended it should do, thus once more illustrating the total depravity of inanimate things. It will shoot through space a distance governed by the ordinary laws of momentum and initial energy. It can point its nose upward and then turn a neat somersault. It can then turn its nose downward and dive like a duck . . . the principal difference being that the duck comes up without ruffling a feather, while the aerodrome stays down in the mud and is dragged up piece by piece. It can do such tricks as these, but it cannot fly. That seems to be settled as well as the likelihood of any further congressional appropriation for the costly experiments.

Not finished with Langley, the *Tribune* asserted that "even if a machine could be made to fly, no one would wish to fly with it," then signed off with the usual theological declaration: "Nature has fitted us with appliances for getting

"HE WOULD NOT BE CONSOLED."

Clifford Berryman of *The Washington Post* draws "Langley's Folly" short distances through the water, also with appliances for getting over the ground short and long distances, but no trace of a wing can be found or anything that indicates nature intended us to navigate the air. There is little possibility of that until we become angels."

Soon politicians began to have their say, with intentions more serious than one day's sport. Several used Langley's failure as a weapon with which to attack excessive federal spending. "You can tell Langley for me," one congressman told a reporter, "that the only thing he ever made fly was government money."

Despite the onslaught, Langley went back to the Board of Ordnance and Fortification in March 1904 with a plea for more money. The financier Jacob Schiff and other men of wealth had offered to underwrite further experiments. But Langley refused to accept private aid for a project intended solely for the public good. To the board, he insisted the launching mechanism had caused the accident, that "the machine itself is probably well fitted for its purpose," and that "a cessation of these experiments at this point will be unfortunate." With another twenty-five thousand dollars, he could construct "a new launching apparatus, which might possibly be used upon the land rather than over the water," make "some slight change" in the aerodrome, then bring "the experiments . . . to a successful conclusion."

Langley was not reading the newspapers, friends said, but the BOF's members were. They acknowledged his assertions and denied his request.

FOR A TIME, Langley bore up pretty well. An old Boston friend said "his patience and rare philosophy in meeting that phase of his career were among the noblest traits of his character."

The secretary had worked in Washington long enough to know that journalists and politicians are paid to complain loudly and in public, and though he was a poor politician himself, he was enough of one to know there had been little chance of more funding from the War Department. What crumpled his spirit as the months passed was the quiet disdain of his fellow scientists, the only community of equals Langley recognized. Scientists do not conduct their politics as politicians do. Ill will toward Langley, simmering for years, would have been expressed in quiet conversational asides and discreet private letters. John Brashear, his old friend and aide from Pittsburgh, found him one day in a state of despair. The secretary took Brashear's hand in both of his and said, "Brashear, I'm ruined, my life is a failure." Langley took from his desk two small pieces of steel. They were the pieces that had fouled the launch, he explained—the sole cause of his downfall. Brashear stood helpless as Langley "cried like a child." He reminded Langley of his achievements in astronomy, "but he would not be consoled."

According to Cyrus Adler, a number of scientists undertook a quiet campaign to remove Langley from his post as secretary "on the ground that his mind had given way, that he was endangering the fair name of the Institution by a series of foolhardy experiments which could never result in anything." Langley's friend Daniel Coit Gilman, president of Johns Hopkins University, broke this grim news to the secretary one day in his Castle office. At that moment, Adler said, "I shall never forget how Mr. Langley looked. He was a fine looking man and held himself very well. He closed his eyes and said: 'My years in Washington have been the happiest of my life; I have enjoyed the work here and the opportunities; I like the place as Secretary of the Smithsonian, but I shall never fight for it. If anybody is going to fight and have it taken away from me, well and good. I shall not lift a finger.' "

He didn't have to. Behind the scenes, Adler and other friends fought on his behalf and quashed the attempted coup. But Langley was now drained of confidence and energy. In Boston, his aged aunt, Julia Goodrich, perceived his depression.

"Samuel," she said, "don't bother about what they are saying about you now. Some day they will erect a monument to you for what you are doing."

"Aunt Julia," he replied, "what do I care about a monument, if they would only let me alone now."

OCTAVE CHANUTE'S RESPONSE to the flights of December 17 depended on whom he was addressing. His letters suggest warring emotions—admiration, pride, pique, and perhaps at least a touch of envy. To the Wrights, at first, he telegraphed that he was "immensely pleased at your success." But in a letter to Langley, who had asked Chanute for accurate details of the Wrights' flights, Chanute said the Wrights' success had been distinctly limited. "As you surmise, the Wrights have not performed what the pesky newspapers credit them with." They had not flown three miles, but only fifty-nine seconds, and only a few feet off the ground. This was no more than "a bare commencement," Chanute said—"encouraging," but very far from the ultimate goal of an aeroplane with automatic stability. "I think that this success brings out but one type of flying machines [*sic*]," he told Lawrence Hargrave, the Austrialian box-kite experimenter. "There are others which can probably be driven by a light motor." Perhaps the Wrights were now in the lead, but they had done no more than create a dangerously unstable machine, and the true prize was still up for grabs. He said much the same to Francis Wenham, the aged Englishman who had designed the first wind tunnel many years earlier. The Wrights had achieved something new, to be sure, Chanute told Wenham, but "the uses will be limited." They had flown, but only in a kind of dangerous gymnastic trick. (As for Langley's disappointments of the autumn, Chanute signed on to the party line: "I was much distressed at the accidents with your launching gear which prevented you from testing your own machine. If I can be of any use to you as an advisor please command me.")

In their statement to the Associated Press, the brothers had said: "From the beginning, we have employed entirely new principals [sic] of control; and as all the experiments have been conducted at our own expense without assistance from any individual or institution, we do not feel ready at present to give out any pictures or detailed description of the machine." The brothers intended that last sentence to mean that since they had "paid the freight" for their experiments, they "stood on quite different ground from Prof. Langley, and were entirely justified in refusing to make our discoveries public property at this time." But the words "without assistance from any individual" stuck in Chanute's craw. Thinking of the dozens of letters he had exchanged with Will, he wrote: "In the clipping which you sent me you say: 'All the experiments have been conducted at our own expense, without assistance from *any* individual or institution.'—Please write me just what you had in your mind concerning myself when you framed that sentence in that way."

It was the first unpleasant word that had passed between them, and Will responded to it carefully. They had meant *financial* assistance, he said. The point was to guard against any suggestion that they owed anyone a full accounting. This was needed, he said, only because of "a somewhat general impression" in the press at home and abroad "that our Kitty Hawk experiments had not been carried on at our own expense, &c." He restrained any impulse to remind Chanute that he himself was the one who had spread this impression, especially in France.

Chanute let it drop, though he would pick it up again.

His letter to the French experimenter Ferdinand Ferber was strange indeed, coming from a professed friend of the Wrights. Chanute urged Ferber to press on with his own experiments despite the Americans' success, "for it's very possible that the Wrights will have an accident and that it will be you who are destined to perfect a flying machine."

AT A DINNER MEETING of the Aéro-Club de France on February 4, 1904, Victor Tatin, one of the senior enthusiasts and one of the most respected, rose to speak. Because the reports of powered flights in America were so sketchy and contradictory, he said, it made sense to be skeptical. In any case, "The problem cannot be considered as completely solved by the mere fact of someone having flown for less than a minute." The real glory remained to be claimed, and Frenchmen must claim it.

> Must we one day read in history that aviation, born in France, only became successful thanks to the Americans; and that the French only obtained results by slavishly copying them? . . . It is in France that the first journey by a flying machine must be made. We need only the determination. So let us get to work!

EIGHT MILES EAST of the Dayton line, the brothers found a rare open space that was not being farmed. What they wanted was not so much secrecy as peace—the ability to do their work without interruptions and annoyances. Most scientists did their work behind closed doors, but this was not indoor science. The brothers went to the owner, whom they knew—Torrence Huffman, president of the Fourth National Bank of Dayton—and asked if they might rent the property for their experiments. He said they could use it for nothing, as long as they moved his cows and horses out of the way.

They built a shed like the ones at Kitty Hawk, and by the latter part of May they were ready to test their new machine—a slightly heavier and stronger version of the 1903 flyer. They planned a trial and invited family and friends, including Chanute, and reporters from all the Dayton and Cincinnati papers. Chanute was detained in Chicago. Despite rain in the morning, the reporters came out, and Kate and Milton, and Lorin and Netta and the children, and twenty or thirty other friends and neighbors, all of them standing in the wet grass and prickly weeds.

At first there was too much wind, then, quite suddenly, not enough, which was hard to explain to the visitors. And the engine was running poorly. But because people had come a long way, the brothers felt they ought to try. They ran the machine down the hundred-foot launch rail. It slid off the end and bumped to a stop in the soft dirt. The spectators returned to town.

Two days later they prepared to try again but were rained out. By the next day the crowd had dwindled to the bishop and a couple of reporters. With the engine still performing poorly, the machine rose to a height of six or eight feet and flew just thirty feet or so. No one bothered to measure.

The reporters were polite in their brief articles. But they saw no reason to go and watch again. They left the two brothers alone in their field.

## Chapter Eight

# “What Hath God Wrought?”

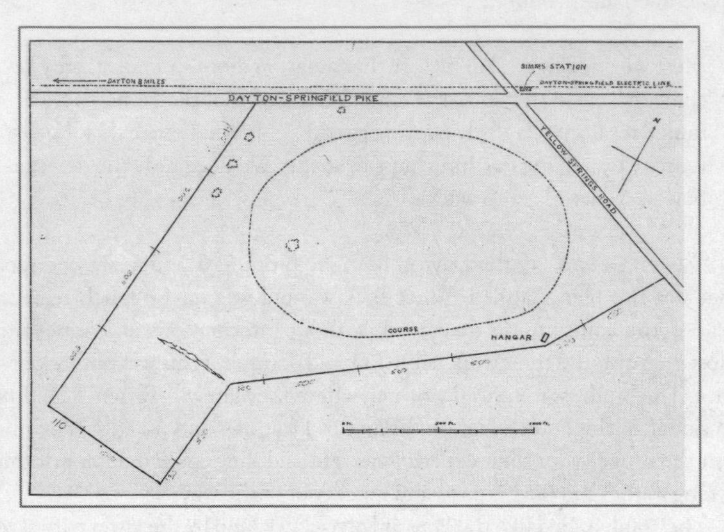


“WE MUST LEARN TO ACCOMMODATE OURSELVES TO CIRCUMSTANCES.”

Orville’s sketch of the flying field at Huffman Prairie

WORK WAS THE Wrights’ play. Their progress on the flying machine had been so rapid in part because they seldom stopped working. Even in their time off, they were forever puttering with a fence in the yard or writing letters or running the lathe or developing photographs or making a toy for a niece or a nephew. Even on Sundays, abstaining from work, they were busy. If they were sitting, they were reading, not to relax but to learn.

The summer of 1904 was thus an exercise in repressed energy. They spent much of it on their rear ends.

Hour after hour, they waited on a makeshift bench at the eastern end of their makeshift flying field, listening to the drone of bees, watching the impossible gymnastics of swallows and wrens, and thinking of the dependable Atlantic winds of Kitty Hawk. They kept an eye on the far end of the field, watching for the distant grass to bow in a breeze that might help them get up in the air. If they saw that telltale ripple, they were up and at the machine in a second. But too often the breeze died, and they sank back on their bench to wait. Or they sat in the shed watching rain drip from the roof. On many days they could make no attempt to fly at all. The sun baked the still air to eighty-five and ninety degrees, and their shirts and trousers clung to their skin.

A RETURN TO KITTY HAWK had been out of the question. Constant work with a homemade gasoline engine demanded machine tools close at hand, and Charlie Taylor's expert assistance. That much had been proven the previous fall; there could be no more emergency runs halfway across the country to fix a broken propeller shaft or any of a hundred other parts. And the wind-driven sand would foul their engine. They needed a new place to fly.

Someone in the family mentioned a particular pasture east of town. Probably it was Orville; he knew the place from school field trips he had made as a boy. Dayton teachers still took children there to see one of the last remnants of genuine prairie left in Ohio, a tiny finger of the grasslands that rolled westward across Illinois to the Great Plains. A layer of clay lay close to the surface of the soil, keeping the ground squishy-wet much of the year. Frost heaves made hummocks all across the field. That meant few trees could grow there. The tract had never needed to be cleared.

Two or three months after their return from Kitty Hawk, Will and Orv stepped off the train at a little depot called Simms Station to look around. The pasture lay kitty-corner from the station, across the junction of the Dayton-Springfield Pike and the Yellow Springs Road. It was about a hundred acres in a seven-sided irregular polygon, full of bristly grasses—big and little bluestem, Indian grass, rough dropseed. Trees stood along the western border, forming a bit of a barrier against the prevailing breezes. The interurban rail line ran along the northern border. To the east and south, farmland lay open and flat. A barbed-wire fence ran around the perimeter, confining some cattle and a dozen horses. A few farmhouses stood off in the distance. Here and there a lonely tree stood in the way, and the ground sloped a little toward the east—a disadvantage for flights launched into the generally westerly breeze. Will thought it looked like "a prairie-dog town." But a heel pressed into the loamy, black soil showed it was unusually soft—not like the dunes at Kitty Hawk, but with enough give to cushion a hard landing. That was one good thing. Better still was the site's convenience. From West Dayton, they could get there on the train in about forty minutes. It was pretty big and pretty open—and very few such properties close to Dayton were not under tillage. Probably it was the best they could find, and "we must learn to accommodate ourselves to circumstances."

The owner of the property was Torrence Huffman, president of the Fourth National Bank of Dayton. No one recorded what the brothers said about their intentions when they asked Huffman if they might rent his prairie. If Huffman had read the newspaper stories, he was not much impressed; he told others he thought the Wrights were "fools."

The brothers promptly erected a duplicate of their long shed at Kitty Hawk, and by late May the shed housed a new flying machine. In size it was nearly identical to the 1903 flyer, with a wingspan of just over forty feet and a chord of six and a half feet. They flattened the curvature of the wing a little more. The machine had a new engine and a bigger fuel tank, and it was nearly three hundred pounds heavier, with the center of gravity farther forward than in the 1903 machine. Then they made their abortive first attempts to fly before witnesses and realized there was to be no quick leap forward to the perfection of a practical powered aeroplane. That job was to be no easier than making a good glider.

AT THE WRIGHTS' SUPPER TABLE, Milton's battles within the Church continued to dominate conversation. Bishop Wright had refused to recognize the legitimacy of the decision to expel him from the Brethren in 1903, sued the White River Conference for wrongful dismissal, and demanded ten thousand dollar in damages. He continued his round of duties, traveling to each of his conferences and meeting with leaders and clergy, many of whom supported him. Other leaders declared Wright's actions to be "revolutionary and productive of a state of anarchy." The editor of the Church's *Christian Conservator*, an opponent of Bishop Wright, said the conflict had grown to "fearful proportions."

Undeterred, Milton and his backers prepared for a showdown in the next churchwide General Conference, scheduled for the spring of 1905.

THE BROTHERS WERE EAGER TO FLY, their eyes on the calendar of the St. Louis World's Fair where flight demonstrations were planned for lighter-than-air craft, and—so hoped the Wrights—their stunning debut. But for weeks neither the weather nor the machine obliged. With neither a slope to run down, nor a steady, strong wind, they had to lay a wooden launch rail some 250 feet long, scything down grass and smoothing out hummocks to make a level path. If the breeze shifted, they had to pull up the track and lay it all over again. In all of June they could make only a few attempts, none of them anywhere near as long as the fourth flight of December 17 at Kitty Hawk. The machine simply would not stay in the air. Unable to find spruce of sufficient length for the machine's leading-edge spars, they had substituted white pine. Thus, when the flyer did manage to labor through the air for a few yards—always with the same troubling undulations of the December 17 flights—it often returned to earth to the sound of splintering wood. Under any sudden impact the pine spars shattered "like taffy under a hammer blow," and each crack-up meant days of repairs. "We certainly have been 'Jonahed' this year," Will told Chanute. And they were still only a few weeks into their flying season.

In St. Louis, many were buzzing about Alberto Santos-Dumont's plan to fly on the Fourth of July. The Brazilian was well aware of the Wrights' claims to have flown a powered aeroplane, and of their hope to challenge him. Chanute informed the Wrights that Santos-Dumont "told me that he was not afraid of you, as he knows how tedious and slow was the working out of a new machine." But then came a reason to suspect that Santos-Dumont was not so confident of his own machine. Late one night near the end of June, someone entered his hangar at the exhibition park and slashed his gas-bag. When the police speculated that the great flyer himself might have done the deed to avoid a humiliating defeat, Santos-Dumont stormed off to Paris to make repairs. In July he sent his regrets, saying it was now too late to finish his repairs in time to return and fly by the deadline of October 1.

Santos-Dumont may have felt no more relieved than Will, who called the incident "a rather strange affair," though "I think I will suspend judgment a while." It seemed "the prospect of a race at St. Louis [with Santos-Dumont] is vanishing into thin air."

A lighter-than-air airship man from California, Thomas Baldwin, replaced Santos-Dumont as the sensation of the St. Louis show. His *California Arrow* dirigible astonished the crowd by making a flight in the shape of an S—an impressive demonstration of control.

FOR MOST OF JULY the brothers bent over the aeroplane, hoping that small adjustments in design might rid the machine of its tendency to bob up and down across the field. To shift the center of gravity backward, they moved the engine, the water tank, and the pilot. When that made little difference, they made more adjustments. On August 6, both of them managed flights of about six hundred feet, but that was still more than two hundred feet short of Will's best mark at Kitty Hawk. Moving under its own steam, the machine would grind along the rail, finally lifting off when it reached a ground speed of about twenty-three miles per hour. But for the engine's thrust to overcome the resistance of the air, they needed an air speed—that is, a combination of speed over the ground against the speed of the wind—of twenty-seven or twenty-eight miles per hour. On these somnolent days at the prairie, that was seldom possible. "It is a pity," Will said, "we cannot trade a few of our calms to Prof. Langley for some of his windy days that used to trouble him so."

Even on good days they could manage no more than four flights, and often it was just one or two—a far cry from the triumphant weeks of 1902, when they could soar dozens of times in a single day.

And they could not stop breaking things. On August 10 Will bounced on the turf and broke the front rudder, then cracked a propeller upon landing. That cost two days for repairs. In the second of four flights on Saturday, August 13, Will finally beat the Kitty Hawk mark with a flight of thirteen hundred feet. But two flights later he broke the rudder again. Three days after that Orville "shot down" and smashed a rudder support, leaving the machine with its tail sticking up toward the sky. On August 23 a sudden gust brought Orv crashing to the dirt. He bruised a hand badly and was "sore all over," and escaped graver injury only because the spar at the leading edge snapped just before it came down on his back. So long as the machine was skimming the ground, its dipping and bobbing would bring perpetual crack-ups. Somehow they had to gain more speed in the launch, to get free of the ground.

IN NOVEMBER 1903, a month before the first powered flights at Kitty Hawk, French aero enthusiasts read a new article by Octave Chanute in the respected *Revue Générale des Sciences*. Entitled "Aviation in America," the text described the Wright gliders of 1900, 1901, and especially 1902 in some detail. A proper description of wing-warping was omitted at Wilbur Wright's request; he knew such publication could invalidate future patents in Europe. But the text was secondary. What struck the Frenchmen like a hammer to the forehead were the ten accompanying photographs of the 1902 glider, including two showing Wilbur Wright actually *steering* the machine in a banking turn to the right. How he accomplished this feat remained mysterious, but there it was, for the first time—evidence that a heavier-than-air machine could be steered in the sky. Otto Lilienthal had kept his balance by shifting his weight, and died in the act. Here was a man controlling a much larger machine by mechanical means. Not a few well-informed Frenchmen understood this photograph portended a new age. If a motor and propellers could be mounted on such a machine, human flight was truly within reach.

Then, only a month later, the same Frenchmen read fragmentary news reports that the Wrights had made a powered flight of three miles.

The early misreporting of the events of December 17—particularly the three-mile claim—did lasting harm to the cause of French aviation. Had the report been accurate, it would have been taken far more seriously. A powered flight of 852 feet would have seemed both marvelous and plausible to knowledgeable French enthusiasts. But they knew enough about progress in gliders to know that any claim of a powered flight of three miles had to be hogwash. The question was little clarified by a report in *L'Aerophile* that Chanute, in response to a query, had cabled back: "Newspaper accounts considerably exaggerated." Either the Wrights were *bluffeurs*, or the American press was unreliable. In January, *L'Aerophile* published the Wrights' own statement to the Associated Press, though the editor appended a skeptical note: The account was "marred . . . by several obscurities." Uncertainty and suspicion remained.

Still, there were those extraordinary photographs in the *Revue Générale des Sciences*, showing the airborne Wilbur Wright in his glider, and the accompanying testimony from the respected Chanute. It could not be denied that the Wrights were doing *something*, and the bewildering combination of fact and fiction sent the Frenchmen of the Aéro-Club into a state of high anxiety.

Men unwilling to risk their own lives, but with the money and clout to induce younger men to do so, issued calls for action. Ernest Archdeacon conceded that "the results obtained are considerable," and said "we must hurry if we wish to catch up." He called for glider competitions to speed the French advance toward powered flyers.

"Aviation is a French science," Victor Tatin declared. Accounts of the Wrights' efforts should be read "with the greatest reserve." By no means could "the problem . . . be considered as completely solved" by a flight of under a minute. There must be no "slavish copying" of the Americans. "We still have in France some men of genius capable of successfully carrying out such work without putting ourselves in tow of foreigners."

In March 1904, Archdeacon and Henri Deutsch de la Meurthe, the oil tycoon whose prize had spurred Santos-Dumont to round the Eiffel Tower in a dirigible, announced the offer of a Grand Prix d'Aviation, soon known as the Deutsch-Archdeacon Prize—fifty thousand francs to the first man of any nation to fly a powered aeroplane in a complete circle of one kilometer.

Ferdinand Ferber, the Army officer who had been the first to try to emulate the Wrights by building gliders of his own, mustered a wan hope that the Americans had mounted a motor out of fears that he would beat them to it. He was soon buoyed by the news that "it is not as wonderful as they say," "not as grand as we thought." Still, "it nevertheless represents a new fact," and a provocation. He went back to work on his latest glider with fresh energy. Ferber gave a lecture on gliding in Lyons, where a twenty-four-year-old student of architecture, Gabriel Voisin, approached him afterward and said: "I have understood the method which you teach, and I mean to devote myself to it." The next day Voisin left for Paris, where he promptly offered his services to Ernest Archdeacon.

THE CHILDREN OF DAYTON were preparing to return to school in the fall of 1904 when passersby on the Dayton-Springfield Pike noticed a new structure out on Huffman Prairie. It was a simple, narrow pyramid about twenty feet tall—four stout poles leaning together and reinforced halfway up. From the apex hung a stout rope with a weight tied to the end. The other end of the long rope ran through pulleys and out to the far end of the launching rail, through another pulley and back along the rail to the flying machine. The brothers had built a catapult. They would haul the weight to the top of the tower. When the operator was ready, he would release the weight, which would plummet sixteen and a half feet to the ground, pulling the rope and thus yanking the aeroplane forward and into the air with enough speed to give the propellers a fighting chance to sustain the machine in the air.

With Kate and her friend Melba Silliman watching, the brothers tried the catapult for the first time on September 7, 1904. The wind was barely breathing. With 600 pounds of weight pulling it, the machine whizzed along the rail—nearly 100 yards in nine seconds—but it flew less than 150 feet. The brothers added 200 pounds to the weight and "almost got a start." They added 200 more pounds. Now Will shot forward and up, and stayed aloft for just over 2,000 feet. After a week of practice, against a little stronger breeze, he doubled that mark. In the next flight, for the first time in nearly a year—since the last flights of the glider at Kitty Hawk in 1903, and for the first time ever in a powered flyer—he shifted his hips in the cradle and warped the wings. The machine leaned into a banking curve. At Kitty Hawk, in the wonderful glider the brothers built in 1902, they had turned to right or left, proving to themselves the machine could be steered. But no turn had been more than a few degrees. This turn did not stop. Will turned and turned until the craft had described a half-circle and the breeze no longer blew in his face but at his back. In just under a minute, he bumped back to the ground, facing in the direction opposite that from which he had begun.

No reporter was present to see this flight, the first ever to describe a half-circle. Since the Wrights' embarrassing exhibition at the beginning of the summer, no reporters had returned to the prairie. Where was the story? A hop across a pasture was nothing compared to Santos-Dumont's grand excursions over Paris. Surely such dirigible balloons held more promise than winged aeroplanes. Had not Langley's spectacular failures put that question firmly to rest?

One of the doubters was James M. Cox, the young publisher of the *Dayton Daily News*, soon to become governor of Ohio and, in 1920, the Democratic nominee for president. Writing some forty years later, Cox recalled: "It is difficult nowadays to understand the incredulity that possessed the public mind." Dan Kumler, Cox's *Daily News* city editor, said in later years: "I guess the truth is that we were just plain dumb." But that was only how it looked later. At the time, it was no more "dumb" to doubt that a machine was flying in Dayton than to doubt that a flying saucer had landed there. Kumler's own brother-in-law was married to one of Katharine Wright's colleagues at Steele High School, where talk of the flying machine was quite common. Kumler heard these reports and discussed them with Cox, his boss. "Frankly," Cox said, "none of us believed it."

Of course, journalists are supposed to check, and checking certainly would not have been hard for Luther Beard, who split his days between his job as managing editor of the *Dayton Journal* and teaching school in the village of Fairfield, just two miles east of Simms Station. He was always running back and forth on the train, where, occasionally, he would see one Wright or the other. He knew what they were up to. But to Beard it seemed the charitable thing not to press them about it. "I used to chat with them in a friendly way and was always polite to them, because I sort of felt sorry for them," Beard said later. "They seemed like well-meaning, decent enough young men. Yet there they were, neglecting their business to waste their time day after day on that ridiculous flying-machine. I had an idea that it must worry their father."

The Wrights chatted with Beard just as amiably, and made no effort to provoke his curiosity about their work. Of course, like any thorough newsman, Beard covered himself.

"If you ever do something unusual," he told Orville, "be sure and let us know."

JUST ABOUT THIS TIME, another man—a man with an outlook on the world quite unlike Luther Beard's or Frank Tunison's—took the trouble to find Huffman Prairie and the brothers at work there. He was a tiny man of sixty-four. He drove a new Oldsmobile Runabout automobile and dressed in well-made touring clothes. His movements were quick and nervous, like a small bird's.

More important, he had capacities with which neither Beard nor Tunison were blessed—capacities for *seeing* a thing for what it really was, for appreciation and imagination, and for wonder. He met the Wright brothers and asked if he might watch what they were doing. They said he might, and over several days they told him a good deal about it.

He was Amos Ives Root. As a friend said, he was "remarkable not in one way, but in many ways," with "a many-sided character, if any man ever had one." He was the leading citizen of Medina, Ohio, a pleasant farm center in the opposite corner of Ohio, near Cleveland. He was wealthy, with vacation homes in Florida and northern Michigan. In his area of expertise—commercial beekeeping, which, in that presynthetic era, produced all the world's commercial honey—he was acknowledged to be the leading authority in the world, and the A. I. Root Company was the leading firm in the industry. His definitive manual, *Bee Culture*, had been translated into several languages.

By the time Amos Root met the Wrights, however, beekeeping was largely in his past. His sons ran his company while he devoted himself to his new passions. These, not necessarily in order, were Christian evangelism and social reform, automobiling, journalism, philanthropy, inventing, gardening, and the close observation of man and his machines.

Born small and sickly on a farm near Medina, Root grew up helping with the family truck garden, not the farm, so he became a master gardener before

"HE WAS ALL QUESTIONS."

Amos Ives Root

he reached adulthood. The garden encouraged an intense curiosity about the natural world and about science, and he read a great deal. He was the sort of person who, when he took an interest in a field, was compelled to learn everything there was to know about it. This was especially true of all things mechanical. "I love machinery," he once wrote, "and have always loved it from a child." As a teenager in the 1850s, he developed a passionate interest in chemistry, then electricity. At the age of nineteen he toured the Midwest giving electrical demonstrations. He also learned how to manufacture jewelry, whereupon he built a jewelry factory and made a good deal of money. He married a local girl, with whom he raised five children.

In 1865, at the age of twenty-six, Root took up beekeeping as a hobby. As he gained in expertise, other beekeepers sought his advice. He became the first to devise standardized beekeeping equipment, and by 1880 he was selling to 150,000 customers around the world. Every two weeks he addressed several thousand readers in *Gleanings in Bee Culture*, a trade journal he founded and edited. It was a small treasure-house of information and advice that went far beyond beekeeping. Topics ranged from "the growing evil of divorce" to the proper method of grafting domestic and wild cherry trees to the effects of "bad air" on the digestive system.

Root was a deliberate contrarian. He liked to be the first to try new machines, even, perhaps especially, if others made fun of him. "No amount of scoffing or ridicule—and he endured it many times—could swerve him from his belief or purpose," a friend said, "and he went straight to his work without faltering or swerving from the path he had chosen." Until it began to pay, people in Medina thought Root's beekeeping was odd. In the 1870s he became the first man in northern Ohio, perhaps in all of Ohio, to own a bicycle. He powered his machinery, including his printing press, with an electrical generator connected to a windmill of his own design. If the wind began to blow hard in the middle of the night, he would rouse his sons to capture the temporary surge in power.

After a long flirtation with the agnosticism of the controversial philosopher Robert Ingersoll, Root became what evangelicals call a professing Christian, and he professed to practically everyone he spoke to, including his employees, who were expected to attend daily prayer meetings on the job. He went about his religion with his congenital zeal, seizing upon all the causes of Christian reform politics. He became a temperance activist, an antitobacco man, and a strict sabbatarian. He was one of the founding benefactors of the Anti-Saloon League and a leader of the Sunday school movement.

All these enthusiasms—gardening, science, beekeeping, technology, Christian reform—were grist for Root's column in *Gleanings in Bee Culture*, which by 1904 he had been writing every two weeks for some thirty years. Each article was a lay sermon based on a biblical text. The column ran under the heading "Our Homes" and developed a devoted following. People with no interest in bees took *Gleanings* simply to read Amos Root. The column was a blend of personal anecdote, useful information, travelogue, news and advice, all delivered to animate Root's core evangelical message.

One of Root's pet causes was to persuade fellow Christians to welcome the technological change that was washing over American society—the telephone, electric power, the phonograph, and especially, of late, the automobile. Since buying his Olds Runabout in the spring of 1903, he had devoted roughly equal time to driving it and writing about it.

In 1903 and 1904, most people outside the cities still depended on horses for local transportation, and the problem of automobiles frightening horses was a lively controversy. Farmers tended to regard automobiles as playthings of the rich, and many did not hesitate to roll out whatever Christian artillery they could muster to denounce the new technology. In this context, Root appointed himself the defender of the automobile and took on the task of forging a reconciliation between the new technology and rural Christians. He believed in the gospel of social betterment through technological progress. He often told his readers that technology was as much a gift from God as was the natural world. After a visit to an Oldsmobile factory in 1903, he pronounced the automobile a "wonderful gift to the children of this age from the great Father above."

Root mentioned the Wrights in his "Our Homes" column of February 14, 1904, telling his readers how he had recaptured the attention of unruly boys in his Sunday school class by describing "two Ohio boys, or young men, rather, [who] have outstripped the world in demonstrating that a flying-machine can be constructed without the use of a balloon." From the details he cited, several of them inaccurate, it was clear he had read one of the misbegotten accounts of the December flights at Kitty Hawk. Still, he had followed developments in aviation closely enough to recognize the significance of the Wrights' claims.

Some months later, in July 1904, Root left Medina on a four-hundred-mile automobile trip through central and southwestern Ohio—a trip taken "with the view of studying humanity, and also of considering the question of automobiles on our public roads." He rolled through some fifteen Ohio towns and cities, including Dayton, and wrote a good deal about the state of the roads, the conflict between cars and horses, and attitudes toward the automobile. Visiting a relative in Xenia, he saw a newspaper story about the flying exhibition planned for the St. Louis World's Fair. Apparently this reminded him of the preacher's boys in Dayton, and he determined to find out more.

It's not clear whether he telephoned or wrote or simply appeared at the door of 7 Hawthorn. But somehow a friendship was begun. Why the Wrights opened up to Amos Root when they were so careful to shut others out remains a puzzle. Possibly they saw him as a future investor, a wealthy man with more than a purely commercial interest in their work, like Octave Chanute himself. More likely they simply responded to Root's curious charm. He was, after all, a man rather like themselves—intensely curious, fascinated by machines (including the bicycle), and inclined to march against the crowd. He was of a certain type of older man whom the Wrights knew very well—a churchman of the old school—and though they did not trust a churchman simply because he was a churchman, they were comfortable with men of Root's type. Too, Root was the antithesis of a hurried reporter looking for a superficial scoop, and he already possessed an inkling of the significance of their experiments. Very few people in 1904 knew enough to recognize the significance of a powered machine that could fly without the assistance of a balloon. Root was one who did know. They did not have to make a case; he already was awed by what they were doing.

For all the Wrights' reticence, we ought to consider how it felt that summer to be unrecognized. They were patient men. Yet even for them, it must have seemed hard, at times, not to crow a little. Amos Root would have been the perfect audience. It is not hard to imagine their feeling a sense of relief, even exhilaration, in telling their news to someone who could appreciate it and wonder at it. Nor is it hard to imagine, given their particular sense of irony, that they relished the prospect of *Gleanings in Bee Culture* scooping the world.

In any case, Amos Root was the Wrights' guest at Huffman Prairie on September 20, 1904, when Will planned to attempt the first flight in a complete circle. He would try to fly like a bird from its roost—to go somewhere and return, all through the medium of thin air, without ground or water to offer the inertia and friction that heretofore had been essential to all human locomotion.

FOR MOST OF THAT FLYING SEASON the brothers had been taking turns, as they had at Kitty Hawk. But since Orville's injury on August 24, Will had been doing all the flying—eight or ten attempts over four weeks, all with the catapult. Charlie Taylor was at the prairie to help.

On his first visit to Huffman Prairie, Amos Root noticed that although the field was in plain view, passersby paid no particular attention. "The few people who occasionally got a glimpse of the experiments evidently considered it only another Darius Green, but I recognized at once they were really *scientific explorers* who were serving the world in much the same way that Columbus did when he discovered America. . . . Nobody living could give them any advice. It was like exploring a new and unknown domain." Root was a careful, thorough observer, and he wrote down much of what he saw immediately or soon afterward, combining his account with schoolmasterly explanations derived from his talks with the Wrights themselves. The Wrights had asked him to publish nothing until they finished flying for the season. But he wanted to draft his account while the scenes were fresh in his mind.

He was all questions. Why was the catapult necessary? Why did the operator lie down on the wing? How fast would the machine fly? When the brothers showed him the action of the propellers, which clearly pushed the machine *forward*, not *up*, he was puzzled. Would the same propellers lift the machine if placed horizontally above it? Not at all, the brothers said; in that position the propellers would not lift a quarter of the machine's weight. Then how, placed vertically, could the propellers possibly keep the machine in the air? "The answer involves a strange point in the wonderful discovery of air navigation. When some large bird or butterfly is soaring with motionless wings, a very little power from behind will keep it moving. Well, if this motion is kept up, a very little incline of the wings will keep it from falling. A little more incline, and a little more push from behind, and the bird or the butterfly, or the machine created by human hands, will gradually rise in the air." It was an astonishing notion. Root had read more than most people about efforts to build a flying machine. But this was utterly new to him.

Will's first flight of the day was a thousand meters, the longest yet, but not a full circle. Then Will launched himself again as Root looked on, spellbound.

"The machine is held until ready to start by a sort of trap to be sprung when all is ready; then with a tremendous flapping and snapping of the four-cylinder engine, the huge machine springs aloft."

The flyer stayed within ten or twelve feet of the ground, unless it was turning. With the Wrights' tutelage, Root grasped the parallel with bird flight: "If you will watch a large bird when it swings around in a circle you will see its wings are tipped up at an incline. This machine must follow the same rule; and to clear the tip of the inside wing it was found necessary to rise to a height of perhaps 20 or 25 feet."

Will was down the field and coming back in less than sixty seconds. Root stood with Orville not far from the catapult, directly in the machine's path.

Wright State University

Samuel Pierpont Langley, the world's leading proponent of the feasibility of manned flight, was the descendant of Puritan divines and Boston merchants. A close observer of his career remarked that "a certain part of Langley was attracted to the spectacular."

Smithsonian Institution

Just as his passion shifted from astronomy to aeronautics, Langley assumed the leadership of the Smithsonian Institution, the most prestigious post in American science. The Smithsonian Building, known unofficially as the Castle, is pictured here in 1903.

Wright State University

Wilbur Wright, shown above in high school (*left*) and at the age of thirty (*right*), once told an irate sister-in-law: "I entirely agree that the boys of the Wright family are all lacking in determination and push." Will's younger brother, Orville (*below, left*), led him into partnerships in printing, journalism, and bicycling. Their sister, Katharine (*below, right*), taught Latin full-time and ran the household. "Without you," her father told her, "we should feel like we had no home."

Wright State University

Library of Congress

Wilbur likely composed his letter to the Smithsonian at the writing desk in the parlor (*above*) of 7 Hawthorn Street, Dayton. By 1899, the household was dominated by adult children. Bishop Milton Wright (*below, left*) was often away on church business, and his wife, Susan (*below, right*), had died ten years earlier.

Wright State University

Wright State University

Journalists seldom referred to Wilbur (*above*) and Orville (*below*) as "the Wright brothers" —indistinguishable partners—until many months after the first powered flights of December 17, 1903. In the early years, the small circle of aeronautical enthusiasts who knew anything about them spoke of Wilbur as the principal experimenter.

Wright State University

As the Wrights' reputation grew, Octave Chanute (*above*), their friend and encourager, could not resist the temptation to imply to others that he also had been their teacher and collaborator. Samuel Langley, needing an engineer, searched for a "young man who is morally trustworthy…with some gumption and a professional training." A friend at Cornell recommended a promising senior, Charles Matthews Manly (*below*), who became Langley's invaluable "assistant in aerodromics."

Rare and Manuscript Collections, Kroch Library, Cornell University

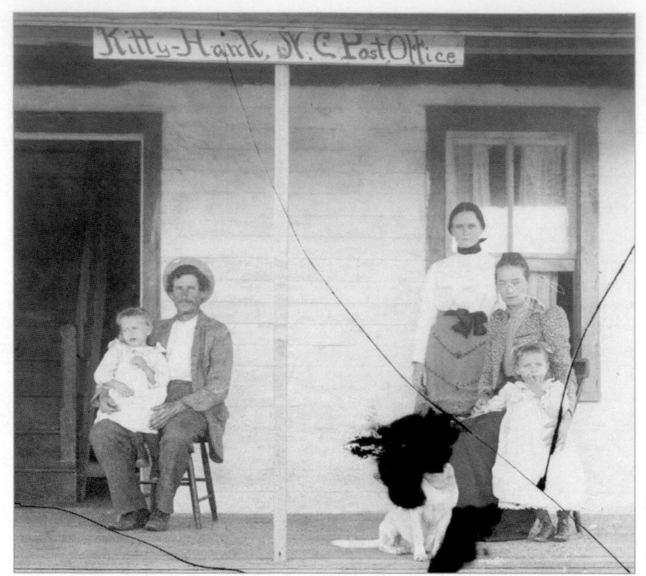

Library of Congress

Bill Tate was the leading citizen of Kitty Hawk, North Carolina. Shown above with his family at the village post office, which Addie Tate served as postmistress, Tate promised Wilbur "a stretch of sandy land one mile by five…not a tree or bush anywhere to break the wind current." The brothers' first view of the distant Kill Devil Hills (*below*) proved that Tate had not exaggerated.

Library of Congress

Library of Congress

The Wrights' second glider (*above*), built and tested in 1901, confounded their expectations. Wilbur nevertheless mounted the machine to make many free glides (*below*). Orville apparently attempted none that year. Wilbur said: "The operator…is so busy manipulating his rudder and looking for a soft place to alight that his ideas of what actually happens are very hazy."

Library of Congress

Wright Brothers Aeroplane Company

The Wrights' wind tunnel (*above*) and lift balance (*below*)—shown here in authentic replicas built by the Wright historian Nick Engler—allowed the brothers to test more wing surfaces at more angles in a single day than they could during an entire season at the Kill Devil Hills. In a few weeks at the end of 1901, they rewrote and vastly extended the entire store of man's aerodynamic knowledge.

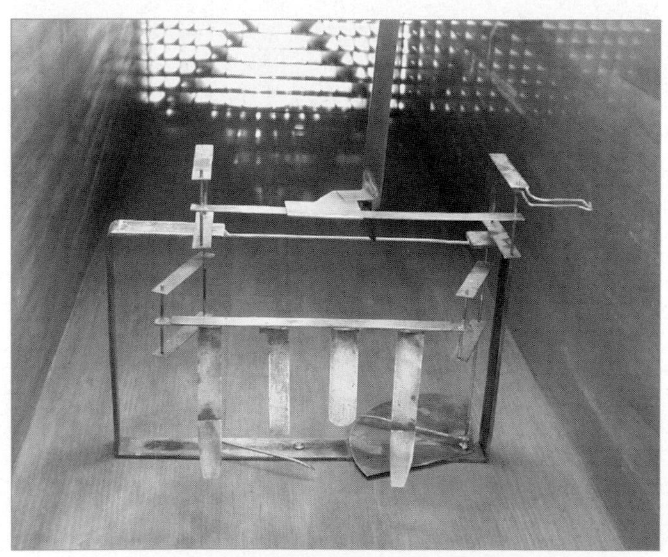

Wright Brothers Aeroplane Company

Library of Congress

The brothers returned to their camp at the Kill Devil Hills in 1902 with a new glider and fresh hope that the data gleaned from the wind tunnel would enable them to master the mysteries of balancing in thin air. Their camp kitchen (*below*) reveals their meticulous attention to detail.

Wright State University

Library of Congress

The 1902 glider (*above*) hoisted its lines (held by Dan Tate and Will) nearly to the vertical—proof of superb aerodynamic efficiency. Lift was maximized, drag minimized, and the goal of true soaring flight brought within reach. The glider in free flight (*below*) was "the largest machine ever handled" with "the longest time in the air, the smallest angle of descent, and the highest wind!!!"

Wright State University

Warm mutual affection and respect developed between Charles Manly and Secretary Langley, who nevertheless allowed Manly to pilot the great aerodrome despite the lack of any provision for a safe return to the earth. Supervising the assembly of the craft atop its enormous houseboat, Manly feared that any gust of wind might send it tumbling off its platform. But he scouted the Potomac's banks "so that I can readily recognize all the landmarks while flying through the air."

Library of Congress

Winter crept into the Wrights' camp in 1903 (*above*) as they raced to fix devilishly unreliable propeller shafts. Orville's twelve-second powered flight on December 17 (*below*) was far shorter than Wilbur's flight of nearly a minute later that day. But Orville's was "the first in the history of the world in which a machine carrying a man had raised itself by its own power into the air in full flight, had sailed forward without reduction of speed, and had finally landed at a point as high as that from which it started."

Library of Congress

Library of Congress

At Huffman Prairie outside Dayton, the Wrights flew close to the ground until they were entirely confident they had mastered their machines. Will said: "Skill comes by the constant repetition of familiar feats rather than by a few over-bold attempts at feats for which the performer is yet poorly prepared."

Library of Congress

Library of Congress

Alexander Graham Bell tested the flying power of his tetrahedral cells in dozens of configurations, from rings to stars to triangles. Kite flying at his Cape Breton estate was both a scientific endeavor and a family pastime. Bell is shown below with a grandson and several assistants, tugging the line of a giant tetrahedral kite.

Library of Congress

Library of Congress

The Aerial Experiment Association's White Wing (*left*), the first aeroplane equipped with working ailerons, was a gauntlet thrown at the Wrights' feet. Glenn Hammond Curtiss (*right*), "fastest man in the world" as a racer of motorcycles, became enchanted by the aeroplane's capacity for speed, and determined to surpass the Wrights as the pioneer and leader of American aviation.

Wright State University

The Voisin biplane of Leon Delagrange was one of several that French enthusiasts declared equal if not superior to the Wrights' heralded but unseen machine.

Wright State University

Arriving in New York in 1909, Wilbur told reporters: "I have not come here to astonish the world. I don't believe in that kind of thing." Days later, his impromptu flight around the Statue of Liberty did precisely that.

Harper's Weekly

For a moment he felt plain fear. He could not move. "The younger brother bade me move to one side for fear it might come down suddenly; but . . . the sensation that one feels in such a crisis is something hard to describe." Charlie Taylor told Root he had felt something similar when he first watched the weight of the catapult plunge and the machine rocket forward—"he was shaking from head to foot as if he had a fit of ague." One's first sight of a human being in flight was downright unnerving, whether out of fear for the pilot or sheer shock at the sight, or both. Yet Root "said then, and I believe still, it was one of the grandest sights, if not the grandest sight, of my life."

He groped for a comparison.

"Imagine a locomotive that has left its track, and is climbing up in the air right toward you—a locomotive without any wheels we will say, but with white wings instead. . . . Well, now imagine this white locomotive, with wings that spread 20 feet each way, coming right toward you with a tremendous flap of its propellers, and you will have something like what I saw." In fact the 1904 flyer looked nothing at all like a locomotive. Perhaps Root had in mind not so much the flyer's actual appearance as the complicated emotional response that it provoked. Like a locomotive, the flyer was technological power in motion. Though both machines obeyed their pilots, they could do things their pilots could not do. Men had made them. But they transcended man.

Root found Will's return to earth almost as enchanting as the flight itself. "When the engine is shut off, the apparatus glides to the ground very quietly, and alights on something much like a pair of light sled-runners, sliding over the grassy surface perhaps a rod or more. Whenever it is necessary to slow up the speed before alighting, you turn the nose [up]. It will then climb right up on the air until the momentum is exhausted, when, by skillful management, it can be dropped as lightly as a feather."

Root asked Will if he could have flown higher. Yes, he was told, "There was no difficulty whatever in going above the trees or anywhere he chose; but perhaps wisdom would dictate he should have still more experience a little nearer the ground. The machine easily made thirty or forty miles an hour, and this in going only a little more than half a mile straight ahead. No doubt it would get up a greater speed if allowed to do so—perhaps, with the wind, a mile a minute after the first mile. The manager could doubtless go outside of the field and bring it back safely. . . . But no matter how much time it takes, I am sure all the world will commend the policy so far pursued—go slowly and carefully, and avoid any risk that might cause the loss of a human life. This great progressive world can not afford to take the risk of losing the life of either of these two men."

Root's appreciation of the automobile was now quite surpassed. If God had been beneficent in providing man with the tools to create a horseless carriage, then His generosity now was beyond reckoning. Root envisioned an extraordinary future. "Everybody is ready to say, 'Well, what use is it? What good will it do?' These are questions no man can answer as yet. . . . The time may be near at hand when we shall not need to fuss with good roads nor railway tracks, bridges, etc., at such an enormous expense. With these machines we can bid adieu to all these things. God's free air, that extends all over the earth, and perhaps miles above us, is our training field. . . .

"When you see one of these graceful crafts sailing over your head, and possibly over your home, as I expect you will in the near future, see if you don't agree with me that the flying machine is one of God's most gracious and precious gifts."

TO AMOS ROOT, the circling maneuver overhead was cause for devotion. To the brothers, it was a fascinating but tricky business, a job of keeping warring aerodynamic forces in balance. Their system of balance and steering—the pilot's hip cradle, the twisting wingtips, the vertical rudder in back, the horizontal rudder in front—was put fully to the test for the first time.

To initiate a left turn, for example, the pilot shifted his hips slightly to the left. The cradle pulled on various cables, causing the right wingtip to shift to a slightly steeper angle while the left wingtip shifted to a slightly flatter angle. At the same moment, the vertical rudder in back shifted slightly to the left.

The craft began to turn. It was one thing to watch it happen, quite another to put it in words for someone who had never seen it. From the ground, the motion looked natural, organic, all of a piece. Every part of the aeroplane was connected to every other, and they all moved in harmony to produce what seemed to be a very simple maneuver—a turn in the air. Will and Orv held a clear image of the aerodynamics in their minds. But to reduce the process to words was invariably confusing to those who lacked the brothers' hard-won familiarity with flying.

The shifting wingtips in a left turn, as Will put it years later, "caused the machine to tilt so that the left wing was lower than the right wing, which, of course, in turn, caused the machine to slide somewhat to the left." Several things were happening. The machine was tipping to the left; thus it began to "slide"—that is, to fall—to the left, just as a sled slides down an icy hill, only this was a sideways slide. The vertical tail in the rear struck the air flat side on, and like a weather vane, tended to twirl the machine around, also to the left. The left wing, with its flatter angle, now enjoyed less lift than the right wing, with its sharper angle. The left wing dipped and slowed, while the right wing rose and accelerated. It was a banking turn, a carefully controlled plunge to one side, an incipient disaster that begins but is never allowed to end. Again, the best comparison is to the turning of a bicycle. The plunge is carefully caught in time and put to work for the operator, who wants not to fall but just to turn. When you saw it happen in the sky, it seemed obvious and natural, because it looked so much like the familiar sight of a bird.

Without seeing the machine, the proponents of "automatic" or "inherent" stability—including Chanute, Langley, and many French enthusiasts—found the Wrights' theory of control not only hard to grasp but frightening. They were loath to risk a pilot's life in a machine that depended on that pilot to manage such a tricky combination of forces. Certainly it took practice. But as Will implied, it was no more risky than the safe assumption that a feathered arrow, if dropped on its side, will strike the ground point first.

You had to see it to believe it. Amos Root saw it and believed. Octave Chanute, for all his intimate correspondence with the brothers, and all his knowledge of aeronautics, visited Dayton that fall only for one day and saw only one meager flight. It was a twenty-four-second, 420-meter hop by Orville that ended in a poor landing—a far less impressive performance than the circle flight Root had witnessed in late September.

Especially in France, Chanute continued to be regarded as the dean of aeronautics. He believed the Wrights' reports of circle flights, and spoke up for them. Yet he had not seen the thing that instantly rendered the state of the art, as he knew it, archaic. For all concerned—Chanute, the French, and the Wrights—things might have turned out better if Chanute had seen what Amos Root saw.

WITH HANDSHAKES and good wishes and promises to stay in touch, Root departed. Through the rest of September and all of October, Orville made most of the flights. But only once could he match Will's circuit of the field.

On Tuesday, November 8, the incumbent Theodore Roosevelt won his first full term as president by defeating his Democratic opponent, Judge Alton Parker, in a landslide. The Wrights approved of the president, an eminent bird-watcher and naturalist, and the next day, Will said, "We went out to celebrate Roosevelt's election by a long flight." Will flew nearly four times around the field for a distance of nearly three miles—by far their best effort.

His father was watching. So were at least two employees of the Dayton, Springfield and Urbana Railroad. The interurban rumbled to a stop at Simms Station, just across the road, about once every half-hour. The brothers knew the schedule and tried to time flights to avoid been seen. But weather and happenstance sometimes made this impossible. The railroad men and probably some passengers caught sight of the machine in the air and watched.

On December 1 Orville matched Will's four-circle flight.

Yet they *still* were not really in control of the flyer. Too often, the machine continued its infuriating tendency to bob up and down, despite their best efforts to fight the horizontal rudder for mastery. Of some forty flights after Root watched them on September 20, only three surpassed the distance Will made that day, and seven flights ended in injuries to the machine. Of the five flights that Will recorded in early December, all were less than one hundred meters long and two ended in crunching accidents. They quit on the ninth of December.

Just before Christmas, Will wrote to Amos Root, telling him he was free to publish his account whenever he wished. He did so in his issue of January 1, 1905. As always, Root introduced his essay with a scriptural text, this one from the Old Testament Book of Numbers, "*What hath God wrought?*"

"Dear friends," the little man began, "I have a wonderful story to tell you—a story that, in some respects, outrivals the Arabian Nights fables."

Beyond the subscribers to *Gleanings in Bee Culture*, no one took notice. Even Root's readers were not uniformly impressed. They objected to Root's characterization of the Wrights' experiments as ground-breaking and historic, citing the flights of Santos-Dumont and many others. Root wrote to the editors of *Scientific American*, offering his account and giving permission to reprint it. In earlier years, the editors had published several of Root's submissions. This one they rejected.

*Chapter Nine*

# "The Clean Air of the Heavens"

"THE GRANDEST SIGHT OF MY LIFE"

Wilbur flying at Huffman Prairie, November 1904

IN 1899 AND 1900, Wilbur Wright had pursued the problem of flight as a diversion, a hobby, a sport, with only a distant glimpse of the possibility of fortune and fame. His brother had joined in for the fun of it. In 1901 and 1902 the hobby became a hermetic scientific quest. Orville joined in in earnest, and all the brothers cared for was to solve a mystery that obsessed them. In 1903 the quest continued, though with the sense now that they stood on the brink of a historic achievement. In 1904, their singlemindedness encountered complications. If they were to fulfill the promise of their infant success in December 1903, they would have to make their problem the sole focus of all their working hours. That would require them to abandon their livelihood, and thus somehow to make their living from their invention.

What they wanted most of all was to continue their pursuit of the science of flight. They had known no better days than the autumn of 1901, when they had devoted endless hours to study, theory, and experiment with their wind tunnel, or the autumn of 1902, when they had tested the fruits of their research at Kitty Hawk. "[Will] and I could hardly wait for morning to come to get at something that interested us," Orville said later. "That's happiness." But full-time research would require financial independence.

To transform their science into business success, they had to sell something to someone. One possibility was to sell the spectacle of flight—that is, to stage exhibitions and charge the public for admission. This they were willing to consider, but a life of exhibitions—"to be montebanks in the montebank game," as Will put it later—was hardly their style, and such a life would leave little time for science. Another option was to sell the pure achievement of flight—that is, to win cash prizes like the one offered at St. Louis. This they were willing to try for, if the conditions suited them. Certainly a lucrative prize here and there could boost them along toward their goal.

The least appealing prospect was to manufacture aeroplanes and sell them to the public, as an automobile maker sold automobiles. Was there a market? Probably, at least among a few wealthy sportsmen. But to succeed at this, they likely would need to defend their patent in lengthy and draining legal battles. Neither brother wanted to spend his days running a business. They had found their life's work. They were explorers, scientists, and engineers. The idea of returning to manufacturing and sales, with stakes far higher and pressures much greater than any they had known in the bicycle business, left them cold.

Toward wealth for its own sake they felt just as indifferent. Both brothers possessed brains and shrewdness enough to prosper in any field they might choose. Instead, their temperaments and longings had pushed them on a path toward the unknown—never the choice of people who wish to pile up a great fortune. Now they wanted money, but only as a means toward a different end. Money would grant them freedom to experiment further. "Perhaps we could go into the manufacture of these flyers and make a fortune," Will told a reporter a little later, "but that would involve a lot of work for us, and that is precisely what we are determined to avoid. We want to get enough out of the sale of the first flyer to make us independent of work, so that we shall have leisure to pursue our investigations without having to be constantly pressed for funds or hampered by the necessity of devoting our time to the control of a great manufacturing business."

Only one sort of customer might offer them this freedom while sparing them the rigors of business-building and patent fights:

Nation-states.

• • •

IN 1904 THE BROTHERS HAD seldom flown much higher than their shed. But they stayed low only to stay safe. Their machine could ascend to any height the pilot wished, and they had glimpsed the extraordinary vistas that would open before a pilot's eyes when he ventured near the clouds. The practical uses of such a long-distance view had been obvious at least since Napoleon's day, when the French had constructed the first military balloons for scouting and signaling. A flying machine with infinitely more mobility than a balloon would be all the more valuable.

Pondering the implications, the Wrights came to believe that a few flying machines could render an army virtually invulnerable to surprise attack. Thus their invention "would make further wars practically impossible," Orville said, since "no country would enter into war with another of equal size when it knew that it would have to win by simply wearing out its enemy." To the sons of Milton Wright, who believed America to be God's instrument in the perfection of mankind and the coming of the millennium, it was no great stretch to imagine their machine might play a role in the banishing of war. Chanute encouraged the hope. "A fast flying machine will render the enemy's disposition of forces so easy of observation, and its directing minds so exposed to destruction, that nations may incline more and more to universal peace."

Of course, this was the long view. The Wrights were well aware that flying machines would likely drop explosives on their journey to the millennium, and they were not pacifists enough to let that prospect keep them from selling the machine to soldiers. "We stand ready," Will told Chanute, "to furnish a practical machine for use in war at once."

THE FIRST MILITARY MAN to take the Wrights seriously was Brevet Lieutenant Colonel John Edward Capper, commander of the British Army's Balloon Section and perhaps His Majesty's most knowledgeable expert on aeronautics. Just past forty, Capper was known to be "a scientific soldier," a capable administrator, a strict and able field commander, and a passionate advocate for the founding of a full-fledged Air Arm. Capper had first learned of the Wrights from Patrick Alexander, the British flight enthusiast who had visited Dayton at the end of 1902. In the fall of 1904, on a trip to America chiefly to see the aeronautical exhibitions at St. Louis, Capper took the advice of Alexander and Chanute to swing through Dayton to meet the Wrights himself.

While in St. Louis, Capper developed a keen skepticism about American claims. "It is of no use whatever pointing anything out to an ordinary American; they are all so damned certain they know everything and so absolutely ignorant of the theory of aeronautics that they only resent it." Before leaving St. Louis, the Englishman spoke with a reporter about the Wrights and expressed "grave doubts concerning the veracity of even the comparatively modest results they were claiming."

Yet only a few days later, the same reporter ran into the colonel in New York and found him "positively enthusiastic" about the Ohioans. Capper had not seen the machine fly. Of the machine itself he had seen only the engine. He had done nothing more than share breakfast with Wilbur Wright and look at some photographs. Yet Capper came away entirely confident that the brothers had done what they said.

The Wrights invariably struck people this way. If it had been an act, one might call it "the Wright treatment." But it was no act. It was merely the impression made by habits of expression and behavior that had developed in the home of Milton Wright. In print, the brothers' claims sometimes looked astonishing and implausible. But in person, the men themselves never seemed to be claiming more than they could prove. They did not say what they could do, only what they had done. There was no grandstanding or strutting. They barely spoke of themselves at all. They spoke only of the machine, and with the air of a proud but strict father who does not want his talented child to get a swelled head. "We call our invention 'the Flyer,' " Orville said once, "which is merely a simple name, as we believe more in practical results than in names. The thing will fly as well under any name, so this is not of any importance." Newcomers found them courteous, reserved, self-effacing, practical, and unusually careful in their choice of words, with a hint now and then of ironic humor. One who met them in 1905 thought Orville had "a face more of a poet than inventor or promoter," while the elder brother was "even quieter and less demonstrative than the younger. He looked the scholar and recluse." They were "very modest in alluding to the marvels they have accomplished—looking more to the renown and glory that will surely come to them as the solvers of the problem of mechanical flight, than any possible pecuniary reward."

With Wilbur especially, something in his gaze and bearing whispered of genius and inspired absolute confidence, even with total strangers. A British acquaintance said, "One could not help being impressed by his absolute honesty, sincerity and self control, as well as by his obvious intellectual powers. . . . It was impossible to evade the thought that he was a man apart."

In his detailed report, Colonel Capper told his superiors the Wrights "have at least made far greater strides in the evolution of the flying machine than any of their predecessors. "I do not think they are likely to claim more than they can perform."

WILL HAD TOLD CAPPER that he and Orville were not quite ready to do business. But just two months later, in January 1905, the brothers made their first approaches to two governments—the American and the British.

On January 3, Will told his story to Dayton's Republican congressman, Robert Nevin, who offered to help. Nevin asked Will to write him a formal letter, which he would then pass to William Howard Taft, President Roosevelt's secretary of war; then Nevin could help to arrange a meeting. Will wrote that letter—which summarized the experiments and asked Nevin "to ascertain whether this is a subject of interest to our own government"—and another letter to Colonel Capper, saying the brothers were ready to offer a deal to the British, too, if the British were interested.

The timing of these approaches was bound up with the Wrights' concerns about privacy and patents. They believed they needed at least one more season to bring their invention to a state of practical perfection. But with their application for a patent not yet answered, they worried about witnesses. Better to turn the secret over now, win their financial reward, and continue their work in official privacy, far from the eyes of competitors and reporters, than to risk more exhibitions out on Huffman Prairie. The time required to perfect the machine and to train military men in its operation, plus "the increasing difficulty of securing the necessary privacy for further experiment," made them think they should, as they told Capper, "bring the matter before military authorities for their consideration." The advantage to the purchaser would be substantial, Will promised: "There is no question but that a government in possession of such a machine as we can now furnish and the scientific and practical knowledge and instruction we are in a position to impart, could secure a lead of several years over governments which waited to buy perfected machines."

Until now, the Wrights had had to depend on no one but themselves and their own employee, Charlie Taylor. At this point, for the first time, they needed someone else. The brothers held a powerful bias in favor of self-reliance. What happened next only seemed to vindicate that view.

Congressman Nevin, as promised, passed the Wrights' letter to the War Department. But the congressman proceeded to get sick, and so could not apply the personal touch that might have secured a meeting between the War Department and the Wrights—the sort of meeting in which Will seldom failed to impress his interlocutors. No doubt to cover his own credibility in a city where "Langley's Folly" was still a fresh memory, Nevin endorsed the Wrights in terms that fell far short of enthusiasm.

> I have been [so] skeptical [he told Secretary Taft] as to the practicability and value of any so-called "Flying machine" or "air ship" that I did not give much heed to the request made . . . by the gentlemen whose letter I attach hereto, until I was convinced by others who had seen their experiments . . . that there was really something to their ideas. I do not know whether you, or the proper office of the government to whom this matter will be referred, will care to take it up or not, but. . . . I am well satisfied they have at least succeeded in inventing a machine worthy of investigation.

Without Nevin's personal intercession, Taft's staff simply passed the letter along to the appropriate office—that is, to the supporters of Secretary Langley, the Board of Ordnance and Fortification.

In the thirteen months since the great aerodrome had made the BOF a laughingstock, nothing had happened to dispose its members any more kindly toward the inventors of flying machines. They glanced over the letter from Dayton and whipped out their form response—sent to Nevin, not the Wrights.

> As many requests have been made for financial assistance in the development of designs for flying-machines, the Board has found it necessary to decline to make allotments for the experimental development of devices for mechanical flight, and has determined that, before suggestions with that object in view will be considered, the device must have been brought to the stage of practical operation without expense to the United States. It appears from the letter of Messrs. Wilbur and Orville Wright that their machine has not yet been brought to the stage of practical operation, but as soon as it shall have been perfected, this Board would be pleased to receive further representations from them in regard to it.

It need not have ended there. That it did—for the time being—must be blamed on the stiff-necked anger of Milton Wright's sons when anyone doubted their good word. Obviously, a bit of clarification was needed. The Wrights were not asking for an "allotment for experimental development."

They were offering a product for sale, a product that was close, at least, to "the stage of practical operation." A little more cajoling from Nevin, plus a few photographs, could have made both of these facts quite clear, and a productive meeting might well have followed. Instead, the Wrights chose to read the letter as no more than an insulting "flat turndown." They turned immediately to the Europeans—a move that caused Chanute to fire off a fatherly note of concern to Dayton.

Will had no apologies to make, no second thoughts, and no intention of applying any high-pressure sales tactics. On the contrary, he thought Washington owed the apology. He wrote Chanute:

> Our consciences are clear. . . . It is no pleasant thought to us that any foreign country should take from America any share of the glory of having conquered the flying problem, but we feel that we have done our full share toward making this an American invention, and if it is sent abroad for further development the responsibility does not rest upon us. We have taken pains to see that "Opportunity" gave a good clear knock on the War Department door. It has for years been our business practice to sell to those who wished to buy, instead of trying to force goods upon people who did not want them. If the American Government has decided to spend no more money on flying machines till their practical use has been demonstrated in actual service abroad, we are sorry, but we cannot reasonably object. They are the judges.

The Wrights' reaction was not quite as peremptory as it looked, for this was not the first time the War Department had spurned them. In December 1903, two rich brothers in Boston, Samuel and Godfrey Lowell Cabot, read news accounts of the first powered flights and took them seriously. They whipped off one letter of congratulations to Dayton and a second letter to their U.S. senator, Henry Cabot Lodge, one of President Roosevelt's oldest and closest friends. Citing the news stories, Godfrey Cabot told the senator, a kinsman, that "this may fairly be said to mark the beginning of successful flight through the air by men unaided by balloons. It has occurred to me that it would be eminently desirable for the United States Government to interest itself in this invention with a view to utilizing it for war-like purposes." The senator passed Cabot's letter to the War Department, where it disappeared into oblivion.

In London, Colonel Capper's personal assurances induced his superiors to take the Wrights' claims at face value. But the result was the same as in Washington—no deal. The brothers' letter passed up the chain of command, and correspondence ensued. The Wrights' price looked awfully high—twenty-five thousand pounds for one machine. More information was needed. The War Office directed its military attaché in Washington, Colonel Hubert Foster, to go to Dayton and ask for a demonstration. But Foster was assigned not just to Washington but also to Mexico City, and that was where he had scheduled himself to spend the spring and summer of 1905. He would not see any correspondence about the Wright brothers until November. Officials told the Wrights they would hear from Foster later; then they let the matter ride.

IN THE FERVENTLY PATRIOTIC Third French Republic, there was no Col. Capper to vouch for the Wrights' character or to argue that they were men to be taken in deadly earnest. Instead, a deep and frantic skepticism was taking hold.

Immediately after the Kitty Hawk flights, Ernest Archdeacon had envisioned catching and passing the Wrights quickly and easily. First he or some other Frenchman would master the Americans' gliding technology. Then, they would leap ahead by taking advantage of France's commanding lead in light automotive engines. So, early in 1904, Archdeacon commissioned a hasty replica of the 1902 Wright glider, based on descriptions and drawings supplied by Octave Chanute. Archdeacon believed it was "exactly copied" from the American machine. But Chanute, as usual, had provided more smoke than light. In fact, Archdeacon's version was smaller, had greater wing curvature, and lacked any device for balance and steering. Piloted by Gabriel Voisin in April 1904, it flew only twenty meters.

A month later, the engineer Robert Esnault-Pelterie tested his own "exact copy" of the Wright glider, this one with wing-warping. It, too, was a failure, and Esnault-Pelterie promptly pronounced wing-warping dangerous.

In March 1905, Archdeacon hitched another glider of the "*type de Wright*" to an automobile and towed it into the air. It crashed immediately.

Yet to Archdeacon and Esnault-Pelterie—and many other French enthusiasts—the performances of these "exact copies" were cause for something like celebration. The Wrights' reported results could not be replicated; therefore, they were liars, and the field remained open to the French!

In a lecture in January 1905, Esnault-Pelterie told the Aéro-Club that his machine had been "exactly like that of the American experimenters" (except for "some questions of construction and detail"); that "the disposition of the controls was exactly that indicated by the Americans." Yet he quickly "proved that no suitable experiment could be attempted with this machine."

Archdeacon drew the obvious conclusion in a letter to the Wrights themselves. Chanute once called Archdeacon "a genial enthusiast whose only fault was that he talked too much." This was certainly one such case. With a patronizing sneer, he dared the Wrights to invite him to "risk a voyage to America to see your apparatus, if I am permitted to see it and to see it operate—unless you yourselves think well of coming to France to show us that which you know how to do."

> The results which you have already obtained, if I am to believe the accounts rendered by yourselves and published in different French newspapers, are absolutely remarkable; so remarkable that they even create in my country a certain incredulity.
>
> That incredulity results from two reasons: First, from the long time which has elapsed without new results since your memorable experiments of December, 1903; and on the other hand from the mystery [with] which . . . you surround yourselves to give you time for securing your patents.
>
> On that point, I have been a little astonished; in the sense that for my part, I do not believe in patents; and that, on the other hand, if you really have in your apparatus any novel devices you have taken a long time in getting the said patents.

Archdeacon concluded with his nose higher still. "If you are our masters in aviation, we certainly are the masters in the matter of light motors. . . . You can find motors with us weighing two kilogrammes per horse power, which you certainly do not have with you." And he challenged them to compete for the Deutsch-Archdeacon Prize: fifty thousand francs for the first flight of a kilometer in a closed circle.

In Dayton, the Wrights found this both hilarious and revealing. Such a missive from a man who clearly could not fly showed them precisely how far behind the French were. If Archdeacon did not believe they had flown, it meant he understood very little about gliding as the true path toward powered flight.

"Mr. Archdeacon will find more compliments than information in our answer to his letter," Will told Chanute, "as we are more ready to congratulate him upon the interest in aviation he has succeeded in arousing in France, than to show our machines and methods." As for the Deutsch-Archdeacon Prize, he said they might get around to it later, "when our other arrangements will permit."

"They are evidently learning that the first steps in aviation are much more difficult than the beginnings of dirigible ballooning, and are skeptical of what others are reported to have done in that line. It is not surprising. They have much to learn."

The Wrights were keeping a close eye on developments in France. They subscribed to *L'Aerophile* (which Orville translated) and saw clippings from other papers. So far, they saw nothing to make them think their lead was in danger. They did worry that Frenchmen would be hurt, though their reading tended to confirm their father's bias against the French as a group. When the aero enthusiast Alexandre Goupil took a potshot at Chanute in print, Will called the gesture "little and contemptible. It displays gratitude for [Chanute's] kindness in a truly French manner." Certainly they had no intention of offering helpful advice. One reason for closely guarding their secrets, Will told Chanute, was "to give the French ample time to finish and test any discoveries of the secrets of flying which any Frenchmen might possess, and thus shut them off afterward from setting up a claim that everything in our machine was already known in France."

They did not care for the increasingly common hints "that in France the Americans are not to be believed upon their mere word." Privately, Will said the brothers regarded "all such intimations with great amusement and satisfaction." He did not mean it in a friendly way.

AS THE BROTHERS SOUGHT to perfect the control of their flyer in the air, their father regained control of his church, and the Wright family put to rest the trouble that had hung over them for more than thirty years.

The General Conference of the Church of the United Brethren (Old Constitution) was to be held in the town of Caledonia, Michigan, 330 miles north of Dayton, in the middle of May. The meeting would decide Milton's fate once and for all. Wilbur set aside his work for a week to accompany his father, and his tract on the Keiter struggle was distributed to the delegates. It may have helped, but Milton's assiduous campaign of letter-writing already had won a majority in his favor, and he knew it. His foes charged that he "dominated the conference with an iron hand equal to that of any political despot." Still, debate was heated. On the third day, Bishop Wright's supporters moved that his expulsion from the Church be declared null and void. The vote in his favor was overwhelming. Milton immediately declared his intention to retire at the close of the conference.

Back at home, Will told Chanute: "It was the decisive battle in the contest of which you have heretofore heard us speak. We won a complete victory; turned every one of the rascals out of office, and put friends of my father in their places. It will be a relief to have that matter off my mind hereafter." As for Millard Fillmore Keiter, he and a few supporters broke away from the Brethren and attracted followers in several states. The followers reversed course when Keiter was found to have embezzled two thousand dollars from one of his supporters and was arrested for land fraud.

The Wright brothers had seen the vindication not only of their beloved father, but of his life's method, as well—to grasp the nettle, no matter how difficult, and not let go. There is a parallel to the brothers' success in aeronautics.

Samuel Langley perceived the problem of balance for twenty years, but held back from wrestling with it. He hoped to circumvent or at least postpone the inevitable confrontation by the application of sheer power. The Wrights perceived the same problem, seized it immediately, and never let go until they solved it, which they now proceeded to do, once and for all, in the late summer and early fall of 1905. As ever, the solution had little to do with their engine and everything to do with their mechanisms of balance. "The best dividends on the labor invested," they said, "have invariably come from seeking more knowledge rather than more power."

CHARLES MANLY, however, refused to give up. Staying on as Langley's aide in aerodromics, he declared himself determined to renew the trials, "if not immediately, at the earliest time possible." He ensured that all the necessary equipment—tools, engines, the launching car and derrick—was safely stored, ready for new uses at any time. Rather than sell the great houseboat, he scouted the Potomac riverfront for a suitable place to moor it indefinitely. He oversaw the careful repair and storage of the aerodrome itself. He considered alterations in design and strategy, concluding the wings should be superposed and trials made over land.

And he worked on his chief's fallen spirits. "I take this occasion," Manly told Langley in June 1904, six months after the disastrous test the previous December, "to again repeat that I have full confidence in the ability of the machine to fly if given a fair trial, that I think the expense of giving it a fair trial over the land will not exceed ten thousand dollars, and that I am determined that in some way or other a fair trial and the success it deserves must be had." Others echoed Manly. Octave Chanute, never reluctant to back more than one horse in a race, believed for many months that the War Department could be persuaded to pay for new trials, for "to do otherwise would be to confess that the Board of Ordnance did not know what it was about in providing funds."

But Langley had had enough. As soon as he learned his appeal to the War

Department had been rejected, he began to gather the wastebooks and diaries and letters that recorded his twenty years of aeronautical experiments. He did not intend to seek out new paths to explore, but to compile a history. It was over.

He maintained his busy schedule of social engagements, often seeing such luminaries as the Bells, Secretary of State John Hay, even the president and Mrs. Roosevelt. His most intimate friendships meant more to him than ever. Eliza Orne White, an old acquaintance, encountered him one day probably about this time and "thought how changed he was, how old, and gray." Yet just a week later, when Mrs. White happened to see Langley again at the home of the popular writer Edward Everett Hale, she watched as the secretary practically bounded up the Hales' steps and, "with his face positively illuminated," announce that Hale had just invited him for an extended visit that summer—"and that if he 'really meant' it, he was coming." Mrs. White said: "I never saw so complete a transformation in anyone; the years seemed to have dropped from him, and the old man of a week before was the embodiment of life and joy."

In March 1905, Charles Manly, "with very great regret," left the Smithsonian to begin a career as a consulting engineer in New York. Thanking Langley for "the many personal kindnesses which you have shown me," Manly, now nearly thirty years old, restated his belief "that the work can be brought to a successful completion. . . ."

> While you have chided me for my optimism, yet I think that if you will review the records or your earlier work with the aerodromes, you will agree with me that it was only optimism and a determination to overcome apparently insurmountable difficulties, which enabled you to achieve the most noteworthy results which you have done in the field of aerodromics.
>
> While the aerodrome itself has been judged a failure by the general public, who are dependent on garbled newspaper reports for their information, yet the data, and the information on *what not to do,* that has been obtained is certainly most valuable, and I feel very certain that were it possible to continue the work, the success which it rightly deserves would be achieved. At the same time whether the work is ever completed or not, the very great thing which you have done in having made it respectable and safe to be reckoned as a believer in the possibility of flying machines, certainly deserves the highest praise, and this I think is recognized even by the misinformed public.

The houseboat was sold. The Army requested the return of items it had loaned to the Smithsonian, including four ladders, six pairs of oars, and fifteen feet of firehose. There was no request for the great aerodrome.

THREE MONTHS LATER, on Saturday, June 3, 1905, N. W. Dorsey, a Smithsonian clerk, came to Richard Rathbun and told the assistant secretary that the Institution's account at the Treasury Department had been reported as overdrawn. Rathbun sent Dorsey to find William Karr, the Smithsonian treasurer, and alert him to the problem. Dorsey found Karr at home, too drunk to say anything intelligible.

First thing Monday, Karr, now sober, said there must have been a mistake at the Treasury. He would see that it was corrected. The next day, Dorsey and a colleague sought out Cyrus Adler and told him something was gravely wrong in Karr's office. Adler went and found Rathbun. Together they went to Langley, and "a hasty examination of certain papers showed that there might be a discrepancy in the Smithsonian Institution account of something over $40,000."

The Institution's attorney was summoned while Rathbun borrowed the secretary's carriage and went looking for Karr. A search of the treasurer's favorite haunts led back to his home, not far from Langley's, some twenty blocks north of the White House. A friend who was already there came to the door and told Rathbun that Karr was threatening suicide. After some parleying, Karr agreed to speak with Rathbun. The two men sat in Karr's parlor. He was calm but still speaking of suicide, "as he could not stand the humiliation." He handed over the keys to his office safe and desk, saying it was "all up" and that "he would soon be beyond any investigation."

Rathbun said it appeared that forty-five thousand dollars was missing. Karr said, "It is nearer sixty thousand." He explained that at first he had meant to pay the money back, "but getting deeper and deeper into it, he saw there was no hope." The assistant secretary asked how long the embezzlement had been going on. Karr said fifteen or twenty years; he could not remember precisely.

It could not have been twenty years, for Karr had been appointed clerk in charge of disbursements only seventeen years earlier, six weeks before Langley was named secretary. Some were led to the inescapable suspicion that Secretary Langley's preoccupation with the pursuit of flight had given William Karr a golden opportunity.

Langley informed the regents he would accept no further salary so long as he served as secretary.

• • •

IF KITTY HAWK IN 1903 had been "too sandy and too far from machine shops," then Huffman Prairie in 1904 had proven "too small and too public. We have thought some of hunting a prairie location in the West." But any field in Kansas or Nebraska that was lonesome enough for the Wrights would lie as far from a good machine shop as the camp at Kill Devil Hills. In the end, they returned to Huffman Prairie for the experiments they planned for the 1905 season.

Once again they had a new machine. Only the engine survived from 1904. The craft was more imposing than its predecessors. The horizontal rudder reached farther out in front of the wings than it had in the 1904 model, and the vertical tail stretched out farther in the rear. The wings stood higher above the skids, and the construction was stronger and heavier, to prevent time-consuming breakages. To the uprights of the horizontal rudder in front, the brothers attached two new vertical vanes in the shape of half-circles. They hoped these "blinkers" or "blinders"—named after the horses' eyewear, which they resembled—would prevent the sideways sliding that had vexed them at unpredictable moments the summer before. Finally, they detached the control of the vertical tail from the wing-warping controls. The pilot would now control three elements independently—wing-warping with the hip cradle; the forward rudder with one hand lever; the tail with another hand lever.

In June and July 1905 they made only nine flights. None was promising, and Orville's attempt of July 14 ended in a frighteningly sudden plunge to the ground nose first. He "was thrown violently out through the broken top surface but suffered no injury at all."

Charlie Taylor came to the field most days. Sometimes they were helped by William Werthner, a colleague of Kate's from Steele High, who would help to re-lay the launch rail or to steady a wing during launchings. "The good humor of Wilbur, after a spill out of the machine, or a break somewhere, or a stubborn motor, was always reassuring," Werthner said later. "Their patient perseverance, their calm faith in ultimate success, their mutual consideration of each other, might have been considered phenomenal in any but men who were well born and well reared."

Heavy rains flooded the prairie. The brothers added still more heft to the horizontal rudder, increasing its surface area more than half again and extending it to nearly twelve feet in front of the wings. Outside the shed, they had to jump from hummock to hummock to keep their feet dry.

It was the last week of August before the field dried out. The new horizontal rudder took some getting used to—"a very comical performance," Will said of Orville's first try—but then, suddenly, things were much better. As they had hoped, the changes reduced the front rudder's hypersensitivity. The sharp bobbing and dipping leveled out, and the frightening darts to the ground were curtailed. On September 7, Will chased a flock of birds in two circles of the field. The next day Orville flew the first figure 8. They altered the shape of the propellers, which improved their performance still more.

"Our experiments have been progressing quite satisfactorily, and we are rapidly acquiring skill in the new methods of operating the machine," Will told Chanute. "We may soon attempt trips beyond the confines of the field." They thought better of that idea. But they decided they were indeed ready to see just how far they could fly.

On Tuesday, September 26, they brought their father with them on the interurban trolley. Will went up and began to circle the field, again and again and again. They lost count of the times; they thought it was sixteen. He glided to the ground only when the gas tank ran dry. He had been up for nearly twenty minutes and covered roughly eleven miles, beating their previous record by a factor of three.

After six years, they possessed "a machine of practical utility."

WITH THE FIRST HINT of autumn in the air, the brothers began to invite guests to Huffman Prairie. Lorin and his wife, Netta, had seen a flight or two. Now they came back with their children. Kate came on October 4, though it was a school day, and saw Orville fly more than twenty miles. Torrence Huffman came again, as did a number of friends and neighbors. Among them were the brothers' landlord on West Third, Charles Webbert, and his brother, Henry, who was Charlie Taylor's father-in-law. Bill Weber, a plumber, came; and Ed Ellis, an old friend from the Ten Dayton Boys club who was now assistant auditor of the city of Dayton; Bill Fouts, a druggist and friend of Orville's, and Fouts's friend Theodore Waddell, an employee of the U.S. Census Bureau who happened to be in town; and C. S. Billman, secretary of the West Side Building Association, who brought his wife and three-year-old son, Charlie.

It's not clear how many of these people came simply out of curiosity, and how many were specifically invited. Ellis, for one, was invited. In any case, several proved in time to suit the brothers' needs precisely—they were respectable, sober citizens whose eyewitness accounts would be taken seriously, yet who knew nothing about aeroplanes, and thus would be unable to give away the technical secrets hidden in the machine.

The flights were now so long that it was impossible to avoid stares from passing trains. Enough talk went around town that reporters began to show up. At one point Will asked William Werthner to stand out at the road and tell any reporters who appeared that cameras were not welcome. Still, the *Daily News* carried a story on October 5, and it was picked up by the *Cincinnati Post*. Finally, the patient Luther Beard of the *Journal* was moved to visit Huffman Prairie, where he learned that his reserved fellow passengers on the interurban had been doing "something unusual" all along.

On that day, Will stayed in the air for nearly forty minutes and flew nearly twenty-five miles.

Then the brothers locked the machine in the shed. They had hoped to make flights for another week or ten days, but they were still waiting for approval of their application for a patent, and "we were frightened at the number of people from Dayton which the flight was beginning to attract . . . ," Orville said later. "Three [trolley] cars passed us that day during the flight. There were more spectators than ever before. We feared that the next day there would be a crowd perhaps of photographers. We had always taken the greatest care to prevent photographs being taken." The machine "had proved the principle necessary to prove. It was only the day previous to the last flight that we had perfected the means of maintaining equilibrium whatever winds might be blowing aloft. . . . When that was proved nothing but the principle remained."

THEIR ABILITY TO CONTROL the aeroplane became rather marvelous. In one of the last flights of the season, the engine became overheated, and Orville had to cut it off while still in the air at the far end of the field. With the engine silent, he glided all the way back to the shed. He came to a stop just outside the door. Ed Ellis was struck by the pilots' obvious control of the craft. "It looked like a monstrous bird in the air," he told a reporter soon afterward, "and flew like one, too, its motion being steady and not in the least wobbly. . . . When it alighted it was as gently as the glide of a bird to the ground after its flight and with as little shock."

The *Daily News* buttonholed the druggist, Bill Fouts, and found him a bubbling fount of enthusiasm—much to the Wrights' dismay. "I wouldn't believe it myself until I saw it," Fouts said. "I went out to Huffman's prairie half expecting to see somebody's neck broken. What I did see was a machine weighing nine hundred pounds soar away like an eagle. When they were about 60 or 70 feet in the air they pointed its nose to a horizontal line and sailed around like a great bird. No wavering. No ducking up and down. It just sailed around that ring as if it had been running in a groove. I told a friend about it that night and he acted as if he thought I had gone daft or joined the liars' club. I got permission from the Wright boys and next day I took him out with me. He groaned and shook his head when he saw the machine, but when it flew up in the air and began to circle gracefully around like an eagle, his grin turned to a look of amazement."

Fouts's friend was Theodore Waddell, the Census man. Waddell was especially struck by the extraordinary means of moving the flyer around on the ground. On the day he was at the prairie, one of the flights ended prematurely at the other end of the field. He asked one of the brothers if he could give a hand in hauling the machine back to the shed. He could come if he wanted, he was told, but he didn't need to help. As Waddell watched, "They walked down the field to where the ship was sitting, poured the gasoline into the tank, and, starting the engine at slow speed, let the machine lift itself clear of the ground and walked it back to the [shed]. It was about the most uncanny sensation I ever experienced, except the other sensation of seeing that machine with a man aboard flying around in circles over our heads."

"It was beyond my comprehension," Charles Webbert said. "I took off my hat and sat down."

HOWEVER DAZZLED, some witnesses still could not foresee a practical purpose. Torrence Huffman, who watched the long flight over his pasture on October 5, asked Will afterward what the machine would be good for.

He got a single word in reply:

"War."

THE ACCOUNTS of these long flights of the autumn of 1905 that one would most like to read do not exist. If Kate wrote down what she saw, or anything about her feelings, the evidence does not survive. Bishop Wright was typically laconic. In his diary on the evening of October 5 he wrote simply: "In forenoon, at home writing. In the afternoon I saw Wilbur fly twenty-four miles in thirty-eight minutes and four seconds, one flight."

The brothers themselves spoke carefully about the sensations of flying. They may have worried that if they reported their experiences too vividly, they might cause thrill-seekers to make reckless attempts to fly. In the coming months, when reporters became desperate to know what this extraordinary new experience was like, they made it a point to answer with maximum understatement—for the fun of it, one suspects, and certainly with an eye toward establishing their bona fides as serious experimenters. "We put many months of study and many more months of practical work on a machine which was mathematically certain to fly, and when we made the test were not disappointed," one of them, probably Will, told the *New York Herald*. "The fact that we had to formulate our own principles as well as work out entirely new formulas gave us the keener satisfaction when we found our ideas were sound in every particular. It was not the surprise to us it was to others when our machine fulfilled all of our promises."

"I am afraid I cannot describe the sensation as I should," Orville told another reporter, "for flight came slowly, at the beginning in short, gliding flights near the ground, then gradually extending when power was added until I became so accustomed to it that I was at last unconscious of being thrilled."

"Is it difficult to guide the machine?" he was asked.

"No more difficult than guiding a bicycle. I could teach any young man within three days."

Occasionally they were more candid, especially with those who shared their obsession. When a reporter asked Will if he had ever gone aloft in a balloon, he said he had not. He doubted it would match the experience of flying a heavier-than-air machine. In the reporter's paraphrase, he said flight had left him "intoxicated," and "after flying once there was little inclination to turn to anything else."

Just before Christmas 1905, in a private letter to the Italian soaring enthusiast Aldo Corazza, Will said: "There is no sport in the world quite equal to that which aviators enjoy while being carried through the air on great white wings. Compared with the motion of a jolting automobile is not flying real poetry?"

The fullest statement about the pleasure of the experience that either of them ever made came from Will a couple of years later, at a time when he was much more eager, for strategic reasons, to establish the brothers' names and accomplishments in the public mind. Clearly, he had thought a good deal about the matter. Writing in *Scientific American* on the subject of "Flying As a Sport—Its Possibilities," he said:

> There is a sense of exhilaration in flying through the free air, an intensity of enjoyment, which possibly may be due to the satisfaction of an inborn longing transmitted to us from the days when our early ancestors gazed wonderingly at the free flight of birds and contrasted it with their own slow and toilsome progress through the unbroken wilderness. Though methods of travel have been greatly improved in the many centuries preceding our own, men have never ceased to envy the birds and long for the day when they too might rise above the dust or mud of the highways and fly through the clean air of the heavens.
>
> Once above the tree tops, the narrow roads no longer arbitrarily fix the course. The earth is spread out before the eye with a richness of color and beauty of pattern never imagined by those who have gazed at

> the landscape edgewise only. The view of the ordinary traveler is as inadequate as that of an ant crawling over a magnificent rug. The rich brown of freshly-turned earth, the lighter shades of dry ground, the still lighter browns and yellows of ripening crops, the almost innumerable shades of green produced by grasses and forests, together present a sight whose beauty has been confined to balloonists alone in the past. With the coming of the flyer, the pleasures of ballooning are joined with those of automobiling to form a supreme combination.

ONE DAY THAT FALL, the brothers saw two unfamiliar men walking in the fields nearby. They thought they were hunters, but they came back the next day—one of them with a camera—and asked if they could watch. The brothers said it was all right, but they asked that no photographs be taken. The man carrying the camera deliberately set it on the ground, far to one side. Apparently he did not introduce himself, or used a false name. He asked to look in the shed. The brothers agreed. Was he a newspaperman? they asked. He said no, though he sometimes wrote for publication. As they talked, he used the proper aeronautical terms for various parts of the machine.

Some time later, Orville saw a photograph of the same man in a New York newspaper. It was Charles Manly. The Wrights never suspected him of any bad motive. Apparently he had visited Huffman Prairie simply because he had wanted, at last, to see a machine fly.

AFTER THEY TOLD THE REPORTERS they were through for the season, the brothers, in fact, planned one last outing. They wanted "to put the record above one hour," and to have Chanute, for the first time, see a really first-rate powered flight. No witness would carry more credibility around the world.

Chanute was all for it. It depended on the weather. On Monday, October 30, the brothers shot a telegram to Chicago: "TRIAL TUESDAY"—Halloween. Chanute caught the train for Dayton that night. But a big storm blew in, and again the brothers' friend missed his chance.

Chanute would continue to talk and write about the Wright brothers with the air of a patron, a teacher, and an inspirer, for a time. Then, increasingly, his tone would turn critical. Yet he still had not seen them fly with their newfound mastery, banking around the field with the confidence of a hawk. And he still did not understand how they did it.

NEWS OF SPECTACULAR FLIGHTS at Dayton reached aeronautical circles around the world. In France there was little faith in the reports. The Wrights summarized the flights in a letter to Ferdinand Ferber. When he told his Army superiors, they treated him as "a mild lunatic," since it was ridiculous to think a French artillery captain would know of such an event when the major American newspapers did not.

The Wrights also sent an account to *L'Aerophile*, which published it. An uproar ensued. Ernest Archdeacon announced that "whatever the respect I feel for the Wrights—whose first experiments without a motor are undeniable and of the greatest interest—it is impossible for me to accept as historical truth the report of their latest tests, which have not been witnessed, and about which they have voluntarily maintained the most complete obscurity," their reports to Ferber and *L'Aerophile* notwithstanding. Such experiments, if they had been made, must be tentative at best, leaving the French plenty of opportunity to solve the problem themselves. "Let two or three men of real ability tackle it, and its final solution will only be a question of months, perhaps of days."

Various newspapers sent correspondents to Ohio to track down the story. Robert Coquelle, of *L'Auto*, was quickly convinced. "The Wright brothers refuse to show their machine," he cabled his editor, "but I have interviewed the witnesses, and it is impossible to doubt the success of their experiments."

IN PARIS, a wealthy American expatriate, balloonist, and Aéro-Club member grew tired of the uncertainty. Frank Lahm sent a cable to his brother-in-law, Henry Weaver, a businessman in Mansfield, Ohio. The cable caught up with Weaver in a hotel room in Chicago. It read: "VERIFY WHAT WRIGHT BROTHERS CLAIM, IF NECESSARY GO TO DAYTON, PROMPT RESPONSE CABLE." Weaver had no idea what any Wright brothers claimed. Then, vaguely, he recalled news reports of flying-machine experiments in North Carolina. He tracked them down and arranged a meeting with Orville, who "promptly told me he would do all he could to satisfy [Lahm] and me that all he reported as accomplished was the truth and nothing but the truth. His very appearance though would disarm any suspicions to the contrary."

Orville told Weaver the whole story; took him to Huffman Prairie; introduced him to several witnesses, who confirmed everything; then took him to 7 Hawthorn, where he met Wilbur. Weaver asked if the brothers were married. "As Mr. Wright expressed it," Weaver told Lahm, "they had not the means to support 'a wife and a flying machine too.' "

"They told me of their correspondence with Capt. Ferber, who I understand is a member of your Aero Club, and laughed over his assertion that there 'was not a man in all France who believed they had done what they claimed.' "

Lahm, like Ferber, did believe them. But he was nearly laughed out of the Aéro-Club de France when he said so.

ONE EVENING IN WASHINGTON about this time, a dinner was arranged to provide Samuel Langley with the comradeship and the talk of science that he dearly loved. Among those attending were Alexander Graham Bell, Octave Chanute, the astronomer Simon Newcomb, and Bell's son-in-law, the botanist David Fairchild. When the conversation turned to aeronautics, Bell got his chance to cross-examine Chanute about the extraordinary claims of the Wright brothers. But the old engineer left no room for debate.

"What evidence have we, Chanute, that the Wrights have flown?" Bell asked.

Chanute replied: "I have seen them do it."

Many years later, David Fairchild recalled that Chanute's words sent a thrill though the company. "To hear from Professor Chanute that man had actually flown made Aladdin's wonderful lamp seem simple child's play," Fairchild wrote. "There are few moments in my life which have compared to that in interest and excitement."

Of course, what Chanute had actually seen the Wrights do and what they now told him they had done were quite different. Chanute did not doubt them. Nor did Bell. But the brothers' claims, far from dampening Bell's enthusiasm for the subject, appear only to have stoked it.

Bell, like Chanute, believed the Wrights' machine might be no more than an exciting step forward—and not precisely in the right direction. Bell feared the Wrights were risking their own lives and the lives of any who followed them in machines that depended on the operator to stay in balance. Yes, their work was a magnificent beginning. But he found no reason to conclude that simply because the Wrights had found one solution to the problem, none other was possible. He believed the future of the flying machine—the true breakthrough that would inaugurate a new epoch of transportation, just as the telephone had forever changed communication—lay in a very different machine.

Once when a reporter was visiting his summer estate in Nova Scotia, Bell pointed upward and said, "See that hawk up there? He has learned to fly, but he has never risked his life. I don't believe there is any necessity in risking human life in learning how to fly heavier-than-air machines. I have always considered the experiments of the Wright brothers very dangerous."

ON JANUARY 20, 1906, near the end of a long evening, Bell stepped to the speaker's rostrum of a raucous banquet at the Waldorf-Astoria in New York.

The occasion was the close of the week-long New York Automobile Show, sponsored by the Automobile Club of America. Behind Bell stood two American flags draped in yellow forsythia. He looked out over a sea of tables festooned with four-foot-tall American Beauty roses. For dessert, balls of white ice cream had arrived on a parade of toy automobiles. The guest list of four hundred was crowded with the names of wealthy New York sportsmen and their escorts, many of them preparing to leave the city that night by midnight train for a week of auto racing at Ormond Beach, Florida, then a second week of racing in Cuba.

These were the elite enthusiasts of the dawning age of the automobile, full of the belief that autos were becoming indispensable, yet savoring the fact that few outside their own class so far could afford to own one. Keen for speed and mechanical novelty, they should have been prime recipients of Bell's message. But they had been imbibing and dining for some three hours by the time he rose to speak, and according to a reporter seated close to Bell, "a lot of promiscuous enthusiasm" made it all but impossible for most people to hear.

"Well, you've got your auto time down to thirty-five seconds for a mile," Bell began. "You're beginning to crawl. The other day I talked to a friend in Chicago and was back in half an hour. You've got something to beat in the telephone."

The joke failed to make much headway against the general noise. Bell gamely plodded forward into a brief review of aeronautical progress in Europe, then said:

"The age of the flying machine is not in the future. It is here. We already have a practical flying machine for the first time developed, and America has done it."

Among people who could hear, this drew some applause.

"The Wright brothers of Dayton, Ohio, have flown 25 miles on their machine carrying a man and a motor . . ."

There was too much noise. The banquet chairman rapped for attention, but he was competing with "bursts of laughter from far corners . . . where someone had told a good story."

• • •

AT ABOUT NOON on Wednesday, November 22, 1905, Samuel Langley was ascending the main staircase of the Smithsonian Castle when he appeared for a moment to lose consciousness. He was helped to his office, where he managed to dictate letters to his aunt, Julia Goodrich, and to his brother, John Langley. Cyrus Adler joined him for lunch, as he often did. Adler noticed that Langley had trouble pouring the tea, and that there was "a slight crookedness in his face." When it became clear the secretary's entire right side was paralyzed, Adler suggested that an ambulance be called to take Langley home. But the secretary said, "I will go home in my customary way," whereupon his carriage was summoned. He soon found it difficult to speak, and it became clear to friends that his memory was impaired. He had suffered a severe paralytic stroke.

Langley's personal physician, Dr. W. B. Pritchard, of New York City, at first saw little hope of recovery. "His usefulness as an executive chief is at an end," the doctor advised Adler, "especially in a publicly official capacity." Richard Rathbun stepped in as acting secretary of the Smithsonian.

But Langley rallied. Fighting melancholy, he soon discovered he could move and exercise his right leg, and he began to write with his left hand. Charles Walcott found that "his interest in all matters pertaining to scientific research and investigation was just as active and just as near to him as ever." Dr. Pritchard, examining him again, was surprised and encouraged. "Mr. Langley will never be our old friend, the stalwart," he told Adler, "but he will walk again and talk again and his thinking machine will do very good work again. His arm will never be of very much use to him, and certain defects of speech and memory will remain, but on the whole, the final net outcome is and will be much better than I had looked for."

Pritchard recommended that Langley spend several months away from Washington in a place where he could be out of doors. So, accompanied by his niece, Mary Herrick, he went to a small hotel in the crossroads hamlet of Aiken, South Carolina. There he rested, wrote to friends, and began a memoir of his childhood.

One day his niece brought Langley a letter from the Aero Club of America, which had endorsed a resolution honoring his contributions to the cause of flight. Its authors included Charles Manly. Knowing of Langley's unpredictable opinions of publicity, it was probably Manly who suggested the resolution be sent to Langley for his approval before the club released the document for publication. The text was all the secretary might have wished. It said "an accident in launching his aerodrome" had prevented "a decisive test of the capabilities of this man-carrying machine, built after his models which flew successfully many times," and it expressed the members' "sincerest appreciation" for "the contributions of Dr. Langley to the science of aerial locomotion."

Langley's niece asked what should be done with the resolution. He said: "Publish it."

• • •

ON FEBRUARY 27, 1906, congressmen on the floor of the U.S. House of Representatives were discussing appropriations to the Army. When Representative Edgar Crumpacker, Republican of Indiana, excoriated the committee on military affairs for wasting millions on such projects as flying machines, its chairman, John Hull, Republican of Iowa, rose in self-defense. It was not his committee, he said, but the House appropriations committee that had given funds for flying machines.

"I regard flying machines as absolutely absurd," Hull said. "I am not a scientific man, but a machine that will fly to the bottom of the river—"

Laughter drowned him out.

LATER THAT DAY, Langley suffered a second stroke and died.

NEWSPAPERS ALL OVER THE NATION reported the story. Langley's achievements as an astronomer and as secretary of the Smithsonian were all but forgotten. The failure of the great aerodrome was recalled only too well.

Like Langley himself, Wilbur Wright was struck by the fact that history seemed to have turned on the malfunctioning of a tiny piece or two of metal. Nothing had changed Will's mind about the aerodrome's ability to achieve true flight. He was sure it could not. But he also believed that even a single straight-line "hop" would have saved Langley's reputation. And as one of the few people in the world who fully understood Langley's deep desire to fly, Will grieved.

"No doubt disappointment shortened his life," he remarked to Chanute. "It is really pathetic that he should have missed the honor he cared for above all others, merely because he could not launch his machine successfully. If he could only have started it, the chances are that it would have flown sufficiently to have secured to him the name he coveted, even though a complete wreck attended the landing. I cannot help feeling sorry for him."

IN WASHINGTON, services were held at All Souls' Church, where Langley's friend, Edward Everett Hale, delivered a brief eulogy. The body was taken by private rail car to Boston, where Langley was buried beside his mother. Among the pallbearers were Richard Rathbun; the presidents of Harvard University and the Massachusetts Institute of Technology; the Harvard astronomer E. C. Pickering, Langley's old friend; Richard Olney, a regent of the Smithsonian and former secretary of state; and Alexander Graham Bell.

Bell stood to pay tribute. He chose not to restrain his bitterness.

"We are parting from one of the great men of the world," he said. "Professor Langley was not simply a man whom we have loved. He was a man whose name is written imperishably on the history of our times. . . .

"He wanted to give man control of the air. A few years ago he led the way for the first time by giving artificial flight to a body one hundred times heavier than the air. I myself have seen his enormous machine making a flight of a mile and a half over the Potomac.

"It fell—yes, it fell. But Professor Langley was only beginning his experiments. The newspapers met those experiments with ridicule. They did not treat Professor Langley with fairness. Ridicule shortened his life. He was not permitted to make his preliminary experiments quietly and in peace. His flying machine never had an opportunity of being fairly tried. His perfected apparatus was never launched in the air. It would have flown. . . .

"Ridicule, I repeat, shortened his life. We have looked upon his face for the last time, but the man and his works will permanently endure."

The elderly Julia Goodrich, Langley's devoted aunt, thanked Bell "with all my heart and soul" for telling the world "the facts that prevented his last ship to be a success. You understand that it cost him his life. . . . I wish most sincerely that you may succeed in all your undertakings . . . and that you may be wholly satisfied with the success you gain for all humanity."

Chapter Ten

# "A Flying Machine at Anchor"

"OH HOW I WISH THAT YOU MAY HAVE SUCCESS AT LAST."

Mabel Bell measures the pull of one of her husband's kites

SOME SKEPTICS SUGGESTED that perhaps the Wrights could fly when others could not because they possessed special gifts in gymnastics, like circus acrobats. The view was that even if they were flying, it was just a kind of stunt, marvelous to see but no good to anyone else. A trapeze artist might astound a crowd, but he did not inspire people to take up the trapeze as a good way of crossing the street.

Of course, there *was* skill involved in what the Wrights were doing, and they knew it. But they believed the skill could be taught to virtually anyone, just as anyone with a little training and practice could learn to ride a bicycle.

"As to our being abnormal in any acrobatic sense," Will told a reporter, "that is the exact opposite of the truth. We are both too nervous to be good flyers. Any mile-a-minute bicycle rider could take our machine and beat us completely."

Of course, no bicyclist could go a mile a minute. He was talking about motorcyclists. In fact, he may have been thinking specifically of the great motorcyclist Glenn Hammond Curtiss, a manufacturer of fine internal combustion engines whose motorcycle sprints had won him the title of "fastest man in the world."

The brothers and Curtiss met in the summer of 1906, in Dayton. Curtiss was there to help his friend, the airship pilot Thomas Baldwin, who was making an exhibition flight. The brothers helped the two men chase down Baldwin's runaway airship, then had a long, friendly talk about engines and propellers and aeroplanes. Curtiss had questioned the Wrights closely. By the time they next met, Glenn Curtiss had taken up aeroplanes in earnest, and Will believed he was trying deliberately and literally "to take our machine and beat us completely."

The Wrights' competition with Samuel Langley and Charles Manly was a contest between gentlemen, conducted at a respectful distance. The brothers and Langley never met, and neither side wished the other ill. Indeed, they scarcely acknowledged that a contest was under way. The Wrights had questioned the secretary's judgment but never his honor, and Langley had respected the Wrights as worthy seekers of the same truths he sought. With Glenn Curtiss it would be different. The contest would be face to face and deeply personal. The stakes would be of the highest order on both sides—for Curtiss, a fortune; for the Wrights, their claim on history. In Curtiss the brothers saw their own Millard Fillmore Keiter—a scoundrel who must be shown up at all costs.

Yet the Wrights and Curtiss were set on a collision course by events that lay quite outside their own lives. These events were rooted in another man's marriage, and in a woman's determination to save her husband from a quiet despair.

MABEL GREENE HUBBARD contracted scarlet fever as a child of five in Cambridge, Massachusetts. The disease left her totally deaf. Her father, a wealthy businessman, could afford the best teachers for her, and eventually he hired Alexander Graham Bell, the son of a Scottish pioneer in the treatment of the deaf and himself widely known as a gifted and devoted teacher. Mabel was fifteen. Bell was twenty-five. Two years later, in 1875, he realized he was "in deep trouble." As he told Mabel's mother, "I have discovered that my interest in my dear pupil . . . has ripened into a far deeper feeling."

His timing was atrocious. Not only were his affections inappropriate, but they added complications to his own life when he least could handle them. By day he was teaching the deaf—and falling in love with Mabel—while by night he was conducting the experiments that would lead to the invention of the telephone. He first wrote to his parents about his feelings for Mabel on the night before he uttered the famous first words over wire: "Mr. Watson, come here. I want you."

At first Mabel was put off by the romantic attentions of this hulking adult. But Bell persisted; she allowed matters to gather steam; and finally she agreed to marry him, saying she loved him more than anyone but her mother. From that dubious endorsement grew an ardent and lifelong love affair and a marriage of unusual mutual dependence.

Their relationship began in the process by which Alec literally taught Mabel to speak, a peculiarly intimate form of pedagogy that continued for many years. To address the problem of her speech—she had trouble making herself understood by people who did not know her well—he examined her mouth and throat closely as she spoke, and directed her to examine his. In the face of his fierce desire to improve her articulation, she alternately resisted and accepted insistent offers of aid such as this:

> I will help you if you will not hate me for doing so. I will work for you—if you will work too. I have been waiting and longing for you to show some wish in the matter—and I think it is growing. Will you let me try to help you—by doing what I advise. A more distinct articulation will open the door of many a new friend for you in the future—and bring you into closer communion with all who love you—Help me to help you.
>
> I have studied your articulation and see very clearly wherein it is deficient. The chief cause of the difficulty many persons have in understanding you is on account of the peculiarity of your voice. This I alone can remedy—you cannot well correct it by yourself.

As the couple matured and raised two daughters amid the fame and wealth that came with the success of the telephone, it became clear that he depended on her at least as much as she did on him. Dividing their time between their handsome home in Washington and their summer estate in Cape Breton, Nova Scotia, they were quite often apart, and he longed for her. "Let us lay it down as *a principle of our lives, that we shall be together,*" he told her once. She was equally bereft during his absences. "You are the mainspring of my life," she told him, "and though when it is gone the other wheels go on by themselves for a time, it is very languidly and more slowly, and I want you back to give me an interest in life."

Yet he was also terribly preoccupied with his work. She could not abide his incurable habit of working late into the night—the only time, he said, when he could "retire into myself and be alone with my thoughts"—and then be unable to rise at a conventional hour in the morning. "I have found by experience that I can only deal with one thing at a time. My mind concentrates itself on the subject that happens to occupy it and then all things else in the Universe—including . . . wife, children, *life itself*, become for the time being of secondary importance."

"I wonder," she asked him once, "do you ever think of me in the midst of that work of yours of which I am so proud and yet so jealous, for I know it has stolen from me part of my husband's heart."

Still, as they grew older, she realized his work was not his mistress but his master, a stern obsession commanding him to prove (perhaps to himself as much as to others) that his early, phenomenal success had been more than a lucky fluke, and that he was capable of giving mankind a second great gift. "I can't bear to hear that even my friends should think that I stumbled upon an invention and that there is no more good in me," he remarked as early as 1879, only three years after he patented the telephone.

All appearances suggested anything but that Bell was a troubled man driven by self-doubt. From the 1880s onward, he was a statesman of science whose views were constantly sought and quoted. In Washington, D.C., his chosen home, he was beloved among wide circles of friends for his warmth, for his support of the arts and education, and for an all-encompassing and contagious curiosity. He dominated any room he entered, not by bombast but by "an indefinable sense of largeness," both physical and intellectual. By 1900, at 53, he looked like a virile Santa Claus. Above a great beard and "extraordinary eyes, large and dark," his forehead sloped to a wavy crown of iron-gray hair. His voice was deep and rich, and as a master teacher of elocution, he spoke, as a friend said, with "clear, crisp articulation" that "made other men's speech seem uncouth." When entertaining, he drew out his guests with thoughtful questions. His interests were infectious. One of his sons-in-law said: "He always made you feel that there was so much of interest in the universe, so many fascinating things to observe and to think about, that it was a criminal waste of time to indulge in gossip or trivial discussion."

One had to watch Bell at work to see clues of his demons. His father-in-law had once commented on "the tendency of your mind to undertake every new thing that interests you & accomplish nothing of any value to any one." If this was harshly unfair—in fact, Bell helped many, and never strayed from his dedication to the deaf—it also identified a troublesome pattern. It showed in his laboratory in Washington, filled with devices of so many diverse types that a visiting reporter was reminded of the U.S. Patent Office. It showed in his preferred bedtime reading material, *Johnson's Encyclopedia*, and what he wrote about it: "Articles not too long—constant change in the subjects of thought—always learning something I have not known before—provocative of thought—constant variety." It showed in his shelves and shelves of lab notebooks, crammed with ideas and sketches that never became more. It showed in his restless movement from one project to another—a device to transmit speech by light waves; a way of duplicating phonograph records and photographs; an effort to condense fog into water, for use on lifeboats at sea; an ancestral version of the iron lung. The telephone had been conceived in a great, concerted rush of work. When a problem would not yield to this approach, Bell had trouble remaining calm and committed to the long haul. He would veer between jubilation ("A new invention has been born") and brooding ("Will not attempt to note ideas for fear of another sleepless night"). Several of his projects were promising, and a couple yielded solid results. But when a new idea failed to develop quickly into a phenomenon to match the telephone, he moved on. He took a certain joy in this intellectual restlessness, but it also gnawed at him. "I have got work to do," he said once, "so much work on so many subjects that I want many more years of life to finish it all."

But Bell's harried search for a focus ended when he fully embraced the problem of flight in the early 1890s. He had flirted with it for years. "From my earliest association with Bell," said Thomas Watson, the gifted machinist who had been the inventor's principal aide on the telephone, "he discussed with me the possibility of making a machine that would fly like a bird . . . I fancy that if Bell had been in easy financial circumstances he might have dropped his [telephone] experiments and gone into flying machines at that time. . . . He made me promise that, as soon as the telephone business became established, if it ever did, I would leave it and start experiments with him on flying machines."

Like many other flight enthusiasts, he had been fascinated by soaring birds of prey as a child, in Scotland. His interest in flight lay dormant for many years while he taught the deaf, worked on the telephone, and waged a long legal war in defense of his patent. Samuel Langley reawakened it. The secretary's aerodromes, even the tiny hand-held versions that stayed in the air only a few seconds, astonished Bell, and he predicted that Langley would confound his critics and learn how to fly. But Bell had no intention of leaving all the fun to his friend. Seizing his notebook just after seeing one of Langley's hand-held models fly, Bell wrote: "I shall have to make experiments upon my own account in Cape Breton. Can't keep out of it. It will be all UP with us someday."

Like Langley, he was well aware that such experiments would expose him to ridicule. But he would not stay silent. The power of his obsession was evident in an exchange with an eminent friend, the great English physicist Lord Kelvin. In a crowded social gathering in Halifax, Nova Scotia, Kelvin—whom Mabel thought a "kindly, loveable, simple man"—remarked to Bell that he regretted the American's foray into aeronautics. When Bell objected, the two "plunged right into scientific talk right there in the midst of the crowd." Afterward, Mabel wrote to Kelvin, insisting her husband's work was not a lark but a serious scientific endeavor. The Englishman replied that he never doubted Bell would proceed "by careful and trustworthy experiment. Even if the result is to demonstrate to himself that a practical useful solution of the problem is not to be found, I am sure what he finds by his observations and measurements will be very interesting." His intent, Kelvin said, had been only "to dissuade [Bell] from giving his valuable time and resources to attempts which I believe, and still believe, could only lead to disappointment, if carried on with any expectation of leading to a useful flying machine." Yet the warning of this man, far more learned in basic science than himself, failed to deter Bell. Two years earlier he had taken his photographs of Langley's models wheeling overhead, and they spoke more loudly than Kelvin's objections.

Bell tested many forms of apparatus related to flight—whirling horizontal rotors; hand-held aeroplanes modeled after Pénaud's toys; gunpowder rockets; propellers. None led to notable success. Shortly after the death of Otto Lilienthal in 1896, he wrote: "I am finding out in the laboratory that a great deal has yet to be learned concerning the best way to combine aero-planes or aerocurves—so as to gain the full benefit of the surfaces."

ON A VACATION in the mid-eighties, the Bells had fallen in love with the remote village of Baddeck, Nova Scotia, a cool, unspoiled place of long vistas and transplanted Scots who reminded Bell of his homeland. So did the twisting, hundred-mile reach of ocean water the locals called Bras d'Or—"arm of gold." On a peninsular headland across the water from Baddeck, the Bells built a massive, turreted house they named Beinn Bhreagh (ben VREE-ah), Gaelic for "Mountain Beautiful." The house could sleep twenty-six, not counting servants, with eleven fireplaces, and in the high months of summer it was filled with relatives and friends. These groups often included Samuel Langley (whom Bell's daughters called "grandfather"), and Simon Newcomb (who once debated Langley for hours on the Bells' great veranda on the question of how a cat, dropped upside down, could land on its feet). Over the years, the Bells built stables, gardens, wharves, guest cottages, a laboratory, and twelve miles of roads. By 1900 the big house was the center of a private little kingdom. For his own castle keep, Bell used a beached houseboat. He would sit there all alone, sometimes for an entire weekend, sketching designs and capturing "fugitive thoughts," unable to set his obsessions aside and rest.

With its long slopes to the water and stiff breezes nearly every day, Beinn Bhreagh was a perfect setting for flying kites, which Bell began to do in 1898. It was a relaxing pastime only at first—and even then, Bell made notes. He had learned in the telephone patent wars never to leave any scientific observation, however casual, unrecorded. When sharp breezes blew off the Bras d'Or, he would spend the entire afternoon on a long meadow that soon became known on the estate as the "Kite Field." Like any heedless hobbyist, he put friends and family to sleep with talk of this new passion. "I suppose Mr. Bell has nothing but kites and flying-machines on his tongue's end," Helen Keller remarked in 1901. "Poor dear man how I wish he would stop wearing himself out in this unprofitable way." The people of Baddeck were baffled. A boatman told a visitor: "He goes up there on the side of the hill on sunny afternoons and with a lot of thing-ma-jigs fools away the whole blessed day, flying kites, mind you. He sets up a blackboard and puts down figures about these kites and queer machines he keeps bobbing around in the sky. Dozens of them he has, all kinds of queer shapes, and the kites are but poor things, God knows. I could make better myself. . . . It's the greatest foolishness I ever did see."

For Bell, the appeal of a kite, especially a big one, lay in its marvelous violation of the law of gravity. "There is a great fascination in watching large structures floating in the air; structures of great weight as well as great size," he said. And once he had made a significant weight float overhead, and remain in his control, the next speculative leap became irresistible. "One cannot help dreaming a little concerning the possibility of a man being carried up in one of these structures, of an engine and propeller being installed, and then cutting loose from the anchor mooring that constitutes the machine as a kite and flying off under its own propulsive force.

"In a large kite I see a flying machine at anchor."

FROM THE LIBRARIES OF WASHINGTON, Bell gathered all the writings he could find, hundreds in all, on the subject of aerial navigation. He told Langley he was "surprised, indeed overwhelmed by the extent of the literature. . . . An enormous mass of material exists, which it seems to me, from the little study I have given it, to be in an utterly chaotic condition." He set out to make an exhaustive bibliography of titles and references, convert the list to index cards, sort it from earliest to latest, "and then study the matter historically."

In these records of tiny progress and much failure, he saw an obvious problem that other enthusiasts and experimenters seemed to ignore: Those who actually attempted to fly tended to die shortly thereafter. In Bell's eyes, the deaths of Otto Lilienthal and Percy Pilcher proved that the aeroplane design, in particular, was simply too dangerous to provide the ultimate answer to the problem. Even Langley's aerodrome looked too risky to Bell. "The great difficulty in developing an art of aerial locomotion lies . . . in the difficulty of profiting by past experience," he wrote. "A dead man tells no tales."

So, with safety foremost in his mind, he sketched designs for a radically different sort of flying machine. First, it must be inherently stable. That is, if wind gusts or an operator's error knocked the machine off kilter, its shape would naturally bring it back into balance. He insisted on this. Second, the machine must be light. The heavier it was, the faster it must go to attain lift, and speed was a killer. Since launches and landings were the points of greatest peril, the craft should be flown over water, a more forgiving medium than the hard ground. The ideal flying machine would rise straight up, and it would be capable of hovering like a kite (but without a line), allowing the operator to drop a rope ladder and step safely down to terra firma. It was an extraordinary image—a man-carrying device floating in one spot overhead, oblivious both to gravity and to the wind.

To gain this end, a kite was the ideal experimental device. The crazy "thing-ma-jigs" glimpsed by the Baddeck boatman actually represented stages in the evolution of Bell's thinking. There were circular shapes and paddle-wheel shapes. Others were stars. One was shaped like an H. Some looked like giant spools of thread. Often Bell imitated the simpler designs of the Australian engineer Lawrence Hargrave, inventor of the box kite. Hargrave was another fervent believer in the cause—"I know that success is dead sure to come," he told Chanute. His box kite was a rectangular compartment, or cell. It was really a twin of the biplane—two plane surfaces connected at their ends and capable of lift when they encountered the wind. In 1894 Hargrave rose to a height of sixteen feet on the strength of four interconnected box kites. He never did better, but his basic design spread through Europe and North America.

What appealed to Bell about the box kite was its stability. If a small box kite flew stably, why not a big one? But here he collided with the rule that Simon Newcomb had cited time and again—any device twice as big as another of the same shape will be not twice as heavy but eight times as heavy.

With hopes of lifting the weight of a man, Bell built box kites with rectangular cells as big as rooms. But the bigger the kite, the poorer it flew.

Bell drove in upon Newcomb's problem. In 1901 he struck upon the idea of joining many small box kites together. "Do not increase the size of the cell, but compound small cells into a large structure," he jotted in his lab book, "and where the two sticks come together omit one, and in this way the larger kites will have less weight relatively to their surfaces than the smaller kites, and yet be equally strong." Next came the insight that even more weight could be saved by discarding the rectangle as the basic shape. Instead, "let everything be built up of equilateral triangles. . . . Whole thing could be built up into a solid compact form of almost any desired shape."

In Bell's mind, a form arose that would be wonderfully strong yet wonderfully light. Its basic unit would be a pyramidal cell—a tetrahedron. That is, in the language of the patent he took out in 1904, "The tetrahedral skeleton or frame may be composed of six bars or rods so connected at their ends as to form the outlines of four triangles." Like a three-legged stool, this was an inherently strong and stable form.

Better yet, something mysterious and wonderful happened when one joined a second cell of exactly the same size and shape to the first. Again, the first cell required six bars. But the two cells together required not twelve bars, but only nine, since the two cells shared one of their triangular faces. The new structure was twice the size of the first, but only one-and-a-half times the weight. And this paradoxical logic applied as more cells were added. The additional weight did not keep pace with the additional size.

That was not all. The simple parts—just rods of wood or a light metal, with connectors between them—could be transported, assembled, and disassembled with ease. Bell soon realized the cells' rigidity and strength made them highly adaptable to all sorts of building projects. In fact, the first structure he assembled with them was a giant wall on the Kite Field to shield kites from the wind while they were being assembled. He built a single-cell shed on the lawn, with one side open for viewing the kite flying.*

Of course Bell's key aim was not structural but aeronautical. He saw that if

* Bell's design prefigured R. Buckminister Fuller's invention of the geodesic dome in the 1940s. Fuller knew nothing of Bell's work. "I was astonished to learn about it," he told an interviewer in 1978. "It is the way nature behaves, so we both discovered nature. It isn't something you invent. . . . You just take two spheres and they just touch one another—that's all. You nest a third one down between the two and you get a triangle. Then you nest another on top and you get a tetrahedron . . . So it seems to be fundamental to nature" (Dorothy Harley Eber, *Genius at Work: Images of Alexander Graham Bell* [New York: Viking Press, 1987], 9–10).

"LET EVERYTHING BE BUILT UP OF EQUILATERAL TRIANGLES."

Two of Alexander Graham Bell's smaller tetrahedral kites (top); the inventor sits in a tetrahedral chair to observe his kites aloft (bottom) he covered two triangular faces with fabric, he had a pair of dihedral wings. This was "the brick, as it were, out of which the flying house must be made." Many such "winged tetrahedral cells" joined together would resemble a great flock of birds flying in unison. And, most important to Bell, the dihedral angles of the multiple little "wings" would guarantee extraordinary stability.

He began to speak of a new theoretical model for manned flight. Langley had chosen the eagle as his model. Bell now looked to the butterfly. A reporter put it this way: "Eagle flight is marked by its strength and speed; the butterfly hovers lightly over the earth, gently rising or alighting, unambitious, but safe. Perhaps, thinks Professor Bell, man would do better to imitate this kind of flight in the first trial of his mechanical wings. An initial velocity of sixty miles an hour cannot be acquired by the new experimenter, without danger to his life and limbs, not to mention his fragile apparatus. Is there not some way by which we may learn to fly more gently?"

A mass of tetrahedral cells did not look like a wing. But it acted rather like one. With Bell watching and assistants holding the lines, the first tetrahedral-cell kite shot skyward in 1902. It whistled as it rose to an impressive height, where it remained nearly as stable in the wind as a three-legged stool on a hardwood floor.

"In multi-cellular kites of pure tetrahedral construction we have forms that seem to possess automatic stability in the air," Bell informed Chanute. "The steadiness of flight in a gusty wind [is] really remarkable." His hopes became faith. "Other kites, in a gust of wind, will dance about, but my kites for some reason are perfectly stable. . . . In aerial machines what we want above all things is automatic stability, and this quality is possessed by these tetrahedral kites."

Soon after that, a tetrahedral kite with two broad banks of cells connected to each other as tandem wings, like Langley's aerodrome, was sailing aloft. A squall hit. The kite yanked two handlers off the ground.

Bell's kite experiments became a cottage industry. Workers from Baddeck began to cross the bay by boat each morning—carpenters to assemble the cells from sticks of wood or aluminum and seamstresses to sew silk triangles to the frames, all by hand. In this sparsely settled back country, any paid work was welcome, and Bell was known as a kindly boss. Seamstresses competed to see who could make the most cells in a week; the winner got an extra day off. When enough cells were ready, they would be joined together, sometimes hundreds in all, to form structures of extraordinary geometry.

The kites were easy to build, so Bell tried designs of endless variety and size, hoping to hit on a great flyer through dogged trial and error. No one had ever seen their like. He chose red silk for most, so they would show up well in photographs, and the color reinforced their unearthly appearance. Long afterward, a Baddeck woman who sewed cells for a dollar a day recalled the daily scene on the Kite Hill in summer. "There were so many kites, so many that never had names. They were experiments. They were really beautiful in the clear sky . . . usually in the late afternoon when the sun was going down.

"He loved to see them fly. 'Beautiful,' he'd say. 'Beautiful.' "

BY 1902 MABEL BELL found her husband "continually more wrought up over his kite experiments than I like." Yet she knew success was the only tonic for what ailed him, and she wanted it as much as he did.

"I do so appreciate all the wonderful, unfailing, uncomplaining patience that you have shown in all your work," Mabel told Alec at one point during the kite experiments, "and the quiet, persistent courage with which you have gone on after one failure after another. How many there have been, how often an experiment from which you hoped great things has proved contrary. How very, very few and far apart have been your successes. And yet nothing has been able to shake your faith, to stop you in your work. I think it is wonderful and I do admire and love you more as the years go on. But oh how I wish that you may have success at last."

She was especially hopeful about Bell's prize kite of 1903. The configuration of its massed cells reminded the Bells of a soaring hawk, so they named it *Oionos*, for the birds of prey the Greeks watched to foretell the future. Mabel was in Washington, where the papers were full of Langley's great aerodrome. But her money was on *Oionos*. "You simply must get that machine to fly," she wrote her husband. "I have been watching a buzzard fly and it looks just like your Oinos [sic], and it goes up and down on the wind without moving it's [*sic*] wings. It's the most graceful thing in the world and we surely, surely must be able to do it. No ugly clumsy balloon thing, but a bird. . . .

"You are working so differently from all the other men, they seem so far ahead of you and yet are they really? I think not. I can not but believe the strong compactness of your machine must win through in the end. It is built according to nature's law. . . . I love you and believe in you and your success."

*Oionos* did fly—so well, in fact, that Bell believed he soon would have another great invention to give the world.

In the fall of 1905, Bell built a man-carrying kite so large it would need a wind of near gale force to get off the ground. Such a wind arrived one day in November, but the weather was so bad that Bell's assistants refused to come over from Baddeck. "He looked gray when he came home," Mabel said, "wrote a short note dismissing the staff and closing the laboratory, turned his face to the wall and never spoke again that day or night."

But he went back to work on a new man-carrier. Bell had hundreds of cells on hand in his workshops. He assembled every one of them into a structure nearly twenty feet across at its widest point, with a weight of just over sixty pounds. Bell named it the *Frost King,* for newlywed friends named Frost in Baddeck, though the name also matched the weather. Bell equipped the kite with a fifteen-foot rope ladder. The ladder would hang by lines some thirty feet below the kite itself. The plan was for a lightweight man to climb on the ladder when the kite was in flight, then ride it higher into the air to prove the kite's lifting power. The rider would not control the kite. But the sheer act of lifting a man well off the ground was a key step.

And he became "more and more desperate," not only to bring the trial off, but to capture a photograph of the scene. Apparently Bell believed that documentation of a man-carrying tetrahedral kite would give his efforts credibility among other experimenters. Certainly he knew that a photograph in such cases was worth more than a sober report.

Every morning the family scanned the water for whitecaps, the harbinger of a good wind. One day the breeze was strong enough for the *Frost King* to lift a Baddeck youngster several feet off the ground. But the Bells' photographer had chosen that day to hole up in the dark room, and no other camera was handy. On Christmas Day, a good breeze pushed the kite some nine-hundred feet in the air, but by the time it was hauled down to pick up a passenger, the wind had died. Two days later Bell induced the *Oionos* to turn a circle in the air like a wheeling hawk—another key step. But now chances for the season were all but gone. On December 28 the Bells were due to catch a steamer bound for St. John's, New Brunswick, where they would board a train for Washington. The steamer was approaching Baddeck when suddenly a "fair sailing breeze" came up, though from the northwest, "the worst kite quarter." Still, it offered one last chance.

With no lightweight male available, Mabel, at 115 pounds, volunteered to fly. So eager was Bell for a successful test that he was willing to send his beloved wife aloft in a freezing wind. They went down to the snow-covered Kite Field together, but the wind fell slack, and Mabel turned back to the house. But moments later, as she was having the trunks hauled down the stairs, Bell exploded into the great house, waving his Kodak camera over his head and crying, "Develop! Develop! We've got him, and if the photos don't come out I'm not going today!"

He had hastily summoned a few men from around the estate, only one of whom was trained to handle the *Frost King.* They managed to get it well aloft, and Neil McDermid, the lightest man in the group at 165 pounds, was chosen

"MY KITES FOR SOME REASON ARE PERFECTLY STABLE."
Bell's *Cygnet I* at Beinn Bhreagh

to take the ride. With others anchoring the kite, McDermid, still on the ground, grasped one line and walked toward the kite until the tension was strong enough to lift him. Then, quite unexpectedly, he was rising—not on the rope ladder, as planned, but suspended by his arms. In a second he was hanging thirty feet off the ground. It was no flight in any sense. But Bell was elated by the demonstration of the kite's strength, not to mention by the sheer spectacle. He clicked his shutter three times. Davidson, the photographer, snapped several frames of his own.

Across the bay at Baddeck, the steamer's whistle blew; the crew was preparing to leave. Bell got McDermid back to the ground, shoved five dollars into his hand, thanked him, and rushed off toward the house. At Beinn Bhreagh, Bell thrust the camera at Mabel and rushed her to the darkroom. She found that "the servants had been there washing it out and it took some time to find our things which increased the excitement." Finally the developing powders were discovered. With Davidson assisting, Mabel carefully and calmly went about the necessary tasks "in spite of Mr. Bell popping in and out the door, jumping about like a boy." Before their eyes three clear images emerged. They showed Neil McDermid "just clear of the ground. However that was enough, and there was rejoicing." Mabel went off to see the trunks out the door, leaving Davidson to develop more images that would show Mc-Dermid at his peak altitude of thirty feet. When Mabel returned she found

Davidson "almost literally tearing his hair and very certainly slapping his thigh, and shaking himself in high disgust and indignation." He had accidentally "spoiled his beauties."

Still, Mabel said, "The power of the kite and its great steadiness and strength is demonstrated." She urged Bell to stay another day and try for better pictures. But for all his ardor, he was afraid to run the risk again. "You've no idea how high up he was," he told her later. "You don't know how high thirty feet is. It's more than twice the height of the telegraph poles, and if the rope had broken or anything on the kite had given away and Neil had fallen he would have been killed or seriously injured."

"Mr. Bell is so happy and so excited, underneath a quiet demeanor," Mabel wrote a close friend. "It means so terribly much to him—it is an achievement he has known he could do, but has not demonstrated for years, and lately it's been a tremendous pull. We've just lived for this moment."

BELL'S ENTRY into the race to conquer the air was coming very late. Nearly three months earlier, the last, long flight at Huffman Prairie on October 5, 1905, had given the Wrights confidence to take their machine to market. As the morning dews on Huffman Prairie turned to frost that fall, they packed away the paraphernalia of experimentation and turned their minds fully to business. They would not fly again for a long time.

Their aim was to sell everything, including their patents, for $250,000. They wanted a contract that both parties would sign before the Wrights made a single demonstration flight. This condition would protect them from window-shoppers who wanted to see the machine with no real intention of buying. The buyers would be protected by a clause saying no money would actually be paid until the Wrights made fully satisfactory demonstrations. This condition was a hard one for governments to accept, as the brothers soon learned. But they would not back down from it easily. They were determined that no detail of their design would be stolen. The only way to secure their treasure was to keep the design secret until their compensation was guaranteed. They had applied for patents but the patents had not yet been granted. And in any case, they believed that the secret of flight, if sold to the right party, would gain more of what they wanted than patents and manufacturing ever could—not just money but freedom, too. Better to get their price for the whole ball of wax—machines, designs, patents, and training—than to burden themselves with a patent war.

> At the present time the most obvious practical use of a flyer is military, [Will wrote Georges Besançon, editor of *L'Aerophile*.] Prompt acquain-

> tance with ways and means of flying is many times more valuable to a government than patents could be to a private commercial company. We are merely carrying our invention to the best market—a market in which patents would reduce instead of increase the value of what we offer. By selling to governments we also avoid burdening our future with business cares and vexatious law suits, without which a valuable patent can not be maintained. We wish to have our time free for purely scientific work.

The brothers believed any nation that agreed to their price for the technology of flight would find itself richly rewarded. If anything, the price was arguably very low, if only the buyer had imagination. The United States and Great Britain remained their customers of choice, but they were prepared to sell elsewhere, too. They approached the entire process as hard-nosed businessmen, confident in their product and fully willing to wait for the customer who would meet their terms. But they did not intend to sign a contract that would entitle any nation to sole possession of the secret. "The idea of selling to a single government as a strict secret has some advantages," Will remarked to Chanute, "but we are very much disinclined to assume the moral responsibility of choosing the proper one when we have no means of knowing how it will use the invention. And then it is very repugnant to think of hiding an invention of such intense human interest until it becomes stale and useless."

THEIR FIRST STEP was to write a series of letters to key officials, contacts, and journalists at home and in Europe, to say their product was now complete and for sale.

To reignite the possibility of a deal in the United States, they swallowed their pride for once and sent a letter directly to Secretary of War Taft: "We do not wish to take this invention abroad, unless we find it necessary to do so, and therefore write again, renewing the offer."

Back from Washington came another exasperating reply, not from Taft but from the Board of Ordnance and Fortification: No "financial assistance" was available for "the experimental development of devices for mechanical flight" until such a device had been "brought to the stage of practical operation without expense to the United States."

The brothers restrained their irritation. "We have no thought of asking financial assistance from the government," Will said. "We propose to sell the results of experiments finished at our own expense." He asked what conditions the BOF would set for the performance of a useful flying machine, and promised that "proof of our ability to execute an undertaking of the nature proposed will be furnished whenever desired."

This letter ran into a wall of incomprehension. BOF members advised staff to tell the Ohioans they did "not care to formulate any requirements for the performance of a flying-machine or take any further action on the subject until a machine is produced which by actual operation is shown to be able to produce horizontal flight and carry an operator."

These responses were "so insulting in tone as to preclude any further advances on our part," Will said.

At this point a little give-and-take, a little compromise, might have saved them a good deal of trouble, and would hardly have stained their characters. Accusations of self-righteousness—so like the ones often aimed at their father—began to be heard, and they had some merit. But the brothers' unbending style as businessmen was also in part the product of a deep-seated aversion to self-promotion—a strange attribute for visionary inventors, perhaps, but nonetheless real in the Wright brothers. It sprang from a certain church-bred humility, a sense of the proper way to do your business and conduct your life. You did the work at hand as well as you could, then offered it. If others did not want it, that was entirely their affair. You did not push yourself upon them. They had learned this ethic as children and were reminded of it now, for their father was determined that as fame overtook them, they must be moral exemplars. He often warned them to beware of scoundrels and false friends, and "amid it all, be men—men of the highest type. Personally, mentally, morally, and spiritually. Be clean, temperate, soberminded and greatsouled. See two worlds & live for both. You can in humility and simplicity have an influence that will bless multitudes. The world is longing for one merely human example."

THEIR PROGRESS OUTSIDE THE U.S. was mixed. The British would not offer a contract until they saw a flight, and the Wrights would not offer a flight until they saw a contract. The brothers' overtures to France and Germany arrived in Europe as the two nations were approaching war over competing interests in Morocco, and the crisis quickened interest in the Wrights in both countries. But hackles rose in Berlin when one of Will's letters—poorly translated into French—referred to Kaiser Wilhelm as being in a "truculent mood," and the Germans backed away.

A tentative deal was struck with a private French syndicate, leading in turn to a secret visit to Dayton by a delegation from the French War Ministry. The brothers showed photographs and asked for another round of testimony from the Huffman Prairie witnesses, whereupon the investigators departed as true believers. But by the time they returned to Paris, the Morocco crisis had cooled, and the War Ministry had begun to think there was no point in paying a million francs for a pig in a poke when French experimenters were promising to overtake the Americans soon.

If, indeed, the Americans had flown at all. When the *Times* of London published a British engineer's letter avowing his belief in the Wrights' claims and his intention to duplicate their efforts, the editors hastened to add a disclaimer: "It is not to be supposed that we can in any way adopt the writer's estimate of his undertaking, being of the opinion, indeed, that all attempts at artificial aviation on the basis he describes, are not only dangerous to life, but foredoomed to failure from an engineering standpoint."

The story about Kaiser Wilhelm caught the eye of some U.S. editors, and not to the Wrights' benefit. The editors of *Scientific American*, who fancied themselves the ultimate arbiters of engineering progress, had ignored the eyewitness account that Amos Root sent directly to their attention a year earlier. But they now published an elaborate sneer at "The Wright Aeroplane and Its Fabled Performances." The article presented a growing and not implausible reason for doubt:

> If such sensational and tremendously important experiments are being conducted in a not very remote part of the country, on a subject in which almost everybody feels the most profound interest, is it possible to believe that the enterprising American reporter, who, it is well known, comes down the chimney when the door is locked in his face—even if he has to scale a fifteen-story sky-scraper to do so—would not have ascertained all about them and published . . . long ago?

The same doubt began to nag at Bell, though he blamed the inventors more than the press. "It seems strange," he told Mabel, "that our enterprizing American Newspapers have failed to keep track of the experiments in Dayton, Ohio, for the machine is so large that it must be visible over a considerable extent of country. . . . This seems to be due to the desire of secrecy. . . . I do not understand how it is that so little attention has been payed to this matter by the American Press. I am now studying carefully the details published."

The brothers knew of such doubts but were disinclined to argue with people. "If they will not take our word and the word of the many witnesses who were present at one or more of the flights," Will said, "we do not think they will be convinced until they see a flight with their own eyes. A few months more time will settle any reasonable doubt."

• • •

IN FRANCE, "*l'incident Wright*" of 1905, whether one believed in aeroplane flights over Ohio or not, at least "had the advantage of shaking our aviators out of their torpor," said *L'Aerophile*. A dozen or more experimenters now were laboring on gliders and powered machines, some with two wings and some with only one. But the most exciting new fact was the conversion of Alberto Santos-Dumont from lighter- to heavier-than-air experiments.

The Brazilian's star had faded a bit in recent years, but it brightened with his announcement that he intended to compete for all aeroplane prizes. He hired Gabriel Voisin and Voisin's brother, Charles, as his codesigners and builders, and for six months he dropped out of the Paris social scene entirely. In the spring of 1906, a very large, white apparatus emerged from his workshop for testing. It looked like, and in a sense was, an assemblage of box kites—two rectangular box-kite wings; a long, boxlike fuselage with a propeller and fifty-horsepower engine; and in front, a box-kite rudder.

The earliest experiments appalled even his friends. First, to test the craft's balance, Santos-Dumont suspended it from an overhead tight-wire, then pulled it along behind a donkey. Next, to test the controls, he attached the machine to the underside of his dirigible No. 14 and attempted to fly this enormous hybrid in a scene that descended into chaos.

But Santos-Dumont and the Voisins kept at it, and in September 1906, in a great field in Paris's Bois de Boulogne, with Archdeacon and many Aéro-Club members watching, his wheels left the ground for a few meters—this was confirmed by men who threw themselves on the ground to see the distance between the wheels and the turf.

In late October Santos-Dumont tried again. A crowd of a thousand stared as he rose two meters, maybe three, off the ground. With no means of balancing, the craft immediately began to lean to its left, thus entering what one reporter called "a graceful curve." Santos-Dumont cut the engine and dropped with a smack, crushing his wheels. Members of the official measurement committee were too flummoxed to do their job, but it was agreed that the machine had been airborne for roughly sixty meters. The crowd erupted in euphoria. Paris's newspapers announced the first flight in history. One declared: "MAN HAS CONQUERED THE AIR!"

Some Frenchmen of the Aéro Club, testy as always about the preeminence of the Brazilian in their midst, remarked that they could do better. One said that with Santos-Dumont's powerful engine a grand piano could have flown for sixty meters. But in the roar of acclaim there was no room for technical quibbles. Santos-Dumont was again the hero of the hour.

• • •

THE WRIGHTS, meanwhile, seemed little closer to a sale by the fall of 1906 than they had been a year earlier when they locked their shed at Huffman Prairie. Their patent applications had been approved in the United States, France, Belgium, and Germany. They had gained representation in the Manhattan offices of Charles Flint & Co., organizer of trusts and facilitator of American arms sales to many nations. Flint's man Hart Berg, a debonair American with high-level contacts throughout Europe, would be their agent across the Atlantic. But they were still without a deal.

Even sympathizers suggested they were being foolish by setting such a high price while refusing to fly. When a reporter asked Orville about it on a trip to New York, he replied:

> It seems to be the opinion of all that we are secretive and making a mystery of our flying machine. That is all wrong. We would be pleased to give a public demonstration of what we have and have tried to do so, but as we are situated that is impossible. We do not know whom we will finally close with, but if it should be some government, then the fact that it is a secret would greatly enhance its value.
>
> We are not wealthy men and have put years of work and study into this machine. We have withdrawn from all other business and given our entire attention to the flying problem, and we have succeeded. We also believe that we are the only men who have succeeded. Now, here we are, with something that is either worth a good price or it isn't worth a penny. The public can call us fakers or even crazy. We don't care. We know what we have got and others know, too.

IN MARCH 1907, they thought of a new way to let lots of people know what they had. The six-month-long Jamestown Tercentennial Exposition was scheduled to open in Virginia on May 1, 1907. It had been planned originally as a celebration of American culture and the "the industries of peace." But the program had been transformed into "the grandest military and naval celebration ever attempted in any age by any nation," "a continuous and varying scene of martial splendor" featuring "the greatest gathering of warships in the history of the world." The eminent social worker Jane Addams and a host of progressive reformers and pacifists were aghast. But as the Wrights looked at the exhibition calendar, then at a map, they saw an opportunity to shake up their stalled sales efforts and make a deal with their own government, as they had preferred to do all along.

At the exhibition's opening, much of the U.S. fleet would steam through Hampton Roads, the great naval crossroads where the broad James River flows into Chesapeake Bay at Norfolk. This point lay only some forty miles from Currituck Sound, at the back door of Kitty Hawk, North Carolina. The Wrights' admirers often praised them for their high-minded rejection of showmanship and feats of daring. But that image would have been forever altered had the plan they now conceived worked out.

One brother or the other suggested that they assemble a flyer on floats at Kitty Hawk, then "take an unexpected part in the parade" at Jamestown, flying over the fleet and perhaps skimming the surface of the harbor at Norfolk alongside some American battleship. This, surely, would show the members of the Board of Ordnance and Fortification—or at least their colleagues in the Navy—that a machine had indeed been "produced which by actual operation is shown to be able to produce horizontal flight and carry an operator."

The brothers went so far as to make tests with floats on the Miami River, in Dayton. But a propeller broke, and the Miami flooded, and the great Jamestown flyover had to be set aside.

And just then, Charles Flint, their agent for European business, asked that one of them cross the Atlantic in hopes of making a deal with the French.

WILL BOARDED RMS *Campania*, bound for London and Le Havre, in the middle of May 1907. Orville stayed home and prepared several machines for demonstration and sale. For a time their hopes were high that a deal was finally in the offing. But the negotiations that followed proved even more frustrating than the failed efforts of 1906. A plan was developed by which Henri Deutsch de la Meurthe, the industrialist who had rewarded Santos-Dumont for circling the Eiffel Tower, would organize a syndicate to manufacture Wright aeroplanes and sell them to European governments, including the French. But the arrangements became ensnared in French politics. For a time, a deal in Germany seemed to be at hand, but that prospect collapsed when Will refused to demonstrate the machine without a signed contract.

Orville joined Will overseas in July, leaving Kate at home to deal with Charlie Taylor, still one of her least favorite men, and with reporters, strangers, neighbors, friends, and her father. "I do hope that you won't have to stay over there any great length of time," she told Will. "What will sister do without brothers?"

In fact, the brothers stayed in Europe for many weeks, and Kate spent the summer in rising exasperation. With her family on the verge of considerable wealth—or, if no deal came, more financial uncertainty—she anxiously awaited news. Yet the papers were unreliable and letters from her brothers were scarce and slow to arrive. "It is desperate to be so in the dark." Another friend of hers got married and settled into "a pretty house on Cambridge Avenue"—one more unwelcome contrast to her own state as an unmarried woman of nearly thirty-three, dependent for companionship on two perpetually absent and absentminded men. She didn't like what she heard of Charles Flint & Co. and believed "the boys" were going to be swindled. Charlie Taylor was supposed to join them in France if flights were imminent, but the brothers had left virtually no instructions for him, so he called constantly at the house to ask her for news she didn't have. And her father was driving her crazy.

Finally, after many consecutive days of being "so worried" and "so out of patience," she blew up:

> Sister is sick enough of this proposition of staying here alone. Every peculiarity that Pop ever had is in full blossom now. I can't leave home even in the daytime without being lectured. The other evening, the Parkhursts invited me over to the Beckel [Hotel] for dinner and you never saw such a scene . . . I never was so much in need of a little company in my life. It is a pathetic state of affairs when going for the cream is treasured up as the chief diversion of the evening! I don't know what I should do if Lorin didn't come in every day for dinner. . . . I am sorry for Pop for he is lonesome, too. But he makes every thing so hard. . . . It will be a marvel to me if I ever get through another year of school, after this miserable summer. Do settle up something or we will all be in the asylum before long.

THE BROTHERS HAD FRUSTRATIONS of their own, though they were interspersed with visits to the Louvre, fine dinners with interesting people, and balloon rides. One of the most irritating of their frustrations came whenever the name Octave Chanute arose, which was often. It was well known in France, Will learned, that Chanute was chiefly responsible for the Wrights' achievements. He heard it said or implied that Chanute had urged them to take up their experiments; funded them; and supplied them with gliders; that they had been in every sense Chanute's pupils, taking his ideas and "putting them in material form." Clearly, Chanute's remarks in France about his relationship to the Wrights, with their subtle implication that friendship had actually been collaboration, had taken root and grown.

But what could Will say? If he corrected the errors with any force, or in public, he would seem petulant and ungrateful. With certain friends, he quietly set the record straight. Otherwise, he kept silent, "though I sometimes restrained myself with difficulty."

THE SURGE IN FRENCH EXPERIMENTATION that had followed Chanute's reports of the Wrights' activities now appeared to be paying dividends. Serious contenders were competing with Santos-Dumont for the Deutsch-Archdeacon Prize—fifty thousand francs for the first official flight of one kilometer in a closed circle. One was Henri Farman, the son of an English newspaper correspondent but entirely French in his loyalties. A bon vivant, Farman was already known around Paris as a painter, bicyclist, motorcyclist, and automobile racer. Another aspirant was Louis Blériot, a manufacturer of automobile accessories, who also bought and began to fly Voisin machines. A third was Léon Delagrange, an accomplished sculptor. All three were now flying machines built by Gabriel Voisin and his brother. These were box kite–like biplanes without wing-warping devices, but Farman began to coax his machine into flights of several hundred meters. Orville was in Paris for one of Farman's demonstrations that fall. He and Hart Berg drove up just after Farman finished one of his flights. They were barely out of the car when Ernest Archdeacon spotted Berg—he didn't recognize Orville—and rushed over, crying "in his loud squeaky voice," "Now, where are your Wrights?"

Berg masterfully turned the tables, sweeping an arm toward Orville and an associate, and replied: "Here they are!"

Archdeacon either didn't believe him or simply didn't get it, and continued to shout, "Where are your Wrights! Where are your Wrights!" as reporters, who *did* recognize Orville, swarmed around, hurling questions and taking photographs. Orville survived this scene to see Farman make several more flights. These were not like the Wrights' flights at Huffman Prairie. Farman's machine skipped across the field like a stone on a pond. "One could just force it through the air," another pilot of a Voisin recalled, and turns were long, long, skidding affairs effected by hauling on the box-kite rudder, with no way to bank. But French spectators were thrilled. When a reporter asked Wright why Farman seemed able to make repeated flights like these, but nothing better, "I told him that it would be in poor taste for us to pass criticisms on another's machine, but that I thought the flights I had seen very 'nice,' and that I admired Mr. Farman as an operator in that he didn't spend his time telling the newspapers of what he was going to do."

MABEL BELL WAS PROUD of her husband's reputation but far from in awe of it, and she spoke her mind on scientific matters. Indeed, she often reined in his tendency to carry a theory beyond the bounds of common sense. Once she heard Bell argue that a propeller should exert the same force in the air as it did in water, since "nature makes no distinctions."

"You can make a boat go with an oar in the water," she remarked, "but you can't make anything go with an oar in the air."

So much for theory.

She perceived and worried over his strange dual tendency to flit from one project to another, yet also to tinker with a thing forever, to sacrifice its usefulness in the pursuit of perfection. The kites at Beinn Bhreagh thrilled her. The tetrahedron as a basis for construction struck her as highly valuable. But she believed her husband would never turn either idea to practical purposes if he continued to work essentially by himself, with only unskilled and indifferent workmen to help him and no one to draw his kaleidoscopic energies down to a tight focus.

Her father had drummed on Bell for leaving so many ideas in his lab books undeveloped. She agreed. Her husband had figured out a way to do away with the cumbersome telephone switchboard, for example. But he told her he had "worked it out far enough to demonstrate its perfect practicability to his own mind, and he did not care to be at the bother of fussing with it further when there remained other problems to be worked out. This in fact is the attitude of his mind towards all the other inventions [he had conceived without developing], and which soon will be his attitude towards the kites and flying machines which [are] now engaging his thoughts."

In the 1870s, she believed, the assistance of Thomas Watson had been essential to the hard details of making the telephone a practical device. In the 1890s, Alec had likewise benefited mightily from his small association of scientists and technicians called the Volta Bureau. Now he needed some such man or group of men to bring the tetrahedral kites to the stage of practical application as powered flying machines, and to handle the business side of their application. He needed engineers and businessmen—men who could turn his visions and inspirations into working hardware.

> For years [Mabel told him] you have struggled along working under tremendous handicap in the want of intelligent assistants who were able to take your ideas and either carry out your instructions carefully and accurately, or help you by the devising of means of putting your apparatus together. . . . Think of how often your kites were broken by the rough handling they received, of the floats that were not water tight because your workmen had no technical knowledge. You have

> had common carpenters and slow unskilled, half educated boys when you should have had the best intellect and highly skilled labor obtainable.

In the summer of 1907, Mabel proposed the formation of a dedicated team of skilled technical men with Bell as their leader and chairman, who would pledge themselves to the goal of developing his tetrahedral structures and especially his tetrahedral flying machines to the stage of practical usefulness and marketability. She offered to bear a year's worth of the costs of such an association herself. "You have tried struggling under the dead load of ignorance and carelessness and indifference, now let's try what you can do harnessed to knowledge, interest and an established position in the commercial world."

The right men were already at hand, she said.

The first was Douglas McCurdy, whose father had been a friend and assistant to Bell for twenty years. One day in 1885, the elder McCurdy—a Baddeck businessman and editor of the village newspaper—had been fussing with a malfunctioning telephone when he noticed a stranger watching him through his storefront window. The stranger came in and offered help. He confidently unscrewed the telephone's earpiece, removed a dead fly, handed the instrument back, and said, "It will work now." When McCurdy asked the man how he knew so much about telephones, he said: "My name is Alexander Graham Bell." It was the beginning of a long and close relationship.

McCurdy's second son, John Arthur Douglas, was two years old when his mother died, shortly after the Bells began to come to Baddeck. By then the McCurdys and the Bells had become close enough friends that the Bells offered to adopt the youngster. The son had stayed with his father, but he became virtually a member of the Bell family, playing with the Bell girls during their long summer stays at Beinn Bhreagh and roughhousing with their father.

At sixteen, Douglas McCurdy enrolled in the mechanical engineering program at the University of Toronto. There he became friends with Frederick Walker Baldwin, four years older and very much Toronto's big man on campus. Baldwin was a golden boy who might have defined the term. Always called "Casey," for "Casey at the Bat," he was "casual, friendly and unworried," and a superb athlete. As captain, he not only led the university's football team to the national championship in 1905, he scored the winning touchdown in the final moments of the game. He excelled in gymnastics, cricket, and baseball, and was a yachtsman accomplished enough to crew on a boat that nearly won the Canada's Cup. Like McCurdy, Baldwin studied engineering at Toronto, and he became interested in aeronautics upon reading Octave

Chanute's *Progress in Flying Machines*. When McCurdy invited Baldwin to visit Baddeck in 1906, he hit it off with Bell immediately, and built a tetrahedral tower for him.

In 1907, another youngster came to Beinn Bhreagh—Lieutenant Thomas Selfridge, a West Point graduate who had distinguished himself in Army rescue work the year before at the site of the great San Francisco earthquake. Selfridge was an associate of Lieutenant Frank Lahm—son of the wealthy expatriate who had investigated the Wright brothers' flight claims—in the small group of aeronautical enthusiasts in the U.S. Army Signal Corps, and he had arranged an official assignment to Baddeck to study and help with Bell's work. He was reserved, smart, and ambitious.

Finally, there was Glenn Hammond Curtiss. He was far from an exact fit with the others. At twenty-nine, with a family and a measure of fame, he was no boy, and he had none of the others' college-boy charm. Growing up in the small, western New York town of Hammondsport, at the tip of Lake Keuka in the Finger Lakes region, he was no relation to the town's namesake—Curtiss had taught himself mechanics and engines and developed something like a compulsion for speed, first on sleds, then on bicycles, then as one of the early makers and racers of motorcycles. With a fierce competitive streak, he won many bicycle and motorcycle races, then began pursuing records for speed. In January 1907, he had gunned a souped-up motorcycle of his own design and

"MABEL PROPOSED THE FORMATION OF A NEW GROUP."
The Aerial Experiment Association: Glenn Curtiss, Douglas McCurdy, Bell, Casey Baldwin, Thomas Selfridge

construction up to 137 miles per hour in a timed sprint at Ormond Beach, Florida, where there were long stretches of hard, flat sand that attracted racers of all sorts. The newspapers said this made Curtiss "the fastest man in the world." They called him "the hell-rider."

Curtiss told reporters: "It satisfied my speed-craving."

His racing renown won him contracts to build engines, and he established a thriving factory at Hammondsport. Motorcycle engines were powerful but light, a combination that attracted air-minded men such as Thomas Baldwin (no relation to Casey), who bought engines from Curtiss and advised him that a huge market was about to open up for engine-makers—in the skies. At the Aero Club show in New York in 1906, he met Bell, who bought an engine for one of his kites and invited Curtiss to visit him in Washington.

At the Bells' townhouse on Connecticut Avenue, then at Beinn Bhreagh, the tough motorcyclist and the patrician Mabel Bell formed an unlikely friendship, partly, no doubt, because Curtiss's sister was deaf, and he had learned to speak clearly for the benefit of lip-readers.

Bell liked all these young men—McCurdy and Baldwin almost like sons—but he raised objections to Mabel's idea for a formal association. They were quite without experience—no more than boys, except Curtiss, who, for that reason, might resist Bell's wishes and want to go his own way. He worried they couldn't pay any of them enough.

She ticked off responses to each cavil: the youngsters might not be the best possible choices, but "they are the only ones you know of," and in fact had "prepared themselves by many years hard and expensive training." Curtiss could make up his own mind. As for salaries, "It is not a question of dollars and cents but of your long experience and need of help against their youth, enthusiasm, practical training and eagerness to help you and be associated with you. What they want is association with you, quite apart from any salary.

"The whole idea is to have your ideas carried out, your embryo inventions brought to completion according to your own desires as quickly and perfectly as possible and these young men are banded together for this purpose."

Finally, Bell agreed to found the Aerial Experiment Association. Bell, in a concession he would regret, agreed that each member would take his turn in overseeing a design of his own choice. Curtiss would be the director of experiments, Baldwin the chief engineer, Selfridge the secretary, and McCurdy the treasurer. The cost was to be underwritten by a one-year grant of twenty-thousand dollars from Mabel herself. The group stated its aim simply as "to get into the air."

The AEA's first project was to build a man-carrying tetrahedral kite. Some thirty-four hundred cells were quickly assembled into a giant wedge called the *Cygnet*, which was towed out on Little Bras d'Or Lake early in December 1907. After an unmanned trial, Selfridge climbed into the small seat for a human passenger. At the end of its tow line, the kite was flown up to a height of 168 feet. But as it came down after a seven-minute flight around the lake, Selfridge could not see how close he was to the water and failed to cut the tow line. The *Cygnet* was dragged through the water and wrecked. Selfridge was unhurt.

Immediately, the younger members of the AEA told Bell they wished to turn to aeroplanes. So the tetrahedrals were stored away. Bell, in spite of himself, was now in charge of a band of young men who wanted to fly with machines he considered foolish and dangerous.

THE WRIGHTS NOW had serious rivals. Fortunately, early in 1908, just as they were prepared to give up negotiations and begin to build aeroplanes for sale to private individuals, the brothers found themselves with not just one but two major contracts.

In the United States, the right combination of levers finally was touched when Selfridge's friend Lieutenant Lahm intervened on the brothers' behalf. Lahm had recently taken charge of the Army Signal Corps' new Aeronautical Section. After a conversation with Orville in France, he wrote to the chief of the corps. "It seems unfortunate," Lahm said, "that this American invention, which unquestionably has considerable military value, should not be first acquired by the United States Army." This led to a meeting in Washington between Will and several key Army officers, which led, in turn, to a solicitation of bids for the construction of an Army aeroplane.

The Wrights had lowered their sights, and the Army was not seeking to buy their patent. Their bid was twenty-five thousand dollars. The Army accepted it, with the condition of successful test flights at the Signal Corps' headquarters at Fort Myer, Virginia, across the Potomac from Washington, D.C. Augustus Herring submitted a lower bid of twenty thousand dollars; the Army, nonplussed, accepted it, too.

In France, Berg had negotiated a more complex deal with a private syndicate, which would give the Wrights a half-million francs, stock in the new company, and payments for additional aeroplanes—also after successful trials.

*Chapter Eleven*

# "A World of Trouble"

"Bell knew the press would pay attention."
Curtiss, Baldwin, and twenty-two others with the AEA's *Red Wing*

Bell's young men worked very quickly. To acquire experience in the air, they built a biplane glider and took it out to a hill. Like Lilienthal, they flew the glider with their bodies suspended from the lower wing, and kept the craft balanced by shifting their weight. They tried no other method of balancing, but they were quick students and fine athletes, so they managed longer and longer glides—about fifty over several weeks. When they reached a distance of one hundred yards, they decided they were ready to build a powered machine.

Selfridge was chosen to direct the design and construction of the AEA's first powered aeroplane, though all the members contributed ideas. Bell himself had to make suggestions from afar, as he was staying in Washington with Mabel, who was ill. Selfridge pored over all the aeronautical information he had assembled and wrote to the Wrights to ask for more. Their polite reply referred him to several articles describing their machine and to their U.S.

patent, under which it would be permissible to make use of the Wrights' devices, though for experimental uses only.

Selfridge told the Wrights he already had a copy of their patent. That became obvious when the AEA machine emerged from Curtiss's improvised "aerodrome shed." It embodied the fundamental structure the Wrights had invented during the period from 1900 to 1903—a biplane machine with a horizontal elevator in front and a vertical tail in back. There were differences, too. The tips of the wings were squeezed together to form a "bowstring" truss resembling "a sidelong pair of parentheses." Baldwin hoped this configuration would help with control. The machine carried one propeller, not two. It would take off and land on sled runners on the ice of Lake Keuka. And there was no device for twisting the wing surfaces. The horizontal surface in front would make the machine go up and down. The vertical surface in back would turn it to right or left—though there was apparently little expectation of attempting any turns. The operator would sit in the middle, next to a powerful Curtiss engine. The surfaces were covered with red silk left over from Bell's kites. So it was named the *Red Wing*.

By the first of March 1908, after only seven weeks of design and construction, the machine was declared ready for a test. But strong winds intervened. They waited.

Bell left Mabel for a quick run to Hammondsport to look over the preparations. There he found Selfridge jittery and ill. Having lost a kidney earlier in his life, the soldier had received orders to report to an Army hospital for treatment of a mild case of jaundice. But he was desperate to try the machine before leaving.

"It is a beautiful machine substantially similar to Farman's," Bell reported to Mabel, "and poor Selfridge is very anxious to try it before he goes off to the hospital in Washington. He . . . really should not be here at all but I sympathize with his desire to try it . . . I see see no reason why his machine should not do all that Farman's has done—but only wish it gave promise of actual stability. . . . Although I do not like to see experiments made with a man on board with a machine of this kind—in which undoubtedly some risk is involved—still I do not think that anything serious could result from an accident."

When high winds continued, Bell and Selfridge both departed, after which the wind promptly calmed down. Ice at the southern end of Lake Keuka was melting, and the rest of the lake would break up soon. Curtiss said they didn't dare wait for Selfridge to return. They had to go now. The others agreed, and Baldwin was chosen as pilot.

On March 12, a good portion of Hammondsport's population gathered to watch Baldwin roar across the ice, rise to a height of ten feet, then slide sideways and clatter back to the ice, crunching a wing. He had flown a little more than a hundred yards.

Bell understood the value of publicity. He knew the press would pay attention to any endeavor to which his name was attached. So he made sure reporters were told about the flight of the *Red Wing*, and he knew just how to characterize the flight to ensure that any coverage would be given prominent play. The reporters leaped at the bait. Front-page items in *The New York Times* and *The Washington Post* included the same key sentence, just as Bell had framed it: "This is declared to be the first successful public flight of a heavier-than-air flying machine in America."

"First successful *public* flight?" To anyone who knew something about the Wright brothers, the words must have caused eyebrows to rise. An editor at the *Post*, for example, inserted this attempt at clarification: "The flight was witnessed by a number of people from Hammondsport." But if the presence of spectators made the flight public, as it did, then surely the presence of spectators at Huffman Prairie had made the flights of 1904 and especially of 1905 public as well.

WITH SELFRIDGE BACK IN HAMMONDSPORT, the AEA boys were eager to fly the *Red Wing* again before the ice broke up for good. So on March 18, still without Bell, they ignored a snow shower and plowed up Lake Keuka aboard the steam tug *Springstead*. Searching for a place to make a landing on solid ice, they passed an endless line of ducks, geese, and gulls perched at the margin of the water. A wind of about twelve miles per hour was blowing. The snow turned to rain, and the fabric of the flying machine was soaked through, adding a good deal of weight. It was now about as bad a day to fly as could be imagined. But these were four young men in their twenties, three of them under twenty-five, and they meant to fly, rain and ice or no. Selfridge and Baldwin tossed a coin, and Baldwin won. Curtiss, in charge of the engine, revved it higher than in the first flight, to compensate for the extra weight of the drenched wings.

The men released the machine. It slid across the ice for fifty feet, then rose. As Curtiss told Bell, "It immediately became evident that we were not proficient enough in designing and handling an aerodrome to handle the machine in so much wind." The machine leaned into a sickening roll to its left, exposing its entire underside to view. The wing struck the ice at a forty-five-degree angle. Again, Baldwin emerged all right, but the machine crumpled into "a shapeless mass."

The ignominious end of *this* public flight of the *Red Wing* went unannounced by the Aerial Experiment Association.

IT WAS JUST AS BELL had feared. Curtiss had spotted it the instant the *Red Wing* began to roll. No doubt all the AEA boys understood it now. Their machine could not keep its balance. Without balance, they would be like Samuel Langley and Charles Manly, waiting endlessly for absolute calm.

An idea occurred to Bell—"moveable surfaces at the extremities of the wing piece." He put this immediately in a letter to Casey Baldwin, and elaborated: "It might be worth while considering whether the protruding ends [of the wings] might not be made moveable and be controlled by the instinctive balancing movements of the body of the operator."

In the same letter Bell said: "In voluntary control by moveable surfaces what we want to do is reef one wing and extend the other."

That was something different than changing the angle of the wing. "Reef" is a nautical term, meaning to reduce the surface of a sail. The idea of reducing the surface of a rising wing so as to lessen its lifting power—and thus restore balance—had occurred to Chanute and others. This was the idea that Wilbur Wright had rejected upon close observation of West Dayton's pigeons.

Bell's proposal of movable wing tips could be adapted to strategies other than reefing. In fact, it's possible that one of the members of the AEA already had conceived of using wingtips to change the wings' angles. For by this time, according to Glenn Curtiss's later sworn testimony, "We were familiar with the warping wing system of the Wrights and we hoped to develop some other system of balance."

This was a sticky intellectual wicket. If Curtiss's testimony is correct, then the members of the AEA understood—at least in some rudimentary way—that a flying machine could be balanced by altering the angles of its wings. If they were to claim an independent solution of the flying problem, they could not simply duplicate the Wrights' system of wing-warping. The answer lay in using Bell's "moveable surfaces at the extremities of the wing pieces" to create small, separate, winglike surfaces whose angles could be changed independent of the wings themselves. In France, the same idea had occurred to Robert Esnault-Pelterie, though he had been unable to put it to good use. The French term for Esnault-Pelterie's device was "little wing"—*aileron*. Eventually, ailerons would become standard features of airplanes of all shapes and sizes.

Recalling all this a few months later, Bell said he was not sure how the AEA youngsters in Hammondsport had happened to develop the device, since he had been in Washington at the time. But "if, as I have reason to believe, their adoption was due to a suggestion of mine that moveable wing tips should be used, contained in a letter to Mr. Baldwin, I may say that this suggestion was made without any knowledge upon my part of anything the Wright brothers may have done. They had kept the details of construction of their machine secret."

Here, without being obvious, Bell may have been trying to make a very fine distinction. The principles of the Wrights' system of balance were contained in their patent, which had been issued in 1906 and available for public inspection ever since. As to the finer *details* of that system of balance, the patent, it is true, was not specific. So Bell may have reasoned that a different method of changing the angle of the lifting surface may have constituted, in Curtiss's phrase, "some other system of balance." But it cannot be true that Bell conceived the aileron "without any knowledge . . . of anything the Wright brothers may have done." Bell surely had read Wilbur Wright's article, "Some Aerial Experiments," which set out the principles of wing-warping. Selfridge, Bell's new protégé, possessed a copy of the Wrights' patent, which also spelled out the idea. Bell knew very well how to read a patent, and he was voracious for information on the means of flight. Selfridge, Bell, and the other members of the AEA met regularly to discuss crucial matters of aeroplane design. What the AEA conceived in the aileron was not an entirely new and different system of balancing, but a new device for putting the Wrights' system into practice. The principles themselves were unchanged. They had been divined in Dayton in the summer of 1899, and made the basis of a practical flying machine over the course of the next six years.

But that might well be forgotten if the Aerial Experiment Association surged ahead of the Wrights in making flights and gaining public attention.

THE WEALTHY BARONS of the Aero Club of America in New York were delighted that a new and attractive group of Americans was pushing forward, and with none of the strange secretiveness of the obscure Ohioans, who had refused to fly for nearly three years now—if they had ever flown at all.

Every month brought fresh news of flights in Europe. The British were reported to be working on their own machines. Henri Farman was flying. On March 21 came news that Farman and Leon Delagrange had boarded Delagrange's machine and made a hop *together*. The distance was only eighty feet, but it was arguably the first two-passenger flight in history. Then Farman bested his own distance in the Deutsch-Archdeacon trial in January, flying nearly two miles in a circular course.

The Wrights' continuing refusal to fly was causing temperatures to skyrocket in the American aeronautical community, especially in the Aero Club in New York. Its members wanted to see someone—anyone—fly an aeroplane, and as soon as possible they wanted to fly themselves. If the Ohioans refused to show what they could do, they would learn the art from the French, or from Bell's young men.

At the offices of *Scientific American*, editor Stanley Beach was now among the Wright believers. He sensed that flight was the coming thing in science and technology, and he wanted to have his magazine strongly associated with it. He began to organize a *Scientific American* Prize competition. The winner would have to be victorious in three separate tests over three years, beginning with a public flight of one kilometer before official witnesses. He hoped this would lure the Wrights out of hiding and bring Farman and Delagrange from Europe—and he was confident the Wrights would win.

"There is a great deal of interest in aviation here at present," Beach told the brothers, "and everybody is interested and anxious to see a machine in the air." Delagrange might soon cross the Atlantic to make a demonstration in the United States, he warned. "It seems to me now that the opportunity presents itself to practically demonstrate that you are capable of flying . . . You should do this and thus forestall a foreigner coming over here and showing us how to fly."

But when Beach's letter arrived in Dayton in the last week of April 1908, the Wrights were not there to receive it. Kate replied on their behalf, telling Beach she would forward his letter to her brothers—at Kitty Hawk, North Carolina.

THEIR CONTRACT IN FRANCE would require the Wrights to make two flights of fifty kilometers, each in less than an hour. The U.S. Army wanted separate demonstrations of speed and stamina. The machine had to show that it could fly at an average speed of at least forty miles per hour over a five-mile course. It also had to make a flight of at least an hour with a pilot and one passenger aboard—seated, not prone—plus enough fuel for a flight of 125 miles. The Army wanted a machine that could be transported in one of its wagons and assembled in under an hour. It had to be able to take off from any sort of terrain "which may be encountered in field service"—a tall order indeed—and "land in a field without requiring a specially prepared spot and without damaging its structure." The Wrights had exceeded the Army's requirements for speed and distance many times. But they had never made a flight with two men aboard, nor had they made flights with the pilot in a sitting position. For this they had to design new controls for operating the wings, elevator, and tail, and "their operation had to be completely relearned." And of course neither brother had flown since the fall of 1905. They needed "a little practice."

In 1905, the attention of reporters and spectators had brought the spectacular flights at Huffman Prairie to a premature end. The brothers believed that once again they needed the privacy they could find only on the Outer Banks. They decided that Will would go first and prepare the camp, with Orville to follow in a week or two.

AT THE KILL DEVIL HILLS Will found a gloomy scene. Since December 18, 1903, the old camp had been standing neglected, subject to wind, rain, ice, and marauding boys. Now it lay "in ruins." One shed was still standing, but without a roof or doors. The other had blown down entirely. Most of it lay under a foot and a half of drifted sand. The ribs of one of the old gliders, stripped of their linen skin, stuck out of the sand like the skeleton of a fish, and as Will worked, he stumbled upon "various relics of the 1901, 1902, and 1903 machines." Most of their old Kitty Hawk friends had scattered, and Dan Tate, their striking workman of 1903, had died. "Kitty Hawk seems to be being deserted."

The surviving shed was unlivable, so Will had to stay at the lifesaving station, where the men were friendly but the accommodations primitive and the food nearly unbearable. Grumpy and harried by delays in the delivery of lumber, he dragged himself through his work amid storms, strong winds, and a spell of diarrhea. "I will expect you to arrive with relief Saturday," he wrote Orville. "I am not sure I can hold out much longer than that."

"I am making every effort to push things as fast as possible but there are many delays," he told his father.

Charlie Furnas, a Dayton mechanic whom the brothers had hired to help out, arrived before there was a place for him to sleep. But a new shed was nearly complete when Orville walked into camp on April 23.

The first reporters arrived about a week later.

ON May 13, in Rome, Leon Delagrange took off in a Voisin biplane and flew nearly eight miles.

THAT AFTERNOON, at the Kill Devil Hills camp, the Wrights saw a man walking across the dunes toward them. He introduced himself as a journalist and told them what he had learned via telegraph about Delagrange's flight at

Rome. Wilbur Wright "manifested considerable interest in this performance but no anxiety."

"We are not worried," Will told him. "We have already tripled the distance made by Monsieur Delagrange this morning. Our confidence in our leadership rests upon the essential difference between our machine and those used in Europe. We have a practical aeroplane capable of flying in the wind."

For several days, reporters had been moving furtively around the edges of the Wrights' flying field, staying out of sight for fear of scaring the notoriously shy inventors back into inactivity. The circumstances now differed mightily from the fall of 1905, when the story struck Dayton's press as incredible. Now, it had not only credibility but the vast advantage of international rivalry and drama. The Wrights were now widely believed to be carrying a secret worth knowing. They were either geniuses or frauds, and either case would make a great story. And the eight or ten men who arrived in the Roanoke Island town of Manteo in May were not the indifferent and undermotivated reporters of the Dayton press corps. They were experienced, resourceful, and determined professionals. One was P. H. MacGowan of London's *Daily Mail*, a veteran foreign correspondent. Another was James Hare, a *Collier's Weekly* photographer who had shot wars all over the world. Gilson Gardner, who had written a competent account of the Wrights' work for *Technical World* magazine as early as 1905, was there representing one of the Chicago papers. The representative of the *New York American* walked into the Wrights' camp disguised as a hunter. Byron Newton of the *New York Herald* stayed clear of the camp, seeking his information from the men at the weather station instead.

On May 14, Arthur Ruhl of the popular *Collier's Weekly* arrived. A skilled stylist, Ruhl had been courting the brothers for more than a year, hoping to write the first comprehensive story of the Wright brothers. Assuring them that "Collier's is deeply interested in your work," he had visited Dayton and persisted despite their polite rejections.

After a long journey by rail, steamer, and mail boat, the reporters had to brave the final leg of the pursuit on foot through soggy woods infested with mosquitoes and razorback hogs. "Geographically, this may be only four or five miles," Ruhl said, "but measured by the sand into which your shoes sink and which sinks into your shoes, the pine needles you slip back on, the heat, and the 'ticks' and 'chiggers' that swarm up out of the earth and burrow into every part of you, it seems about thirty-five." The reporters were convinced that if they showed themselves, the experimenters would simply refuse to fly. So they tried to stay hidden and spy from the border of the woods and the crowns of sand hills. The men at the lifesaving station were apparently their best sources. They enjoyed watching the Wrights through field glasses, and were happy to pass along information about their doings, substantiated or not.

On the day Ruhl arrived, the reporters crept around the dunes like a ragtag band of sunburned Boy Scouts, debating whether to stay together or move singly; whether to spy from behind a hill or to construct a screen of underbrush. The sun glared from a bright blue sky. A steady wind blew at about fifteen miles per hour, kicking up a haze of sand from the tops of the dunes "like faint smoke from a chimney."

Finally the reporters crept up to a point about a mile from the camp. Peering across the "open ground and heat shimmer" at the base of the Kill Devil Hills, Ruhl could see "two busy little dots" moving here and there around "a rectangle of hazy gray lines, with a white streak at the top, which might have been taken for the white line a receding wave trails along the beach."

After a very long wait, the men glimpsed a flicker of motion. "Two whirling circles appeared, and across the quiet distance came a sound like that of a reaper working in a distant field. The circles flashed and whirled faster and faster, then the white streak above tilted, moved forward, and rose. Across the flat, straight for the ambush, it swept, as fast an an express train. It . . . swerved and tilted slightly, righted itself, dipped and rose, now close to the ground, now thirty or forty feet above it. It had come perhaps half a mile when the operator

"THE SAND, THE HEAT, THE 'TICKS' AND 'CHIGGERS'"

Reporters wait for the Wrights to fly at the Kill Devil Hills, May 1908 saw, for the first time apparently, a dead tree trunk directly in his path. He swerved, but had to alight, coming down easily with a slight splutter of sand." More dots swarmed out to the machine—men from the lifesaving station, Ruhl guessed. They inserted carts under the wings, restarted the propellers, and jogged along beside the machine, guiding it back to the launch rail.

The reporters doubtless had seen artists' sketches of flying machines like this one. But the real thing left them struggling to make sense of it. Large machines of their experience all moved narrow-end forward. Trains did. Ships and boats did. Automobiles did. To see this thing move wide-side forward was strikingly odd. Gilson Gardner could think only of "a runaway street car moving side forward."

Now the reporters saw the "quaint bird" rise again and come toward them. Hare, the photographer, spotted his chance. He raced out into the open and aimed his camera. MacGowan, remembering his war service, called, "Don't shoot till you see the whites of their eyes!" The reporters watched as the machine banked to the right and "swept grandly by," then flew on toward the beach, where a few cows grazing in the long grass "threw their heads upward, and whirling about, galloped away in terror, ahead of the approaching machine. It swept on far above them indifferently," then rose over the swell of a low dune and headed back toward the shed. The reporters heard the distant rattle of the engine stop suddenly, and watched as the machine slid smoothly toward the ground and "alighted lightly as a bird." They looked at their watches. The aeroplane had been aloft for nearly three minutes. They estimated the distance of the flight at about two miles.

According to Ruhl, the reporters now wrestled with deep emotion and warring impulses—to stay hidden and file their stories, or to run across the plain and "tell two plucky young men how much they admired them."

They chose professional duty. With just enough time to file that day, they walked the four miles to the weather station at Kitty Hawk, where they cabled the news "to the world waiting on the other side of various sounds and continents and oceans that it was all right, the rumors true, and there was no doubt that a man could fly."

Ruhl left the Outer Banks the next morning. On his way to Norfolk via Elizabeth City, he dashed off a letter to a young lady friend who was vacationing in the town of Edenton, North Carolina. He had hoped to visit her, but now there would not be time. He was so worn out from his trek across the damnably soft dunes that he could "hardly write an ordinary sentence." Yet he had seen "a wonderful sight," he said, "and I felt that I was . . . present at something almost as extraordinary as the first trip of the first locomotive."

• • •

THE WRIGHTS were only partially aware of the surreptitious presence of the press nearby. A couple of reporters had ventured into their camp. Others had been seen in the distance, and local friends told them that reporters were "hanging about in the woods and crawling through the marsh to get sight of us in flight." A report in one of the Norfolk papers that they had flown ten miles out to sea appeared before they had made a single trial, reconfirming their belief that some journalists simply made things up. But when reporters appeared and introduced themselves, the brothers behaved cordially, obeying Kate's admonition to "be as civil as you can to every one. It pays to make friends of as many as you can." At the same time, they conveyed the impression that if the newsmen tried to come close during the flights themselves, the brothers would close up shop. As Will told the family, "It is a good thing sometimes to have a fierce reputation, like a schoolteacher."

They had wanted to test the new steering mechanism during calms, but the weather had failed to cooperate, and they regarded their performance so far as only so-so—certainly not the sort of showing they wished to make in France and Washington, before official witnesses. In three days they made eighteen trials. Most covered less than a hundred meters. The up-and-down course of the first flight "beat any roller coaster about Dayton all to pieces." The new steering controls were effective, but "it will take some practice to get used to them." And there was little time for that. Will would have to leave for New York and Europe in just over a week.

Perhaps for the first time in their careers as experimenters, they were rushing.

On the morning of May 14 they started especially early. Before 8:00 A.M., with skies clear and the wind steady, Will took Charlie Furnas up for a trip of about six hundred meters. A little while later Orville and Charlie went up again, this flight lasting just over four minutes and describing a complete circle. At one point the wind gusted to thirty miles per hour, and for a few seconds the machine stood nearly motionless in the air, like the gulls poised over the surf nearby.

The wind shifted to the west, and Will made several false starts, catching wingtips in the sand before he could leave the track. After the midday meal he tried again and "got off nicely." At an altitude of about thirty feet, he passed one landmark after another—the West Hill, the Little Hill, the tree they called the "umbrella tree," the dried-up ponds. He passed the camp and went into a second circle. He was traveling with the wind at his back at a speed of more than fifty miles per hour. With no warning at all—no sudden turbulence, no catch in the engine—"the machine suddenly darted into the ground."

A mile away, Orville was watching through field glasses. He saw the tiny figure of his brother pitch forward as the machine struck the ground nose first and went into a somersault. Then he could see only a cloud of sand. "Somewhat excited," he and Charlie Furnas stared across the blank expanse. Thirty seconds passed. Then they saw Will stand and wave.

REPAIRS WOULD TAKE far too much time for any more trials, with Will due in Europe in a couple of weeks. After only five days of flying, they had no choice but to quit. They had flown sitting upright and they had carried a passenger, and the new steering apparatus had worked well enough. But with the machine facing its crucial public tests, they certainly would have wanted more practice. And Will was not at all sure why the machine had dived and crashed.

"I do not think the accident would have happened if we had not been so rushed and overworked," he confided to Kate. "Such a thing as overexerting ourselves before mounting the machine will not occur again."

Will left for New York, where he would meet with his agents at Flint & Co., then cross the Atlantic to prepare for the tests in France. He would fly the machine that Orville had shipped to France the previous summer for the exhibitions that never had occurred. Orville would stop in Washington to inspect the flying grounds at Fort Myer, then return to Dayton to build a new machine for his own exhibition to the Army.

IN WESTERN NEW YORK STATE, the Aerial Experiment Association was preparing its new machine, and "all Hammondsport seems to be alive to the occasion." Boys were cutting school each morning and running out to Harry Champlin's racetrack at Stony Brook Farm, awash in sun and stiff spring breezes. All day they would hang around the AEA's big tent, watching the men as they tested their roaring engines and fitted strips of wood and sheets of linen into unfamiliar shapes. They were "fine bright-looking boys," Bell said approvingly, "and I only hope that their teachers will be merciful to them when their hour of reckoning arrives."

As the new machine neared completion, grown men appeared at the tent, too—reporters and flying-machine inventors and officials of the Aero Club of America, all of them hovering nearby, even skipping meals in town, "for fear the wind would go down during their absence, and they might be absent at the critical time."

Everyone gawked at the solid, genial figure of Alexander Graham Bell, who moved through the little crowd every day, rapping the ashes from his pipe and "waiting for that blessed wind to go down." Having missed the *Red Wing*'s flights, Bell felt "very anxious" to see the performance of this new craft, the *White Wing*. At stake was his own idea for achieving stability with movable surfaces at the tips of the wings. The lives of his young associates were at stake, too, and Bell was on edge. "The machine is distinctly of *the dangerous kind*," he confided to Mabel, "requiring a great deal of skill in the operator. It is unfortunate that such a machine has to be operated—at first—with an unskilled man on board. I have no doubt that it will fly—but whether it will come down safely without injury to the man on board, is a problem which can only be settled by experiment." The Wrights and Farman and Delagrange had flown safely in such machines, he knew, "so . . . this fact is encouraging. For my own part I should prefer to take my chances in a tetrahedral aerodrome—going more slowly—and over water."

He remained just as avid for the tetrahedral form as ever. When the breeze continued to blow across the racetrack at ten miles per hour every day, he grabbed the chance to send three kites up for observation. He planned more full-scale tetrahedral experiments at Beinn Bhreagh later in the summer, and he was still chafing at his young friends for their polite lack of enthusiasm. They complained of the kites' lumbering resistance to the wind, and they seemed no longer willing even to try a tetrahedral device with an engine and man aboard. This offended Bell's lifelong loyalty to the principle of experimentation. "The bug-a-boo of 'increased head-resistance' should not prevent the experiment from being made . . . ," he grumped to Mabel. "We can *try it*." And even if the machine failed to fly as Bell hoped, "the kite," unlike the dangerous aeroplane, "won't come down."

As the controlling partner, he could have ruled out the aeroplane experiments. But he could not bring himself to crush his friends' hopes, or his own. "I have not the heart to throw a damper over the ambitions of the young men associated with me. I feel that in a very few days now America may be ringing with their praises—and they certainly deserve success, and will obtain it too—I have no doubt."

When the breeze finally fell to a murmur, the members of the AEA paid the price for their zealous hurry to get into the air. Rolling the machine out onto the racetrack on May 13, they found they had neglected to take certain crucial measurements. The *White Wing*, larger than the *Red Wing*, could not move even a few yards without bumping its wings against the elevated walls of the racetrack. They tried to run it over the grass inside the oval, but the wheels broke on the muddy ruts. They raised the machine higher above the wheels for another try on the track. But without a device to steer the craft while it remained on the ground, it blundered from side to side.

On May 18, just before supper, Casey Baldwin nursed the *White Wing* up to a height of ten feet off the ground and flew ninety-three yards in a straight line. The reporters crowded around Bell. "We had the first very promising spring into the air today, showing that the machine will fly," he declared. "It was not very much in itself, but a very great thing as showing what it may do in the future." The next day Selfridge made two short flights, also in straight lines.

Now Glenn Curtiss was to have his first turn at flying. He was not long past an attack of the mumps, but he was eager for his first try, and he appraised the machine with the practiced eye of one with a great deal of experience in managing vehicles at high speed. To bring the machine's nose down, he shifted some of the machinery forward, and to better see the ground, he removed the cloth shield from in front of the pilot's perch. He added a fourth wheel under the tail.

On his first run down the backstretch of the racetrack, Curtiss could not make the machine rise. The engine was found to be short of oil. On the second run he sped down the track at close to thirty miles per hour, pulled on the elevator control, and felt the odd sensation of suddenly riding upon air, not earth. The right wing dipped and Curtiss leaned left, activating Bell's wingtips and restoring balance. "Very much elated," he steered too high, then too low, and the wheels brushed the ground. ("As is usual in any balancing act," he said, "the novice overdoes matters.") But he bounced upward again and found himself staring straight at the vineyard beyond the track. He steered left and missed it. He steered right and resumed his course. Close to the ground, he cut the engine and glided down into a farmer's field. He had flown 340 yards in nineteen seconds. On a windless day, Curtiss had kept a flying machine balanced with ailerons. Thomas Selfridge fired off a cable to the Associated Press in New York.

A day later, barely in the air, Douglas McCurdy felt a "puff" of wind from his port quarter. The right wing dropped and caught on the rough ground, and the *White Wing* executed a somersault, dropping McCurdy "gently and without any jar whatever on the ground." The machine smashed into pieces beyond him.

WITH A SINGLE FLIGHT, Glenn Curtiss had proven himself the most accomplished of the AEA fliers and the canniest among them as a student of the flying art, and now it was his turn to supervise the design and construction of an aeroplane. He went at the task exhilarated and determined. In the air over Champlin's track he had not traveled even half as fast as he had on a motorcycle. But he had felt in his sinews the electrifying sense of new possibilities in the realm of speed. For here was a medium without friction and a racetrack without limits. The sky beckoned to him. He was already pining for the enormous silver *Scientific American* trophy, waiting at the Aero Club of America's headquarters in New York for the man who would claim it. Curtiss decided to try for it. His colleagues agreed, and they began to build a new machine.

WHILE BELL'S BOYS were flying the *White Wing* in Hammondsport, Wilbur Wright was preparing to leave North Carolina for New York. As he passed through Manteo, reporters accosted him with a new urgency. They blurted that Henri Farman, in France, had just issued a public challenge to the Wright brothers: Would they compete with him for twenty-five thousand francs to see who could be first to complete a public flight of five kilometers?

Will rewarded them only with a dry crinkle of a smile and a courteous refusal to be provoked into saying a single word more than he preferred to say. Well, couldn't he say *something?* They got another smile. "I will talk to you on any subject except aeroplanes or my plans for the future."

IN NEW YORK a *New York Herald* man, acting on a cable from his colleague in Norfolk, met Will's train, and from then until his ship departed forty-eight hours later, Will enjoyed no peace. He learned that for a week or more, the flights at Kitty Hawk—no better than ordinary by the standards of 1905 at Huffman Prairie—had been the subject of breathless, day-to-day coverage, not only in the great New York dailies but all across the United States and Europe. The little band of writers and photographers sneaking around in the marshy woods had been only the vanguard of a press corps that was now voracious for news of flying machines.

In fact, a tectonic shift in the collective mind of the Fourth Estate had occurred, thanks to an extraordinary convergence of events—the flights of Farman and Delagrange in France; the surprising emergence of American contenders associated with the magical name of Bell; the widely published photographs of men in flying machines that were actually *off the ground;* and amid all this, the electrifying word that the mysterious Wright brothers had at last emerged from their midwestern redoubt and were preparing to fly again. Suddenly the realization that epochal events were under way crashed into the consciousness of the agenda-setting New York papers, and for the third week in May 1908, they flayed the story for all it was worth. The rest of the country's papers followed suit. As the editors of the *Dayton Daily News* proclaimed—with a private twinge of embarrassment, one hopes—"THE WORLD IS TALKING ABOUT WRIGHT BROTHERS."

The writers of headlines delighted in the developing story. After the first day or two of breaking news—"WRIGHT AEROPLANE SUCCESS" . . . "WRIGHT BROTHERS MAKE FLIGHT IN WAR AIRSHIP"—they put their interpretive powers to work. Some chose the historical angle: "THE WRIGHT BROTHERS HAVE AT LAST SOLVED THE GREAT PROBLEM OF SAILING IN THE AIR." Others chose metaphor: "CLOUDLAND TRIP MADE IN AIRSHIP" . . . "LIKE A GIGANTIC BIRD IT SAILS THROUGH SPACE." Some even became blasé: "FLYING EASY IF YOU KNOW HOW." Then the pleasant fact that each day brought a new mark for distance called forth the sportswriter's impulse, and though the longest flight was only eight thousand meters, or 4.97 miles, that was no impediment to the newspapers' claims: "WRIGHTS FLY 2 MILES" . . . "WRIGHT AEROPLANE GOES THREE MILES". . . . "BIG WRIGHT AIRSHIP FLIES EIGHT MILES" . . . "WRIGHT AIR SHIP FLEW 32 MILES."

Then came a delicious plot twist—Wilbur's accident, translated as: "BUNGLING" . . . "DISASTROUS WRECK" . . . "COLLAPSES IN MIDAIR"—followed by the prospect of out-and-out international competition: "AERONAUTS OF FRANCE ARE CONFIDENT" . . . "MR. FARMAN'S CHALLENGE UNANSWERED BY WRIGHTS" . . . "ORVILLE WRIGHT DESCRIBES AIRSHIP HE SAYS WILL WIN."

The Wrights' desire to work in private contributed to wild errors—the most outrageous was the report that they had flown ten miles out to sea and back—but it also whetted the press's hunger for juicy details. "They are so afraid that somebody may seek by illicit means to discover their ideas as embodied in the aeroplane," one reporter whispered, "that they never leave it unguarded and sleep by it with a rifle at hand." (This was probably true, as the Wrights had brought firearms to Kitty Hawk before—though for shooting game, not thieves.) "On account of the mystery attached to the Wright machine, the public appears vastly more interested in it than in Farman's or Lagrange's." If that was not true of the public, which was only beginning to learn of all this, it certainly was true of the American press.

Through it all came the sense that the Wrights were hardly rogues or charlatans but quite real and legitimate figures who, it now seemed, must be precisely what they claimed to be—the inventors of the first practical flying machine.

Of course, this epiphany did not tend to shed glory on the nation's newspapers, most of which had entirely ignored the brothers for five years. So, without quite neglecting to mention the brothers' earlier and more impressive flights, the press managed to convey the understanding that these few trials at Kitty Hawk were somehow the Wrights' first *real* flights, if only because they had been seen by some few of the men designated to record and recognize the notable events of America's public life. On May 14, the likable Byron Newton of the *New York Herald* explained the new sense of things this way: "Today's performance, while not equal to hundreds of flights the Wrights have made, will place them on a new footing before the world, because it will be the first time that a considerable number of disinterested outsiders have ever seen them in any flight and doing what they had told others they could do. There is no longer any chance for questioning that they stand at the head of the world's inventors and operators of a dynamic flying machine."

THE FRENCH were in no mood to agree. In Paris, newsmen besieged the Aéro-Club, buttonholing airmen for comments on the reports from America. Most maintained a discreet silence. But Ernest Archdeacon could not let such an opportunity pass. "In my opinion," he said, "the brothers Wright have certainly made serious flights, but that they have accomplished more than we have in Europe I am not prepared to admit. My own opinion, or if you prefer, my theory, is this: The brothers Wright, being Americans, are essentially business men. For some years they have been experimenting with an aeroplane, and have undoubtedly obtained certain results. What those results are I am not prepared to say. In fact, it is upon this point that I believe the Wrights have bluffed." As "business men desirous of finding a market," they had said over and over that their machine was practical. Yet look what had just happened in North Carolina, he said. "In the first time that it apparently made a serious attempt at flight it met with an accident." And this was quite like the accidents the French airmen had suffered. Therefore it was clear the Wrights were no closer to a final solution than Farman, Delagrange, and the others. Their patents could not be upheld and they knew it, Archdeacon ventured—why else the cloak of secrecy?

WHEN THE NEW YORK REPORTERS cornered Will in the lobby of the City Club, where the Flint men had put him up, or outside the offices of Flint & Co., he alternated between pure taciturnity, a little fun, and a point or two he had been wanting to make in public.

What was his purpose in New York?

"Business."

What response would he make to Henri Farman's twenty-five-thousand-franc challenge?

None.

Why had he and his brother left Kitty Hawk after so few flights?

"A horde of newspapermen drove us out."

What was his opinion of the report that Casey Baldwin and Thomas Selfridge had just made flights of nearly one hundred feet in Hammondsport?

"I think that is very plausible." (Did the newsmen catch the irony?) But then: "It will not do to believe what one reads concerning airships. I find frequently that the published accounts of flights made by others are fully as inaccurate as some of the accounts said to have been made by my brother and myself."

Were the brothers, as rumored, about to publish a full description of their machine?

No, but, "I will say for the aeroplane we have perfected that it is practicable. It is the only machine perfected as yet, in our opinion, capable of doing the work expected of it."

And what did he think of Mr. Augustus Herring's claim to have built an aeroplane engine that weighed only one pound per horsepower?

"I would rather have one that weighed five pounds to the horse power. A motor for flying machine purposes has to be built to stand the work. In fact this whole question resolves itself into the making of a machine that will do what is required of it against the buffetings of the wind.

"You can cross a pond in a tub, but you can't cross the ocean in a tub. If the question were one merely of buildng an airship that would behave well in calm weather it would be easy. How best to overcome the action of strong winds, which always must be expected and guarded against, is the real question. Our theory is that an extremely light motor cannot be built sufficiently strong to withstand the strain that must come on it in practical operation in strong winds."

IN DAYTON, people were stopping Kate and Lorin and the bishop on the street to ask for the latest news. Stories of Wilbur's accident inflamed the uproar. (The family had learned to be wary of alarming reports. "We were worried although we did not believe that the smash-up was as bad as it was reported to be," Kate said.) They all scoured the papers, local and out-of-town, to pick up details. The *Daily News* and the *Journal* had turned downright proprietary about "the boys," and were quick to defend them against all criticism, real or anticipated. Kate was especially tickled by one local writer's analysis of the crash. "To read what Anderson had to say," she told Will, "one would almost imagine that you intentionally slammed into the ground, just to see if you could!" She relished the daily work of searching the mail and the papers for new friends and new enemies. One writer would win her affection for "a very nice letter," while another "makes me sick. . . . He is the worst little whipper-snapper."

Milton was astonished and delighted when a woman friend called to say that her minister had delivered a sermon the day before on the achievement of the impossible, citing the examples of Columbus, Robert Fulton, "and the Wright Brothers."

As letters and cables and callers arriving at 7 Hawthorn increased by a factor of ten, it became more and more clear to the family that the two brothers were no longer private creatures of West Dayton, Ohio, but de facto citizens of the world, which was pulling them into a vast realm of affairs far beyond any of their imagining. The family had long ago moved beyond any misgivings about the worth of the inventors' strange work. But still, this was new.

Kate was thrilled, and adapting quickly. For Bishop Wright, reading article after article about "the Wrights and their Flyer," the experience brought sheer wonder and a bit of unease. Proud as he was, he could not help pondering his sons' estrangement from the Church. Amid this astounding new fame, could he believe that they were nonetheless the true representatives of his faith? He sought grounds for reassuring himself. In a letter to Orville, he reported the particulars of an eastern newspaper's profile. "It says that you are 'slim, sedate and placid,' 'the very antithesis of one's idea of what an airship sailor should be', 'nothing daring, nothing devilish about them,' 'They look like a pair of clerks in a village hardware store, whose pleasure it is to attend the Wednesday night prayer meeting,' &c. &c.

"How sadly they miss the devotional tendency! Yet largely you owe your training and standing to the church."

WHAT WILL HAD TOLD the New York reporters about an impending article by the Wrights themselves—that "we have, neither of us, authorized the publication over our signatures of any matter pertaining to our aeroplane, nor have we written anything about it"—was technically correct. But Will meant to see that this would not be so for long. The rush of publicity and the consciousness of rivals in hot pursuit had reinforced his conviction that it was high time to end their public silence with a bang. In the buzzing confusion of news reports, even a conscientious and fair-minded reader would now find it impossible to judge which of the airmen, if any, had a clear claim to priority in the matter of invention; which flights deserved close attention and accolades; and most important, which machines actually could fly like a bird. The brothers' entire claim to being the true inventors of the practical flying machine was at stake, and thus so was the value of their patents. Among the reporters, Will had shrugged off the news of the AEA's flights in the *White Wing*. But in his own mind he was troubled by the particulars.

On his second day in New York he broke away from his other obligations to visit the office of *The Century* magazine, whose editors, like others, had been courting the brothers. He told them that although he was going abroad and would not have time to write the story of their work, his brother might do so. That night he sent a stack of press stories to Orville and urged him to start writing.

> We need to have our true story told in an authentic way at once and to let it be known that we consider ourselves fully protected by patents. One of the clippings which I enclose intimates that Selfridge is infringing our patent on wing twisting. It is important to get the main features originated by us identified in the public mind with our machines before they are described in connection with some other machines. A statement of our original features ought to be published and not left covered up in the patent office. I strongly advise that you get a stenographer and dictate an article and have Kate assist in getting it in shape if you are too busy.

Inside the Flint offices, Will spoke with his agents about the coming exhibitions in France, and the question of what impact the flights of Farman and Delagrange might have upon the brothers' contract in France. He spent a few minutes with General Nelson Miles, the war hero who had approved the appropriation of federal funds to Samuel Langley. The old Indian fighter had toppled from power years earlier. But it couldn't hurt to win his goodwill before their flights for the Army in Washington.

Charles Flint himself had made arrangements for a quick automobile trip to the West Orange, New Jersey, laboratory of Thomas Edison. But a heavy rain swept in, and the meeting had to be canceled.

Will wrote a last stateside letter to Kate, apologizing for the reporters' hectoring antics in Dayton and teasing her for her haste in packing his things for the overseas trip. "I do sometimes wish . . . that you had raised the lid of my hat box, which was not locked, and put some of my hats in it before sending it on. However, a man can buy hats almost anywhere. . . .

"Good bye to 'sterchens' for a few months."

On the morning of May 21 he boarded the French liner *Touraine* and steamed out of the harbor, bound for Le Havre, "in a fog so dense that we could scarcely see the vessel's length."

AS WILL MADE HIS seven-day crossing of the Atlantic, the press storm shifted from the dailies to the popular weekly magazines. *Collier's* published Arthur Ruhl's funny narrative of the reporters' pursuit of the brothers at Kill Devil Hills. Orville, delighted with the story, called it "the most interesting thing I have ever seen concerning our experiments." Ruhl wanted to write more, and wrote to Dayton to say so. And he had another piece of business, he told Orville. His publisher, Robert Collier, wanted to purchase a flying machine.

Responding warmly, Orv noted Ruhl's regret that the *Herald* had broken the story of the Wrights' design, and he observed the odd fact that no reporter had thought to write the same story many months before on the basis of their American patent, which had been lying undisturbed in a file at the U.S. Patent Office since 1906. Indeed, the ways of the press perplexed Orville. "I have never been able to discover exactly what is the 'secret' of which the newspapers so often talk," he said. "The great mystery surrounding our work has been mostly created by the newspapers. They have told so many contradictory stories, that people are inclined to doubt all of them."

As for Collier's desire to buy and fly a Wright machine, Orv said the pressure of current business would keep the brothers busy for several months. "If Mr. Collier is then interested, we would be glad to furnish him a machine. I do not think he would have any more trouble in learning to operate it than he had in learning the automobile. As a sport, I think flying is far superior."

ON THE *Touraine*, with "few distractions of any kind," Will brooded. He worried about finding competent assistants to help him assemble the flyer that had been sitting in storage since the previous summer. He worried about Orville's ability to juggle the crucial tasks he faced at home—inspecting the flying field at Washington; writing the article for *The Century*; filing new patents; building a new machine for the Fort Myer trials. He worried about his brother's tendency to take on too much, to take too long, to keep matters to himself. He fired off instructions and entreaties. From New York he had warned Orville that "it always takes more time to do things than is expected." Now he reminded him: "Do not attempt more than time will allow." And he asked Kate to watch for trouble. "If at any time Orville is not well, or dissatisfied with the situation at Washington, especially the grounds, I wish you would tell me. He may not tell me such things always."

And he thought again and again about the bad moment at the Kill Devil Hills, when the machine had inexplicably darted downward toward the sand. "The more I think of the circumstances immediately preceding the accident at Kitty Hawk the less I can account for it," he confided to Orville. "I cannot remember that there was any indication of the approach of any disturbance such as I had noticed at other times. I do not think there was any upward turn such as had preceded some of the darts at Simms [Huffman Prairie]. On the other hand, I feel pretty sure the trouble did not result from turning the rudder the wrong way."

On Will's sixth day out of New York, Leon Delagrange made a flight of two thousand meters in Rome. Three days later, at a field near Ghent, France, Henri Farman offered Ernest Archdeacon a seat beside him and flew for a kilometer before a great crowd.

ARRIVING IN PARIS, Will was distressed to find no news of Orville, either in the papers or in letters from home. "The fact that the newspapers say nothing of a visit to Washington leads me to fear you did not stop there in returning home [from Kitty Hawk]. It is a great mistake to leave a personal inspection of the grounds go till the last minute."

Will went immediately to work. He met with Hart Berg; toured the countryside to find a suitable field for demonstrations; made arrangements to work at the automobile factory of a new friend, Leon Bollée. There was still no word from Orville.

Will began to seethe. He himself was writing to Dayton nearly every day. "Does he not intend to be partners any more?" he asked Kate. "It is ridiculous to leave me without information of his doings and intentions." Anticipating this complaint, Kate said "Bubbo" was doing his best, though with so many visitors and so much to do, "He doesn't [write] as much as he would otherwise." She promised to pitch in with correspondence.

By the middle of June, Will was ready to open the long wooden crates that contained the flying machine that Orville and Charlie Taylor had packed in Dayton the previous year and sent to France in expectation of an exhibition. Yet when he lifted the lids, instead of finding smaller boxes inside, each with its own cargo of parts carefully cleated down and wrapped to prevent damage in shipping, Will found an ugly jumble of metal, wood, and fabric. Inside the assembled wings he found ten or twelve ribs cracked and the white linen smeared and "torn in almost numberless places." The radiators were "badly mashed." He pulled out a broken magneto, damaged coils, a bent axle, and squashed tubes. The seat was broken. The mess would cost him many days of extra work, and he let Orville have it.

"I am sure that with a scoop shovel I could have put things in within two or three minutes and made fully as good a job of it. I never saw such evidences of idiocy in my life. Did you tell Charley not to separate anything lest it should get lonesome?" He catalogued the damage for his brother and issued strict instructions for next time: "Hereafter everything must be packed in such a way that the box can be dropped from a height of five feet ten times, once on each side and the other times on the corners. . . . Things must be packed at least ten times as well as they were last time."

Will was forced to take one wing apart and rebuild it almost from scratch. When several days of "regular French weather" caused the fabric to shrink, he and some helpers built a fire under the wings to dry the cloth. It caught on fire. He worked his fingers raw with sewing, and the muscles of his hands wore out from the effort of stretching the cloth tight on the wings. Then he discovered that the French mechanics hired to prepare the engine had bungled the job so badly that he had to spend most of two days on repairs. "They are such idiots! And fool with things that should be left alone. I get very angry every time I go down there."

As he went through the crates and assembled the machine, he concluded that Orville and Charlie Taylor not only had done a poor job of packing, but had assembled some parts shoddily and left others out altogether. Screws and wires were the wrong length. Nuts were missing. Bolts were lost. He remained furious at Orville for days. With every lost hour and every scrape of a knuckle he fumed at his little brother a little more. "I have had an awful job . . . ," he sputtered. "In putting things together I notice many evidences that your mind was on something else while you worked last summer."

French officials hovered, waiting with increasing impatience. On the Fourth of July, Will went as usual to Bollée's shop and worked all day on the engine. In these tests, a device for cooling the roaring engine by circulating water through its chambers was used. Because Orville and Charlie Taylor had failed to send the proper rubber tubing, it was necessary to use a poor substitute that was too large for its purpose. Just as Will and Bollée's mechanics were about to quit for the night, a section of the tubing ruptured, and a spout of boiling water struck Will on the forearm and side. For two or three minutes he was in agony. Bollée quickly applied picric acid, containing the danger of infection. But a blister twelve inches long arose on the arm and another the size of his hand on one side. An incompetent doctor made matters worse by coating the injury in oil, and the problem became worse for several days before it began slowly to improve.

Though the injured arm was useless for the time being, Will continued to work, knowing it made more sense to stay in bed and try to hasten his recovery. But he was now so far behind in preparing the machine that he could not stay away from the shop. The result was to irritate the injury and bring his frayed nerves to the brink of disintegration. Again he unloaded on Orville.

> If you had permitted me to have any anticipation of the state in which you had shipped things over here, it would have saved three weeks time. . . . If you have any conscience it ought to be pretty sore. I do not know of a single thing that I have not had to practically make over either because it was injured in shipping, or made defectively, or only partly finished and the hard part left undone. . . . If you had finished the parts, and left the assembling alone it would have saved me a world of trouble.

With an ocean protecting him from his brother's wrath, Orville responded with a meager excuse and mild regrets. As for the disarray of the machine in its crates, the missing parts, the work left undone, he said only, "The trouble comes at the Customs house." And, "I hope you are getting over the burns."

AT ROUGHLY THE SAME MOMENT that Wilbur Wright was scalded in France, Glenn Curtiss was preparing to fly his new aeroplane. Bell, fascinated by the flying insects that swarmed through the Finger Lakes region as spring turned to summer, had named the machine in their honor—*June Bug*.

In Hammondsport, a crowd of several hundred, including many reporters, turned out to see the great motorcyclist attempt to fly one kilometer to claim the *Scientific American* trophy. Augustus Herring was strutting through the crowd, bragging of "having a marvelous new machine up his sleeve, an aeroplane so small and efficient that he could pack it in a suitcase."

And Charles Manly was there. He had established himself as an independent engineer in New York, and was taking an active role in the affairs of the Aero Club.

The Bells had been unable to attend, but their daughter Daisy had come to represent the family—she even climbed into the pilot's seat before Curtiss was ready—as had her husband, the botanist David Fairchild, who watched with a scientist's appreciation of fine detail.

After a day of clouds and showers, the sun had come out. Curtiss made one try at suppertime but had to come down after only a few hundred yards; the tail was out of proper adjustment. Reporters and photographers waited in knee-high clover. The Fairchilds stood with Mrs. Curtiss at the edge of a patch of potatoes. David Fairchild felt "an anxiety lest something happen, lest [one] should be on the point of seeing a tragedy . . ." He thought of Otto Lilienthal falling to his death; "it haunted me as I studied the situation."

About 7:30, with dusk approaching, the engine was restarted. The machine rolled across the bumpy field, gathered speed and left the ground. It passed over the spectators at a height of twenty feet. Fairchild noticed how slowly it seemed to move. "Thirty miles an hour in an auto seems fast going where fence posts and wayside flowers mark the speed," he said, "but in the air with nothing but the distant hills to go by the passage of this giant flying thing seemed leisurely and graceful." They watched the wings pass down the valley.

From the air, Curtiss spotted a photographer whom he thought had "seemed to be pleased rather than disappointed" when the first flight fell short of the mark. So he flew on past the red flag that marked the kilometer, contin-

"VISIONS OF GREAT FLEETS OF AIR SHIPS"

Glenn Curtiss at the controls of the *June Bug*

uing "as far as the field would permit, regardless of fences, ditches, etc," before coming down in the fields.

"In spite of all I had read and heard," Daisy Fairchild told her parents, "and all the photographs I had seen, the actual sight of a man flying past me through the air was thrilling to a degree that I can't express. We all lost our heads and David shouted, and I cried . . ."

On the train bound to Washington, Fairchild wrote a long note for his father- and mother-in-law, recording the events of the day and his own emotions.

> The thing is done. Man flies! All the tedious details of perfecting a practical passenger carrying machine are forgotten. Even the previous successes of which you have seen reports mean nothing and with one leap the imagination builds on this positive fact which your eyes are seeing, a whole superstructure of world locomotion. You think of the plovers that hatch their young in the summer of the Arctic Circle, teach them to fly in Labrador and spend the winter with them in the Argentine. . . . You remember the flights of homing pigeons that cover 500 miles in eleven hours and these suggest strange visions of great fleets of airships crossing and re-crossing oceans with their thousands of passengers. In short we cast aside every pessimism and give our imaginations free rein as we stood watching the weird bowed outline pass by.

*Chapter Twelve*

# "The Light on Glory's Plume"

"A CRACK LIKE A PISTOL SHOT COMING FROM OVERHEAD"

Fort Myer, Virginia, September 17, 1908

WILBUR WRIGHT was confident that in all of nature there was only one way to fly. The birds knew it. He and his brother had discovered it. The essence of flight was the coordinated action of horizontal and vertical surfaces as they encountered the air, and it was embodied in the Wrights' patents.

So in Will's mind it was a simple syllogism—if the *June Bug* had flown as the newspapers said it did, then it infringed on the Wrights' patents. He wasted little time in issuing instructions to Orville about what to do. Sixteen days after the Independence Day flight at Hammondsport, Orville warned Glenn Curtiss that the *June Bug* and any airplane like it constituted intellectual theft. He did not say so in those words, but the implication was clear and threatening.

"We believe it would be very difficult to develop a successful machine without the use of some of the features covered in this patent. . . . If it is your desire to enter the exhibition business, we would be glad to take up the matter of a license to operate under our patents for that purpose."

Curtiss ducked. The newspapers were mistaken, he said. He intended no exhibition flights—not for money, that is. His work was experimental only.

IN FRANCE, Wilbur slept beside the flying machine, first at the factory in Le Mans of Leon Bollée, the automobile man whom he had befriended the year before, then at a shed near Les Hunaudieres, the horse-race track where he had arranged to make his flights. Several French workmen were detailed to help him. "The shed is not nearly as good as those we had at Simms and Kitty Hawk, but while the warm weather lasts I will be reasonably comfortable." The family at the farm next door provided milk and water. He fixed café au lait for himself each morning on a little alcohol stove. For lunch and dinner, he and the workmen walked down a lane to the inn of one Mme. Pollet, who offered a simple but excellent French table. He made friends with a six-year-old neighbor boy—"a truthful little chap"—who spoke English and German, and with a half-starved stray dog whom he rid of his fleas, fed, and named Flyer.

Will had two principal allies at Le Mans. One was Hart Berg, who made himself indispensable in the dual role of press agent and bodyguard, shooing away nearly all those who came to gawk and question. (Berg sometimes lent a hand with the mechanical work, too, but, "I always tremble when the engine is mentioned," he wrote Orville, "as I am responsible for the high tension ignition & if there is ever a missfire—well you know Wilbur well enough to *know* that I occasionally hear of it.") The other was Leon Bollée, who not only put his shop and his workmen at Will's disposal but offered friendship with no strings attached. Mme. Bollée, who was about to give birth, became a friend as well. Will promised her he would make his first flight in France on the day her baby was born.

The airmen of France could not resist the lure of the American they had read about for so long. Henri Farman wangled a lunch date—"he is a pretty nice sort of fellow and disposed to be friendly," Will thought—and Louis Blériot sent a respectful request to be allowed to watch the demonstration flights. "On my part," Blériot added with pride, "I should be delighted to give you the pleasure of seeing my monoplane apparatus fly, when you wish it."

The Aéro-Club men and their admirers among the French press were not so solicitous. The furtive flights at Kitty Hawk in the spring had only inflamed their frustration over the Americans' secrecy. Georges Besançon, editor of *L'Aerophile*, spoke for many when he said that although he believed the Wrights had flown, he could not fathom the reason for their years-long refusal to do so in public, which had "caused them to be despised in Europe, and had certainly spoiled the market for the sale of their secret."

When Will suffered his scalding, and it was announced that tests would be postponed during his recovery, one of the Paris papers said: "*Le bluff continue*."

A few reporters suspected Will never would fly. But others feared he would take to the air the moment they strayed away, so they kept up a steady vigil. "I think they have had their patience about as nearly tried as mine has been," Will said. "I have told them many times that they were wasting their time, but they do not believe me." He assured the newsmen there would be nothing spectacular to see at first, that he planned merely to "potter around" until he was sure the machine would work properly. But expectations were rising. "When it became known that a shed had been erected on the racecourse," said the *Paris Herald*'s man, "and that within the Bollée factory the mysterious machine was gradually taking form, attention became focused on the man. Everyone knows who he is now, and there is a very general feeling that in flying near Le Mans Mr. Wright is about to fix the general attention of the world on the town."

As he worked one day, an Englishman arrived. This was Griffith Brewer, a London patent attorney and balloonist who hoped to meet this prospective client. Brewer handed his card to a mechanic, who passed it to the American. Brewer watched as Wright glanced at the card, looked for Brewer among the onlookers, nodded, and returned to his work. Brewer hung on until the day's work was done and Wright had disappeared inside his shed. The Englishman was about to give up when Wright emerged and said, "Now, Mr. Brewer, let's go and have some dinner." Over supper at Mme. Pollet's, Brewer carefully avoided all mention of flying. Instead the two men talked far into the evening "on topics of mutual interest," chiefly "American life and habits." Brewer perceived that "this complete change must have been welcome after answering the continuous stream of questions about the aeroplane." Brewer's tact and good fellowship that evening launched a lifelong friendship with the Wright family.

IN NEW YORK, Will's hasty departure for France had caused outrage at the Aero Club of America. Most Americans were still digesting the recent news reports from Kitty Hawk, and only beginning to understand their import. Yet Wright was leaving! Departing American shores and taking his machine with him, to make his first public demonstrations not in the United States but in France! Furious, Aero Club organizers decided that if the Wrights were not patriots enough to make public flights in the United States, then someone else would, and they invited Henri Farman to cross the Atlantic. Hasty preparations were made at the racecourse at Brighton Beach, in Queens, and plans were announced for flights no later than the first week of August.

Thomas McMechan, secretary of the Aero Club, took charge of managing Farman's appearance in America. In spite of the Wrights and Bell's Aerial Experiment Association, McMechan insisted on promoting Farman's approaching feats as "the first public flights in America."

As workers hurried to put coverings over the ditches in the infield at Brighton Beach, so that Farman would have a level field, McMechan blasted the Wrights at an impromptu press conference. "We are disgusted with the conduct of some of the American inventors," he said.

> They have been hard at work, chasing the almighty dollar, and have given secondary attention to solving the problems of flight in a heavier than air machine. Whatever they have discovered they have kept secret. They have guarded against their friends as they have against their rivals. It may be that they have been able to make money from this secrecy—I assume that they have—but they have not advanced the science of flight by it. The American inventors have not been fair to America by this policy. The Aero Club of America sought to induce the Wrights to enter public tests, but they would not. . . . It is on this account that we induced Mr. Farman to come here to make a public test of his machine. We know America and the quality of her mechanics. We know that as soon as we can arouse public interest in the problem of flight and get our inventors at work on it that sooner or later some one will develop a practical flying machine.

Just then Farman came up.

"What is the trouble between you and the Wrights?" a reporter asked.

"There is no trouble," he said. "The Wrights are piqued—that is all."

"What do you think of the proposed experiments at Fort Myer?"

"Nothing will come of them. The conditions imposed by the government's engineers are impractical. The requirements are five years in advance of the present state of affairs."

BRIEF SYNOPSES of the Wrights' work—both written by the same authority—appeared in *The Independent* magazine and the leading German aeronautical publication, the *Illustrierte Aeronautische Mitteilungen*, in June.

> It is now generally conceded that the Wright brothers have accomplished the extraordinary performances claimed by them. . . . They inaugurated negotiations for the sale of their invention to various governments for war purposes, asking, it must be confessed, very high prices. Being somewhat opinionated as well as straightforward, they made two mistakes: the first that the principal market for flying machines would be for war purposes (where cost is no object), instead of for sporting purposes, as more correctly judged by the French, and the second that contracts could be obtained for a secret machine contingent upon making a flight of thirty or forty miles within one hour. Two years were therefore spent in fruitless negotiations.

The writer said the Wrights' Army trials likely would succeed, "if no disastrous accident intervenes," but that the military uses of the aeroplane were being vastly overestimated. Dirigible airships would do better in war, he said.

The author of this faint praise of the Wright brothers was Octave Chanute. Orville read it with the same capped frustration that he and Will were growing accustomed to in all dealings with their old friend, whom Orville now believed was "endeavoring to make our business more difficult."

"I think I will write him," he told Will.

But he did not. The prospect of a quarrel apparently seemed too painful, and any attempt at a public response to Chanute's assertions and insinuations would make the brothers look both ungrateful and ungenerous. So they continued to keep quiet.

CHANUTE WAS FAR FROM the only irritation at 7 Hawthorn. Orville, Kate, and Milton were worrying about everything: patents, contracts, competitors. "The Bell outfit" was offering *Red Wings* for sale at five thousand dollars each—"They have got some nerve." The brothers were being mistreated by the editor of *Aeronautics* magazine—"a fool or a knave," according to Kate. Farman was about to beat Will to the punch at Brighton Beach—"If you don't hurry," Orv nagged his brother, "he will do his flying here before you get started in France." Without their older brother's calming confidence, Kate and Orville were inflaming each others' nerves. "We miss you more and more," Kate told Will. "I am so tired of your being away all the time."

Will professed to be unconcerned about Farman or the others.

"If the trials are successful," he replied, "there will be no trouble with anyone."

But he used the word "if" advisedly. In fact, though he kept it hidden from his family, he was deeply worried. "While I was operating alone," he told Milton later, "there was the constant fear that if I attempted too much and met with a serious accident we would be almost utterly discredited before I could get the machine repaired, with no materials & no workmen. The excitement and the worry, and above all the fatigue of an endless crowd of visitors from daylight till dark had brought me to such a point of nervous exhaustion that I did not feel myself really fit to get on the machine."

And if he was feeling this way among the French, wouldn't Orville, with perhaps less self-discipline, face the same distractions in his flights for the Army? Will sent more cautions across the Atlantic:

"You should get everything ready at Washington as soon as possible. . . . If you have trouble of any kind . . . cable me, and I will come over. . . . Be cautious about throwing over the screws [propellers] always. . . . Do not attempt to go up without an absolutely reliable means of stopping the motor. . . . Make sure that everything is as perfect as it can be made. . . . Be exceedingly cautious as to wind conditions and thorough in your preparations. I wish I could be home."

LATE ON THE NIGHT of August 6, 1908, with the reporters and the hangers-on elsewhere, Will and his assistants hoisted the machine onto a cart, hitched

"GENTLEMEN, I'M GOING TO FLY."

Wilbur's flyer en route to the flying field at Le Mans. Transportability was a condition of the French contract.

the rig to an automobile, and towed it over rutted roads to the Hunaudières racetrack. When the press discovered the move the next day, word went around that a flight was planned. By the afternoon of August 8, a small crowd was lounging in the grandstand. Estimates of the number vary, but it was probably no more than a hundred. Most were from Le Mans and the countryside nearby, but some had made the 125-mile trip from Paris. Will saw the usual reporters, two officers of the Russian Army who had been waiting for weeks, Louis Blériot, and other members of the Aéro-Club, including Ernest Archdeacon himself, who was telling people why the Wright machine would disappoint them. Everyone was watching the American to see what his decision would be about making a flight that day.

The air was calm, the day "the finest we have had for a first trial . . . for several weeks." The raw spot on Will's arm still pained him, and the machine was not quite the way he wanted it. But there were the Frenchmen, watching him closely, the hopeful and the skeptical, and so "I thought it would be a good thing to do a little something."

He intended to make only a very short flight to ensure the untried machine was in proper shape for longer trials. For several hours he worked as the Frenchmen waited. He attached the propellers from the 1904 machine and tested them. He walked around and around the machine, checking each surface and cable. He tested the launching derrick. He surveyed the oval field: It was roughly eight hundred meters long by three hundred meters wide, with a belt of trees around the perimeter and other trees dotting the infield, which was bisected by a ditch. He set the launching rail at an angle precisely perpendicular to the gentle wind and checked it carefully. He gave precise instructions to the men who would run along at the wingtips at the launch, to make sure they stayed level. He directed the weight to be raised to the top of the derrick. Hart Berg recalled being struck by Wright's utter lack of flamboyance—the gray suit, the plain cap, the starched collar.

Dusk was approaching when he reversed his cap and said quietly, "Gentlemen, I'm going to fly."

Men hauled the rope that lifted the launching weight to the top of the derrick. The propellers were spun and the engine started.

Wilbur lowered his hand to the clip that released the weight. It plunged. The machine leaped forward so suddenly and accelerated so quickly that one of the men assigned to run along at the wingtips failed to take a single step. In four seconds the aeroplane was off the ground.

One of those watching from the grandstand was François Peyrey, a young journalist who had followed the development of aviation with studious devo-

"THE WIND DOES NOT SEEM TO TROUBLE HIM."

Wilbur over Les Hunaudières, August 1908

tion as aeronautical editor of *L'Auto*. He had some grasp of the theoretical problems of flight and he was an acute observer. He knew the French aspirants. He had watched Farman and Delagrange manage their long takeoffs across the field at Issy les Moulineaux; had seen the French machines inch upward off the ground and make their wide, wide turns in the air, slipping sideways, straining to stay aloft.

Now Peyrey stared as Wilbur Wright rose immediately and confidently to a height of thirty feet and "began with the most delicate of all maneuvers in aviation—namely circling." As the machine rose higher still, Peyrey could clearly see the American ease his control shaft to the side, and the instantaneous and extraordinary response of the wings, one tip twisting downward, the other upward, and the vertical tail, angling in perfect synchrony with the wings—and the equally instantaneous response of the machine as a whole, rolling promptly and smoothly upon its longitudinal axis and sweeping into a tight and graceful turn—not in an alarming way, but under the obvious and complete control of the operator. "To behold this flying machine turn sharp round at the edge of the wood . . . and continue on its course is an enchanting spectacle," Peyrey reported. "The wind does not seem to trouble him."

The truth, obvious and stark before Peyrey's knowledgeable eyes, was that this man was the first to fly precisely as a bird flew, his machine a "great white bird" with "perfect lateral stability." "To deny it," he declared, "would be childish." He could see that the horizontal rudder had to be carefully controlled with constant small adjustments, but he grasped the essential fact that, "as in the case of a bicyclist, the movements necessary to maintain equilibrium probably soon become instinctive."

"The spectacle was marvelous and delightful."

And it was immediately clear to Peyrey that no one could any longer question Wilbur Wright's claim to have flown hundreds of times in America, and at distances of up to twenty-five miles. Clearly, he could do as he pleased in the air.

It was over in less than two minutes—the first tight half-circle; the race down the backstretch of the track, high above the steeplechase hurdles; a second half-circle; the straight course back to the starting point; then "the descent with extraordinary buoyancy and precision" and a smooth skid along the grass.

All around Peyrey, people were shouting and cheering. He ran close to the machine and looked hard at the flying man to catch his reaction. "I saw the man who is said to be so unemotional turn pale. He had long suffered in silence; he was conscious that the world no longer doubted his achievements."

Perhaps Peyrey really did see some dramatic change in Will's countenance. Perhaps it was wishful thinking. In either case, Peyrey doubtless was right in perceiving that the scene struck Will with some force. He had believed that his first successful flights would erase all doubts in France. But he could not have been prepared for the emotional uproar that greeted him in the little grandstand. The French spectators were not just convinced. They were simultaneously shocked and thrilled. Even those who had believed the Wrights' claims were transfixed by the difference between expectation and reality.

Paul Zens, who, with his brother, Ernest, had tested a powered biplane of their own design just a few days earlier, had been waiting all day. He blurted to a reporter, "I would have waited ten times as long to have seen what I have seen today. Mr. Wright has us all in his hands." Another reporter grabbed René Gasnier, who was building another biplane. "The whole conception of the machine—its execution and its practical worth—is wonderful," he said. "We are as children compared to the Wrights."

Asked for his opinion, Louis Blériot replied that he was "not sufficiently calm" to express it.

For a moment, patriotic loyalty withered. Even Ernest Archdeacon lowered the French tricolor long enough to make a brief concession. "For a long time, for too long a time, the Wright brothers have been accused in Europe of bluff—even perhaps in the land of their birth. They are today hallowed in France, and I feel an intense pleasure in counting myself among the first to make amends for that flagrant injustice."

THE NEXT DAY WAS SUNDAY, so there was no flight, but the Paris newspapers reported an extraordinary event—"a revolution in the scientific world," said *Le Figaro*—and on Monday, August 10, two thousand spectators engulfed the grandstand and stood in chattering crowds. Any Aéro-Club leaders who had missed the Saturday flight were now there, including Henri Kapferer, one of the minority in the Aéro-Club who had supported the Wrights' claims, and Leon Delagrange.

In the first flight Will found himself too close to a row of trees and landed after less than a minute. But in the second flight the machine described a figure 8, and the spectators were now simply beside themselves. When a reporter caught up with Delagrange as he prepared to catch his train back to Paris, the flier said simply: "*Nous sommes battus*." "We are beaten."

On Wednesday, before a crowd of three thousand, Will made one flight of seven minutes and two shorter ones in "violent gusts"—another feat that caused jaws to drop. On Thursday it was eight minutes. But in landing, he mishandled the controls and cracked a wing and a skid. By now most Frenchmen were past the ability to find any fault at all in his performances. "Wright is a titanic genius!" exclaimed a dirigible man who had been a vocal skeptic of the Wrights' claims. "The little accident is nothing. . . . The broken wing is the punctured tire of the automobile. . . . You see how easily he kept his seat through it all. The machine is beyond criticism—that is to say, I defy anyone for the moment to say how it can possibly be improved."

These flights were the first in which the Wrights controlled the vertical tail independent of the wing-warping controls. Will told Orville he "could turn very short curves with the new arrangement, and as soon as I become familiar with it, we will have full control."

After nine flights at the cramped racetrack, Will now sought permission to move to a site he had preferred all along—a vast, open field at a nearby French Army post called Camp d'Auvours. The French were now eager to comply with any request, and Will made preparations to transfer flyer, tools, and supplies several miles to the east. Taking time off from his repairs at the end of the week, he wrote his first long letter home since the flights had begun. In his public remarks that week he had been all courtesy and modesty. Now, in the privacy of a letter to his brother, he allowed himself to savor the victory just for a moment.

After the figure 8 on Monday, he told Orville, "Blériot & Delagrange were so excited they could scarcely speak, and Kapferer could only gasp and could not talk at all. You would have almost died of laughter if you could have seen them."

HENRI FARMAN WAS IN NEW YORK, where he made a few short, straight hops at Brighton Beach. Soon after he learned of the flights at Le Mans, he boarded a ship for home. In Paris, not having seen Will fly, he said: "I believe that our machines are as good as his."

THE GREAT EUROPEAN NEWSPAPERS and the aeronautical journals alike fell over each other in their scramble to praise the Wrights. All testified to the extraordinary similarity to bird flight, to the machine's utter obedience to the will of the operator, to its "facility," "dexterity," "versatility," "virtuosity," and "grace." In London, the *Times*, reporting "immense enthusiasm" in France, said the machine appeared "almost as manageable as if it were a small toy held in the hand," while the *Daily Mirror* referred to Wilbur Wright's "perfect control" in flights that showed "the consummate ease and grace of a swallow." *L'Aerophile* and *Le Figaro* were at his feet. Without wishing to denigrate the French pioneers, the reporter Frantz Reichel told the readers of *Le Figaro*, "One is obliged to recognize that there is a whole world of difference between their machines and the Wrights'. . . . It is enough to have seen it fly, just once, to be convinced by it."

Clearly, a new and practical form of transportation had been born. Yet it was not a sense of the usefulness of flight that sprang to the minds of witnesses who saw that first, shocking figure 8, but rather an appreciation of its "incomparable beauty. . . . Nothing can give an idea of the emotion experienced, and the impression felt, at this last flight, a flight of masterly assurance and incomparable elegance."

Will told Orville: "You never saw anything like the complete reversal of position that took place after two or three little flights of less than two minutes each."

• • •

HART BERG and the commander of Camp d'Auvours put their heads together to arrange a system of special passes to control the crowds. But Will slowed down for a couple of weeks, partly because of bad weather, "partly because I have not felt like doing much hustling."

He wrote to Kate: "You can scarcely imagine what a strain it is on one to have no one you can depend on to understand what you say, and want done, and what is more no one capable of doing the grade of work we have always insisted upon in our machines. It compels me to do almost everything myself, and keeps me worried."

THE U.S. ARMY POST OF FORT MYER, Virginia, stood on the heights overlooking Washington from the Confederate side of the Potomac River—not that Confederates ever had occupied this hallowed piece of ground. The property, called Arlington Heights, originally had belonged to the adopted son of George Washington. It passed from him to his only child, Mary Anne Randolph Custis, the wife of Robert E. Lee; then to soldiers of the Army of the Potomac, who moved in after the Lees fled to Richmond when Virginia left the Union in 1861.

In 1863, General Montgomery Meigs, a Georgian loyal to the Union and quartermaster general of the Army, looked over the property. Meigs and Lee had been friends. Many years earlier, as Army engineers, they had built an ingenious jetty in the Mississippi River that saved the commerce of St. Louis from suffocation by sandbars. Now, deeply bitter toward Lee, Meigs decided the Custis-Lee estate was just the right place to bury Union dead whose families were too poor to have their bodies brought home. By the end of the war, sixteen thousand soldiers were buried on the gentle slope.

On the heights above, the temporary fortifications became a permanent Army post—Fort Whipple, renamed Fort Myer in 1881. Its high elevation made it the natural home of the U.S. Army Signal Corps. Soon the expanse of open ground attracted the eye of General Phil Sheridan, who made it a showcase for his cavalry regiments. This was why, in 1908, Fort Myer was the only Army property in the vicinity of Washington where there was enough open land to accommodate trials of flying machines.

Charlie Taylor arrived from Dayton on August 19, followed the next day by the crates that held the unassembled pieces of the machine and the unwieldy weights for the catapult—six iron blocks weighing a total of twelve hundred pounds. Orville came that evening, Charlie Furnas the next morning, the engine a day or two later in a crate of its own. Orville and Taylor found the parts "in perfect shape," and Orv seized the opportunity to tell his brother, a little tartly, that "they were packed exactly as were the goods sent to Europe."

The post and the city were buzzing about the achievements of Thomas Baldwin in his dirigible, the *California Arrow*—a sixty-six-foot-long "car" slung under a 253-pound bag of silk and rubber pumped full of twenty thousand cubic feet of hydrogen. Baldwin had his own contract with the Army and was making trials at the same time. The bag had taken several days to inflate, and there were delays for high wind, but then Baldwin—flying with Glenn Curtiss, the designer and manager of the *Arrow*'s engine—came close to satisfying all the terms of his Army contract. The two men flew a four-mile speed test at just under twenty miles per hour and a two-hour endurance test at just under fourteen miles per hour. The only mishap occurred when horses of the Thirteenth Cavalry spooked at the sight of the rising dirigible.

Brigadier General James Allen, chief of the Signal Corps, was much impressed by Baldwin and doubtful that Orville Wright could do as well. But Baldwin himself, haunting the enormous balloon shed where the Dayton men were assembling the aeroplane in preparation for their own tests, watched with delighted curiosity.

"Would you really think a thing like that could fly?" he asked a reporter. "It's a wonderful thing to me. I have been so long and so intimately associated with balloons and dirigibles that they are no longer any more than commonplace, but that—why, that is a wonder."

Between Baldwin's flights, Lieutenant Benjamin Foulois, an enthusiastic young recruit to the new Aeronautical Division, watched Wright and his mechanic work. "They paid no attention to anyone else . . ." he remembered. "They talked only to each other, as if they were on a desert island miles from civilization."

The third entrant in the Army trials was nowhere to be seen, but a grand entrance was foretold. Augustus Herring had informed the press in New York that he would not bother to have his aeroplane shipped by train. He would simply fly it from New York to Fort Myer—though he needed a few more weeks to prepare. The Army granted him a month's extension. Orville told Will: "The Signal Corps does not hesitate to express its skepticism."

Glenn Curtiss was not at all skeptical of Herring. Indeed, he had heard talk that made him think Herring might possess a crucial advantage over the Wright brothers. "You have probably seen the photos and description of the Wrights' [machine]," he wrote to Bell. "They do not seem to have anything startling, but I cannot say as much about Mr. Herring; I believe he employs [a] gyroscope, and I think there are real possibilities in this line. I see no other solution of automatic stability."

ORVILLE FOUND the Signal Corps officers friendly and hoping for his success. But he immediately conceived a dislike for Thomas Selfridge, who was present as an official Army observer. "He makes a pretense of great friendliness," he told his family, but "I don't trust him an inch. He is intensely interested in the subject, and plans to meet me often at dinners, etc. where he can try to pump me. He has a good education, and a clear mind. I understand that he does a good deal of knocking behind my back. . . . He is endeavoring to do us all the damage he can."

With the news of Wilbur's successes in France, reporters assigned to Fort Myer now champed at the bit for their own versions of the Wright story, and he spoke to them with new candor and perfect courtesy, even friendliness. And he allowed photographers easy access to the machine. Following instructions from Wilbur, Orville was now in fact eager to get the details of their design firmly associated with the name of Wright. "We haven't any secrets now," he said. "The only ones we ever had were about construction details not then patented." He explained the machine's features in detail. The press rewarded his candor with admiring references, and Orville, accustomed to Will being the public spokesman, grew a little proud of his ability to handle himself in the spotlight. "The reporters seem to think I am not in the least uneasy about fulfilling our contract," he told Kate. "They say that I do no boasting of what I can do; but that they can get but little out of me as to what I expect to accomplish, but that I have the air of perfect confidence!"

By day, at Fort Myer, scientists waited to have a few words with him. Reporters were a constant presence. People from the neighborhood wandered by to gawk and ask questions.

"Do you mean to tell me that that bunch of twisted wire and old sheets is going to fly?" one man asked. "I say it won't."

"What do them whirligigs do?" a boy wanted to know. "Cool the engine?"

"Would you mind telling me," a young woman asked one of the mechanics, "why you have so many wires about it? I should think it would look much better if you left all those things off and you just had wings, like a real bird."

During evenings at the Cosmos Club, the Army-Navy Club, and the home of Alexander Graham Bell's daughter, he was meeting "stacks of prominent people . . . who are very friendly." And "I am meeting some very handsome young ladies!" Several approached him with letters of introduction from some mutual friend or acquaintance, while "most of the others I meet in bunches, and I will have an awful time trying to think of their names if I meet them again."

"Work goes slowly," he conceded in a letter to his brother, "with a crowd of people standing about the whole time."

This was precisely the sort of thing Will wanted him to avoid. One last time, Will warned him about the dangers of losing his concentration and allowing himself to be rushed.

> Don't go out [flying] even for all the officers of the government unless you would go equally if they were absent. Do not let yourself be forced into doing anything before you are ready. Be very cautious and proceed slowly. . . . Let it be known that you . . . intend to do it in your own way. Do not let people talk to you all day and all night. It will wear you out before you are ready for real business. Courtesy has limits. If necessary appoint some hour in the day time and refuse absolutely to receive visitors even for a minute at other times. Do not receive any one after 8 oclock at night.

A FOUR-POLE LAUNCH DERRICK, thirty feet tall, was erected a short distance from the Fort Myer hospital. Orville positioned the launch rail so that it pointed southeast down a long, gentle slope. This way, as he took off in the machine, the ground would fall away in front of him, leaving a little extra room for error as he tried to get up and away. The field was a carpet of coarse, spindly weeds. The eastern boundary was marked by a low wall of reddish-brown stone and a tall line of oaks. There was room for Orville to fly ovals of about a half-mile in length. This was substantially smaller than the field at Huffman Prairie, and it meant that in the test of endurance, Orville would have to be "turning the entire time," or just about—a requirement that would force him to "be continually adjusting the tips and the resistances and the levers and be on the keen watchout every minute."

For several days he and Taylor and Furnas labored over the engine. He refused to attempt a flight until he was certain it would run without misfiring for extended periods. The separate test of speed, unlike the endurance test, was to take place in a long, straight flight, which meant he would have to fly over uneven ground, with trees and other obstructions. If the engine failed and he was forced down, the machine was unlikely to survive.

C. H. Claudy, the *New York Herald*'s man, a reporter whom Orville had come to respect and like, asked how the machine could be useful to the military if an unplanned landing might wreck it. With the patented Wright self-assurance—and, Claudy said, a smile—Orv answered: "I didn't say I was going to land. If I thought I was going to break my machine up in a tree top, I wouldn't go. . . . It is not a probability, only a chance, but it is the chance which makes the speed test the more difficult of the two, in spite of the greater individual strain put on me in the hour test. I expect to go through both without difficulty."

ON THE MORNING OF SEPTEMBER 3 Orville was ready to move the machine from the big balloon shed to a smaller tent at the west end of the parade ground. Twelve men were needed to hoist it onto an Army wagon, which an automobile towed gingerly along the dry, rutted road. The sky was cloudless, the temperature comfortable. But the wind was blowing at a steady fifteen miles per hour. He could fly in these conditions, but with only one machine on hand, and so much riding on its success, he wanted to take no chance of suffering a crippling accident before the official trials.

Several hundred onlookers waited with him—"all curious, some skeptical"—including Curtiss, Selfridge, and David Fairchild, Bell's son-in-law; General Allen and Major Squires of the Signal Corps; Augustus Post and Albert Zahm of the Aero Club of America; Gifford Pinchot, head of the U.S. Forest Service; a number of government scientists; and most of Fort Myer's officer corps. Theodore Roosevelt, Jr., the president's son, was there. Only the Bell contingent and a few others who had seen Henri Farman's brief exhibitions at Brighton Beach had seen an aeroplane in the air. Signal Corps men herded all of them into a semicircle behind the machine and back a little ways.

As suppertime approached, the fort's snapping flags began to sag. The spectators saw Orville Wright and his mechanics go into a huddle, then speak with the soldiers. Several of them surrounded the ghost-colored machine, found their handholds, and on a signal, hoisted the behemoth off the ground, shuffled in a slow and ungainly maneuver to the launch rail, and set it gently on its little truck. Six men grasped the long rope, turned their backs to the launch tower, and hauled in unison. The iron weights rose to the apex of the derrick, swinging a little, then stopped.

Orville Wright and the mechanics fussed with the engine. Then Wright took one propeller, a mechanic the other, and they pulled in unison. The engine roared . . . and died.

After a moment, one of the mechanics broke out of the huddle. The crowd watched as he hustled down to the tent at the far end of the field, then returned at a jog with some tool or part. More fussing with the engine ensued.

Spectators shifted from one foot to the other, hungry and tired. Several started talking again about bluffs. The only diversion came when the press photographers asked if they could snap the machine from in front as it tried to rise. Orville Wright came over and spoke to them too quietly for the crowd to hear. Afterward they said he had told them that anyone standing in front of the machine ran the risk of being killed, and he couldn't concentrate on what he had to do if he also had to look out for photographers. So they staked out positions to either side.

After nearly an hour, things appeared to be ready again. Chatting in the crowd drained away as the battery was attached to the engine. Again Wright and the mechanic yanked the propeller blades, and "a regular and steady bang-bang-bang-bang" filled the field. Horses started and jostled. Spectators could see the whole machine shaking slightly. The engine seemed awfully strong for a structure so light. A moment earlier, the propeller blades had looked oversized and ungainly, but now they were ominous, circular blurs.

The mechanics backed away and Orville Wright by careful steps climbed into his seat.

Two movements occurred at once—the black weight falling, the machine lunging. The motion was smooth and fast. Spectators leaned to see—was it off the rail? Yes—because the rail was behind it now—but the skids were clipping the heads of the weeds. The machine seemed to lose speed for an instant. The momentum of the catapult was spent. But the machine did not sink. The slope was dropping away and the machine was skimming straight ahead, clear of the weeds and now beginning to rise.

The reporter from *The Washington Star* heard "a long, in-drawn breath from the crowd." Then came only a faint cheer and scattered clapping of hands. The *Star*'s man thought this relatively quiet response was odd until afterward, when he heard one man ask another: "Did you cheer him?" And the second man said, "Nope, too busy thinking."

By the time the machine was halfway down the field it was twenty feet off the ground or higher. At the far end, the engine noise distant, it passed over the balloon tent and the broad wings tilted, the left tips pointing to the ground as it swept into the same tight circle that Will's audience had just seen in France—"exactly as a bird's wings tilt when it is flying in a circle"—and then it was leveling out and coming back toward them, following the line of the cemetery trees and the low stone wall. Its course was far from perfect. It dipped and rose as it came on. When it got close they could see Orville Wright moving the levers and the machine "responding as does a boat or an automobile to the slight touches on one or another of the three levers." The operator was leaning back in his seat, looking as casual and calm as a motorist in an automobile.

Now the machine was approaching them and sweeping into another tight half-circle, and people took quick steps backward as if to get out of the way, though it was as far above their heads as a flag on a tall pole.

Down the first stretch it went again, past the artillery sheds, but it was closer to the ground now and spectators thought it was going to hit the balloon tent. They saw it angle sharply downward. There was a burst of dust. The engine noise stopped. People were alarmed and some started to run. But then they saw Orville Wright getting up and out; he was fine. A skid had been broken in the landing but nobody cared. Stopwatches said the flight had lasted seventy-one seconds. Reporters surrounded the flying man, and a couple of them had wet cheeks.

"I pulled the lever the wrong way," he said. He had come down to avoid hitting the balloon shed. With "the same quiet smile that has characterized him since he became a familiar figure at Fort Myer," he reminded them that it

"A LONG, IN-DRAWN BREATH FROM THE CROWD"

Orville over the parade ground at Fort Myer, ca. September 1908 was his first time flying this new machine, with new controls, and apart from a few minutes in North Carolina the previous spring, he had not flown at all in nearly three years.

The reporters turned for comment to the Signal Corps officers. These sober souls were now "enthusiastic beyond the limits of ordinary official conversation," C. H. Claudy said. "It might not be quite fair to quote all they said in the heat of the moment, but it can be confidently stated that they are more than pleased with this first performance and believe the machine showed better control in the air than the Farman aeroplane."

Men mobbed Orville, reaching to shake his hand and all talking at once. The machine was getting jostled, so soldiers were detailed to get a rope barrier around it. A woman pushed through and said: "I don't know how to thank you for the glory of those seconds. They were wonderful and to be remembered forever."

Claudy, the *New York Herald* man, asked Glenn Curtiss what he thought.

"Couldn't have been better," he said. "If he hadn't made the mistake and had to come down he would have made several more circles. His turns are fine."

FOR SEVERAL DAYS the two brothers generated growing headlines on both sides of the Atlantic. It was an extraordinary display of their prowess, though Orville made the greater sensation. On September 9 he made flights of fifty-seven and sixty-two minutes with several Cabinet secretaries watching. "Have I anything to say?" said Major George Squier, of the Signal Corps. "Well, I should say so. It is just splendid. I hope the news . . . got over to Berlin and Paris right away. We lead the world now in aeronautical supremacy. It is just splendid." General Nelson Miles was in the crowd; he pushed through to Orville and offered fervent congratulations. The *Dayton Journal* led the next day's paper with an enormous double-deck headline: "Orville Wright Conquers Air in Amazing Flight." The *Herald* called upon the city to honor the brothers in such a way as to "make them feel until their dying days that the people of Dayton really and truly are proud to call them sons and brothers . . . Dayton may crumble in dust, but the name of the Wright brothers will endure as long as earth endures."

On four consecutive days Orville set four consecutive world records for endurance in the air. At long last, Octave Chanute was present. With Lieutenant Frank Lahm, Jr., son of the expatriate American who had pleaded the Wrights' cause in Paris, Orville set the record for a flight with a passenger. And these were practice flights; the official trials were still to come.

At Camp d'Auvours, Will read the stories from the United States with pleasure and relief. His newfound friends showered him with another round of congratulations. "Well, it was fine news all right," he told Kate, "and lifted a load off of my mind." To Orv he sent a two-word cable: "*Très bien*."

Another telegram came to Orville:

ON BEHALF OF AERIAL EXPERIMENT ASSOCIATION, ALLOW ME TO CONGRATULATE YOU UPON YOUR MAGNIFICENT SUCCESS. AN HOUR IN THE AIR MARKS A HISTORICAL OCCASION.

It was signed, "Graham Bell."

THE FAMILY was now subscribing to a news-clipping service, so they were able to keep track of nearly everything written about the brothers in the major eastern and midwestern newspapers and magazines and some in France as well. Bishop Wright seems to have read virtually every word. When he learned of Orville's first record flight at Fort Myer, Milton pondered all that had happened. "They treat you in France as if you were a resurrected Columbus," he wrote to Will that evening, "and the people gaze as if you had fallen down from Jupiter."

Kate thought the bishop was eager to go to Washington. "What would you think if Pop and I came on to see the official trials?" she wrote Orv. "When I suggested it to Pop today he said 'What good would it do?' but looked tickled to death at the idea. Do you suppose we could scratch up the cash? Daddy has about a hundred dollars. Anyway he ought to go. . . . You write and ask him to come. It really isn't right for him not [to] be there. He wants to go very much."

They did not go; Milton was scheduled to attend a Brethren conference. But he continued to search for every scrap of news. Pride in his sons and the cautions of his faith apparently battled in his mind. He held a low opinion of the Old World in general and the French people in particular, and thus was pleased, as he told Will, that "your capture of France and Europe . . . seems a strong *coup d'état*." At the same time, he could not help but think of lines by the Irish poet Thomas Moore, from which he quoted to his son:

The world is all a fleeting show
For man's illusion given . . .
And false the light on glory's plume,
As fading hues of even;

And love and hope and beauty's bloom
Are blossoms gather'd for the tomb—
There's nothing bright but Heaven.

Yet he believed history's verdict on his sons was now certain. "You and Orville are . . . secure of a place with Fulton and Morse and Franklin in the temple of fame. 'Conquerers of the Air.' Its extensive results are, as yet, uncomprehended and undreamed of, even by yourselves. Did Fulton have any vision of an ocean greyhound, or Franklin of wireless telegraphy?"

FOR SEVERAL DAYS after the string of records at Fort Myer, the winds were too strong to permit flights. But Orville now felt he was ready to make the official trials. He had learned the new machine's idiosyncracies and reaccustomed himself to flying. "I do not think I will make many more practice flights," he told Kate. "I have now had enough to enable me to complete our contract."

LATE IN THE AFTERNOON of September 17 the breeze faded to about six miles per hour, and Orville indicated he would attempt the first required trial with Thomas Selfridge as his passenger. Selfridge's Army superiors were apparently unaware of the brewing patent conflict between the Wrights and the AEA—which made Selfridge's assignment awkward if not downright improper—and Orville raised no objection.

The machine was prepared for flight and all was ready to go a few minutes after 5:00 P.M. Orville had thought the propellers were warping, so he had replaced them with a new pair the day before. The light was beginning to fade, and a skim of fog was gathering.

Orville gave his last instructions to Selfridge and said: "I don't have to tell you to keep your nerve. You've been up often enough to know how to do that. Just sit tight, and don't move around any more than you actually have to."

AFTER THREE SMOOTH ROUNDS of the field at an altitude of something over a hundred feet, Orville heard a light tapping behind him. He glanced back and saw nothing wrong, but decided to cut the flight short. Then came two loud thumps, and the machine shuddered.

A PHOTOGRAPHER NAMED W. S. CLIME had been to the parade ground to shoot several earlier flights. On this day he arrived late at the south end of the field, where a press tent had been set up. He was told that Wright and Selfridge were about to fly, and he decided to stay where he was, so as to photograph them from the front as the machine rose. A moment later he did so, just as "it swept by with the grace and ease of a soaring bird." As the machine circled the field once, twice, three times, at about thirty-eight miles per hour, Clime did not keep his eyes fixed on it, "the novelty having worn off to such an extent that one only gave a glance upward when it went directly overhead." On the machine's fourth time overhead Clime looked up, and from a distance of roughly a hundred feet, he could see Orville, "hands on levers looking straight ahead, and Lieut. Selfridge to his right, arms folded as cool as the daring aviator beside him."

Clime made another exposure. He turned and began to walk toward the shed when he heard "a crack like a pistol shot coming from overhead." He looked up and saw a long piece of material falling from the machine, off to the right.

About a quarter-mile away, where a crowd of some twenty-five hundred was gathered, the sharp report that Clime heard was indistinguishable from the bang-bang-bang of the engine. Only a few people noticed the piece of falling debris.

Clime saw the machine swerve to its left. Then it appeared to slip backward, then to right itself for an instant, then to pitch forward in a nearly perpendicular course toward the ground.

ORVILLE CUT THE ENGINE and struggled with the levers. Selfridge uttered a soft, "Oh! Oh!"

SOME IN THE CROWD thought the machine nearly recovered itself. Then it struck the ground.

Clime, the photographer, sprinted toward the billowing dust—only two soldiers on horseback beat him to it—and found the machine "an inconceivable mass of wreckage." He heard moans. He dropped his camera, grabbed a wing surface, and pushed upward as hard as he could until it gave way. Underneath he found Orville Wright suspended from a cable and a strut, his arms hanging limply, his shoes barely touching the ground. Blood streamed down his face. Selfridge was underneath Wright, lying on his back, his knees drawn up, apparently unconscious. There was blood all over his clothes.

Clime and the two horsemen were able to get hold of Orville and carry him out of the wreckage, but Selfridge was pinned by cables and debris. Clime looked across the field for more help, and "a weird spectacle presented itself. Horsemen were galloping madly across the broken field in our direction, a picture of my idea of a cavalry charge in actual battle, and in their rear a mass of humanity blended together by the twilight into a low black line, and approaching with ever increasing rapidity."

CHARLIE TAYLOR was among those who reached the wreck next. He helped to get the machine off Selfridge. He stood with Orville for a moment, then walked over to the machine, leaned against it, and sobbed. Charlie Furnas was so disturbed by what he saw that he would never work for the Wrights again.

Dr. J. A. Watters, a New York City physician, pushed through the cavalrymen. Orville said: "It's my leg, doctor, and my chest. They hurt me fearfully."

A half-dozen Army orderlies arrived on the run. With the crowd surging around, they carried the injured men to the post hospital at the north end of the parade ground.

Dr. Watters told Charlie Taylor that he believed Orville would be all right, and the mechanic began to gather up the wreckage. Parts of it were in pieces. The fabric was torn all over. He examined the engine closely and concluded that Orv had switched it off and tried to glide the machine safely to the ground.

In the hospital, the orderlies carried Wright and Selfridge up creaking stairs to the second floor. Chanute was among those waiting, as was Charles Flint. The doctors came down after a time and said Wright was in shock but able to talk. Speaking very softly, he had asked them to telegraph his family in Dayton and his brother in France.

Selfridge had suffered a fracture at the base of his skull. Surgeons operated to remove large fragments of bone. Five minutes after they completed the procedure, Selfridge died.

KATE BOARDED A TRAIN for Washington that night.

Will was not told of the crash until the next morning. Stricken, he called off his own flights for the time being and searched the newspapers for particulars that might allow him to diagnose the exact cause of the accident. He concluded immediately that the fault lay principally with himself; he felt his worst fears had been confirmed. His brother lacked the fierce discipline that permitted Will to freeze out unwanted guests, and it must have been these distractions that caused him to overlook whatever mechanical flaw had led to the accident. He wrote to Kate:

> I cannot help thinking over and over again "If I had been there it would not have happened." The worry over leaving Orville alone to undertake those trials was one of the chief things in almost breaking me down a few weeks ago. . . . A half dozen times I was on the point of telling

> Berg that I was going to America in spite of everything. It was not right to leave Orville to undertake such a task alone. I do not mean that Orville was incompetent to do the work itself, but I realized that he would be surrounded by thousands of people who with the most friendly intentions in the world would consume his time, exhaust his strength, and keep him from having proper rest. When a man is in this condition he tends to trust more to the carefulness of others instead of doing everything and examining everything himself . . . I cannot help suspecting that Orville told the Charleys [Taylor and Furnas] to put on the big screws instead of doing it himself, and that if he had done it himself he would have noticed the thing that made the trouble, whatever it may have been. . . . Tell "Bubbo" that his flights have revolutionized the world's beliefs regarding the practicability of flight. Even such conservative papers as the London *Times* devote leading editorials to his work and accept human flight as a thing to be regarded as a normal feature of the world's future life.

"I will never leave him alone in such a position again," Will told Milton. As for Selfridge, he said, "It is a great pity, a great pity."

MABEL BELL was distraught for many days. "I can't get over Tom's being taken," she wrote her husband, who left Baddeck for Washington immediately upon receiving the news. Selfridge had become a special favorite of hers, "a knightly boy," during his months at Beinn Bhreagh. She treasured his deep reserve, his "quiet fun and good humor," his solicitous regard for the older women who lived and worked on the Bell estate, his habit of seeing to such small needs as bringing a screen to her side to block a draft.

"Isn't it heart breaking? Yes and yet it is better for him than to die as poor Langley did. He was so happy to the very end."

She grieved for Bell "in this breaking of your beautiful Association." But she asked him not to substitute another man for Selfridge in the AEA. "It was beautiful and the memory of it will endure:—'Bell, Curtiss, Baldwin, Selfridge, and McCurdy.' . . . Do anything you think best, but let the A.E.A. be only those to the end, and then take some other name . . . let's hold tight together all the tighter for the one that's gone."

KATE WAS MET BY ARMY DOCTORS, who told her Orville's injuries were serious but not life-threatening. He faced a long recovery and could not be moved home for several weeks.

She found him "looking pretty badly." His face was gashed and his badly broken left leg was in traction. "He was looking for me, and when I went in his chin quivered and the tears came to his eyes but he soon braced up again. The shock has weakened him very much, of course. The only other time that he showed any sign of breaking down, was when he asked me if I knew that Lieut. Selfridge was dead." Chanute was there, and "can't be nice enough," and Lieutenant Lahm, of the Aeronautical Section, who became the Wrights' near-constant attendant and, soon, a good friend.

She was missing school. Despite the brothers' extraordinary achievements, this nevertheless posed a financial stringency for the family. But "school can go and my salary, too . . . Little bubbo shall not be neglected as long as I am able to crawl around."

Will, knowing of the strain she already had been under that summer, offered an apology for upending her life with a flying machine. But he had no qualms about her ability to represent the family in the crisis.

> I am awfully sorry that you have had to pass through so much trouble of a nerve-wracking character this summer. However I am sure that as soon as Orville is well on the way to recovery you will enjoy yourself immensely at Washington. Orville has a way of stepping right into the affections of nice people whom he meets, and they will be nice to you at first for him and then for yourself, for you have some little knack in that line yourself. I am glad you are there to keep your eagle eye on pretty young ladies. I would fear the worst, if he were left unguarded. Be careful yourself also. . . .
>
> I presume that poor old Daddy is terribly worried over our troubles, but he may be sure that, like his Keiter trouble, things will turn out all right at last. I shall be not only careful and more careful, but also most careful, and cautious as well. So you need have no fears for me. I promise you that I will be as careful of myself as I was in 1900 when I gave you a similar promise. It is a pity Orville is not with me as he was then.

When she was not nursing her brother, she was conferring with doctors and Army officers, fending off reporters, sorting through stacks of telegrams and letters, and receiving Orville's visitors—General Nelson Miles ("a pretty fine old fellow, socially"); a Navy admiral ("I was disgusted with [him]"); Charles Flint, her brothers' financier, and his wife ("fearful blowhards"); and many others. Among other items of business, she arranged for an extension of her brothers' contract with the Army until the following year. And she pretty clearly fell at least half in love with Lieutenant Lahm. "I should have died if it hadn't been for him. . . . To look at him you would never imagine how kind and sympathetic and thoughtful he can be."

Managing all this and more, and nearly going without sleep, she nonetheless received a stern reminder of her father's exacting standards. In this case, she had failed to keep her promise to write every day. "The natural inference," Milton wrote, with hurtful sarcasm, "is that you are down with typhoid fever. . . . We are ashamed to tell the many inquirers after Orville's condition, that for four days (to-morrow) we have no word from you, so we have to say, that we suppose you are sick. But if you are down sick, news might disturb you. So I will close. Your father, M. Wright."

DESPITE THEIR SHOCK and grief, none of the Wrights thought Will should stop flying in France. Four days after the crash, he flew for more than an hour and a half at d'Auvours, breaking Orville's records for distance and duration.

He flew all that fall and into the winter, taking up a long string of officers, statesmen, and journalists with him, and a number of women, as well. On the last day of 1908, to win the twenty-thousand-franc Coupe Michelin for the longest flight of the year, he flew two hours and eighteen minutes, some of it through freezing rain and sleet.

"I am sorry that I could not come home for Christmas," he wrote Orville, "but I could not afford to lose the Michelin Prize, as the loss of prestige would have been much greater than the direct loss. If I had gone away, the other fellows would have fairly busted themselves to surpass any record I left. The fact that they knew I was ready to beat anything they should do kept them discouraged." His closest competitor had been Farman, with a flight of forty-four minutes in October.

IT WAS NO LONGER POSSIBLE to deny the Wrights' achievements. But the achievements could be defined in such a way that the Wrights seemed odd and quirky—certainly not the practical, efficient, business-minded men who would be required to bring the new age of flying beyond this difficult infancy.

In an essay circulated within the AEA only a week after the crash at Fort Myer, Glenn Curtiss sketched his view of aviation beyond the Wrights. "The airship which, within ten years, will carry men and freight from place to place, will be a natural evolution of the aerodromes of today and not the semi-accidental discovery of a genius. It will be the work of a man who is thoroughly familiar with the laws of fluid movement; with the effects of wind currents and the means of overcoming the numerous difficulties which are encountered in the air. It is in the practical application of the scientific knowledge at hand that the solution of the problems of aerial flight will be found."

As Curtiss wrote these words, it seems not to have dawned on him that he was precisely describing the approach that had led, at Kitty Hawk and Dayton, to the creation of the aeroplane. Perhaps, in his zeal to muscle in on the founding of an era, he had to interpret the Wrights' work as "the semi-accidental discovery of a genius." That would leave the field open to him, the practical man who would bring aviation to its true fulfillment.

*Chapter Thirteen*

# "The Greatest Courage and Achievements"

"THEM *IS* FINE."

Orville watches as Wilbur and Kate prepare to fly at Pau, France

SOON AFTER SELFRIDGE'S DEATH, Mabel Bell gave ten thousand dollars to keep the Aerial Experiment Association alive for six more months, until March 31, 1909. Her husband, still reeling from the tragedy at Fort Myer, had seen enough of aeroplanes. In the time remaining to the AEA, he wanted to return to his original purpose—"placing a tetrahedral structure in the air." In the face of the Wrights' patents, he felt it made more sense than ever to create an altogether different form of flying machine. Winning patents for the AEA aeroplanes would be "extremely doubtful," Bell now believed. Yet they would have "no difficulty in securing good patents upon aerodromes embodying tetrahedral structures." Trials of Douglas McCurdy's new *Silver Dart* would go forward. Otherwise, Bell wanted to concentrate on his kites.

So as autumn moved toward winter at Beinn Bhreagh, the seamstresses of Baddeck fashioned twenty-two hundred new cells of silk. Carpenters joined these to the five thousand cells left over from 1908 to create what Bell at first called Drome No. 5, soon to be renamed *Cygnet II*. Like its namesake, it was a behemoth of geometrical intricacy clothed in brilliant red silk. In fact it was larger than the first *Cygnet*. But Bell had it built without interior cells—a concession to Baldwin and the others, who argued that interior cells added weight with no increase in lifting power. The original *Cygnet* had been only a manned kite. Its successor would carry an engine, propellers, and a horizontal rudder in front, like the Wright and AEA aeroplanes. A powerful boat was to pull the structure into the air with a man aboard. Once aloft, he would start the engine. If the engine were strong enough, the tow line would go slack. The passenger would suddenly be a pilot. He would cast off the tow line, and Bell would have his triumph—a flying machine no longer at anchor, a tetrahedral aerodrome flying through the air under its own power, serene and safe even in turbulent winds.

Yet the *Cygnet II* would embrace the wind with little more aerodynamic grace than a brick wall. To overcome its stubborn tendency to resist the wind, it would need an extremely powerful engine. For that, Bell needed Glenn Curtiss, who "has only to look at the engine to get it to run well." But in the months after Thomas Selfridge's death, Curtiss became an increasingly elusive partner in the quest for Bell's dream.

In fact, all three of the remaining AEA youngsters soon sidestepped Bell's plea for more work on tetrahedrals. In Hammondsport, where Curtiss was supposed to be building a new engine for the *Silver Dart* and the *Cygnet II*, he and McCurdy were spending nearly all their time on the *Silver Dart*, and on a plan for aeroplanes that would take off and land on water. At Baddeck, with Bell, Casey Baldwin, too, was spending most of his time on water-based craft.

Bell, waiting week after week for news from Curtiss, became exasperated. "What we want to know from Hammondsport," he declared, "is the answer to the question: 'What are you doing?' . . . Silence does not give us any information. . . . The delay in completing the new engine affects us all, for it is needed at Beinn Bhreagh as much as at Hammondsport." When Curtiss did write, he spoke only of malfunctions and delays. He confessed: "We have read so much of the Wrights and others flying, not to mention the fact that we should have been through here long ago, that we are getting very uneasy."

Bell scheduled a full meeting of the AEA for January. With its charter due to expire at the end of March, he wished to make important decisions about the future, and it was essential that all members be present to make applications for patents. But Curtiss begged off, saying the directors of his own company were about to meet. Besides, he said, he had "other important business."

When Douglas McCurdy arrived at Beinn Bhreagh alone, carrying Curtiss's proxy for the patent decision, Bell wired: "YOUR PRESENCE NECESSARY TO DETERMINE THE NAMES TO BE SIGNED TO THE APPLICATION FOR A PATENT. NO PROXY WILL MEET THE CASE. PLEASE COME IMMEDIATELY AFTER YOUR DIRECTORS' MEETING IF POSSIBLE." Bell heard nothing for several days. Then Curtiss's reply came—by leisurely letter, not by wire. He would come, he said, when the engine for the *Cygnet II* was ready, a reason well calculated to soothe Bell's impatience.

When Glenn and Lena Curtiss finally got to Baddeck at the end of January, the Bells staged a convivial "old home week." But when the talk turned to business, Bell could not have escaped a growing understanding that his disagreements with Curtiss now went beyond the issue of aeronautical design.

Bell was all caution. Even if they could acquire patents on the distinctive features of the AEA aeroplanes, they would have to wait years for approval. If they sold aeroplanes in the meantime, the Wrights would sue. Orville Wright's letter left no doubt about that. Bell's memory of his own patent war remained vivid, and he feared another, especially with his own fortune at stake. "I am very much averse to attempting to make money under our present organization," he said, "or under any organization that would throw the financial responsibility on me alone, for I am the only member of the Association that could be touched in the matter." Since before Selfridge's death, he had been thinking about converting the AEA into a nonprofit association to encourage the development of aviation in the United States, giving grants to poor inventors and sharing information broadly—an institutionalizing of the work of Octave Chanute. Working in this way, they might proceed unhindered by the Wrights and advance the cause of flight as a great collective endeavor. In the meantime, he thought they might raise money by designing and selling flying toys. He was also exploring the possibility of selling aeroplanes through the Canadian government to the British empire.

Curtiss had his eye on more immediate gains. Government contracts in the United States were unlikely, he said, as "the Wrights will no doubt be in on the 'ground floor.' " But prize offerings were mounting. In London, the *Daily Mail*'s one-thousand-pound prize for the first flight across the English Channel still stood unclaimed. The publisher James Gordon Bennett, Jr., of the *New York Herald*, was offering five thousand dollars to the machine that reached the highest speed in a grand aviation meet to be held in the summer of 1909 in France.

Other aeronautical prizes were being planned by the organizers of a great exhibition much closer to home, in New York. This was the Hudson-Fulton

Celebration, a two-week pageant to commemorate Henry Hudson's entry into New York Harbor in 1609 and Robert Fulton's voyage up the Hudson in 1807. And clearly there was money to be made by flying the AEA aerodromes for the sensation-seeking public. "These cash prizes looking very alluring," Curtiss told his colleagues. "I think prize chasing and exhibition work should go hand in hand." But that would be merely spadework to prepare for large-scale manufacturing and sales. "Probably by another year the machines will become more standardized and a certain amount of business may be expected from private parties for machines for sport," he said. "Perhaps an Aerial Development Company could be formed to look after Government contracts, prizes, etc., and to get in shape to handle the large volume of business which is bound to come later."

Bell said there was no point in organizing a manufacturing company in the United States without patents for protection. Investors would come nowhere near it. As an alternative, the AEA might start a company with money they contributed themselves and raised from a few friends. Curtiss could manage the concern. When the Patent Office acted on their applications, they could decide how to proceed. Meanwhile, with the risk of patent litigation spread around, Bell was willing to allow flying exhibitions by the *June Bug* and *Silver Dart*. Money would be set aside for "a moderate amount of litigation."

As the AEA's discussions were going on at Beinn Bhreagh, Curtiss was exchanging private telegrams with New York. Not enough evidence survives to allow one to know how much about these exchanges, if anything, he related to his AEA comrades. He may have mentioned that Augustus Herring had presented him with a major business proposition. Or he may have kept it to himself.

ICE COVERED the Bras d'Or Lakes "like a flat tabletop of gigantic dimensions." That meant, of course, that no boat could tow the tetrahedral *Cygnet II* into the air. Still, Bell itched for a trial before spring. So the experimenters attached sledge runners to the underside of the machine. McCurdy's *Silver Dart* was prepared for trials, too.

Neither Alec nor Mabel felt much faith in Curtiss's new engine for the big kite. Bell always professed to be undaunted by the prospect of a failed experiment, saying a failure could teach as much as a success. "Alec is not discouraged," Mabel once wrote her mother, and added, "I don't believe he ever will be." On that day, at least, she did not feel so resilient. "It is a great disappointment that the new engine is too heavy and not powerful enough to do much for his aerodrome. . . . Probably nothing will happen and then there will be the old round of experimental drudgery to be gone over again." Still the promise of stability seemed strong, and the lovely geometry of the form evoked faith. Alec knew full well the tetrahedral design lacked the lifting power of an aeroplane, Mabel said, "but it is so beautifully steady in the air. You can't conceive what a beautiful object these great big red silk kites are in the air. They are so solid and substantial and seem just glued to one place in the sky while all round the wind is whirling past making tall trees bend. I am sure it will come out all right in the end and there's no question but that this is the steadiest and strongest form."

On the ice, Bell, all in fur, sat off to one side with his pipe, utterly still as younger men fussed with his machine. He had issued strict instructions. The idea was only to see if a manned machine would rise, and if so, if it would maintain its balance. Then the operator was to come down immediately. On the first attempt, with McCurdy squeezed into a V-shaped spot for the operator, the *Cygnet II* got away from its attendants and skittered along for a hundred feet or so before the engine sputtered out. On the second try the gas line broke. The men tried again and the propeller twisted off its shaft and shattered. "Although at times it seemed about to do so," Mabel said later, with all the faith she could muster, the *Cygnet II* never raised itself so much as an inch off the ice.

When McCurdy flew his aerodrome the next day—the first flight ever in the British dominions—Mabel thought Bell "almost as pleased . . . as if the Silver Dart was absolutely his own machine and so proud of Douglas." Yet she admitted to her daughter that "Papa feels down this evening although he knew the chances of success were of the slimmest.

"The drome is such a beautiful sight it goes to Daddysan's heart."

Bell immediately began to supervise changes to the *Cygnet II*, including new safety features to protect "the aviator."

DOUGLAS MCCURDY made many fine flights in the *Silver Dart* over the next several weeks, "circumdroming" the lake, as Bell put it, for distances of up to twenty miles. In the meantime, the little AEA convocation broke up. With affectionate farewells all around, the Curtisses, bound for home, were put on the steamship at the little Baddeck pier.

Several days later, the Bells learned that their friend Curtiss, Augustus Herring, and Courtlandt Field Bishop had appeared at the Aero Club of America on East Forty-second Street to announce the formation of the first company in America to manufacture aeroplanes for the commercial market.

• • •

COURTLANDT BISHOP had made efforts to broker a deal between the Wrights and the Army. When the efforts fell through, he gave up in disgust—more at the Wrights, for their stubbornness, than at the government—and went off to race balloons. Now Bishop realized the Wrights were no longer the only game in town. Augustus Herring, Bishop's fellow member of the Aero Club, enjoyed the credibility that came with holding a U.S. Army contract for a flying machine—one still to be delivered, it was true, but the contract alone made Herring's a bankable name for the time being. More important, Herring was saying he held patents (or perhaps just patent applications—he was never quite clear about this) that not only predated the Wright patents but also would guarantee the great prize of automatic stability in a flying machine. He said he had invented gyroscopic devices that would keep an aeroplane safe even in strong winds.

Curtiss made an obvious target for the entrepreneurs. Thanks to his flights in the *June Bug*, his name as an aviator was now well known, and he had manufacturing experience and a factory to throw into the bargain. If he could bring the great name of Bell with him, so much the better.

Herring used each man to bait the other. To Bishop, he confided that he had Glenn Curtiss in the bag, and to Curtiss he said Bishop was in, with other wealthy friends on the way.

The negotiations began in New York and continued via telegram while Curtiss and his wife were staying at the Bells' home. Herring's wires brought heavenly promises to the upstate mechanic: "BIG FUTURE" . . . "BEST POSSIBLE BACKING. SMALL COMPANY, FIRST. WAY CLEAR TO MILLION EACH." Curtiss was salivating. "PROPOSITION AGREEABLE," he replied. When his suitor asked if Bell could be brought in, Curtiss tossed cold water on the idea. "I do not think Mr. Bell would consider making any connection with this company as he has a plan for a big organization, and I think best not to mention this at present; however, if I should come with you I think the other scheme would be given up."

Curtiss weighed his options. The Wrights already had invented a practical aeroplane and patented it. If he wished to sell his own aeroplanes without paying the Wrights for the privilege, he had three choices. He could use the Wrights' system and instantly incur a lawsuit for violating their patent rights. He could adapt or improve the Wrights' system—this was what the AEA claimed to have done, but with no assurance of escaping patent problems. Or he could find some wholly new system of controlling an aeroplane. This was what Bell hoped to do with tetrahedrals, and what Herring said he could do with gyroscopes. Only this option offered an escape from the shadow of the

Wrights. Curtiss had watched Bell's lumbering kites straining on their ropes, overwhelmed by the forces of wind resistance, and he was convinced they represented a dead end. When Herring came to Curtiss with promises of an alternative system of balance, the younger man glimpsed long vistas of speed, fame, and wealth.

And none of these could be won as long as he was in league with his old comrades. Herring said later that Curtiss regarded the AEA members as "un progressive, particularly Dr. Bell, and that he couldn't see that there was much of a future for him in the A.E.A.; that he thought there was more money outside of it, operating independently."

Arriving in New York, the Curtisses were met at Grand Central Terminal by Captain Thomas Baldwin, the dirigible man, who was in on the scheme, too. Baldwin offered a new enticement: The Aero Club would sponsor Curtiss in a bid for the sterling silver Gordon Bennett Trophy and five-thousand-dollar purse at the great aviation meet to be held that summer in Rheims, France.

When he had heard the final details, Curtiss jumped in.

"America has taken the lead in the development of aviation," Bishop told the press. "We have lost the Wright brothers, and we do not intend to let the foreigners take every one else of prominence in developing aerial flight. If Congress will offer no incentive to inventors to remain in their own country, the next best thing is to keep them here by private enterprise. This has now been done. . . .

"This matter of air navigation, especially with aeroplanes, is no longer a fad or a joke, and wide awake men in New York with means realize that fact."

HOW MUCH, if anything, Curtiss had told his AEA colleaguges about Herring's proposition is not clear. The evidence suggests he said something to McCurdy. But he must have said little if anything to Bell, for soon after hearing the news, Bell wired Curtiss: "PLEASE WRITE FULLY CONCERNING YOUR ARRANGEMENT WITH HERRING AND HOW IT AFFECTS YOUR RELATIONS WITH THE A.E.A." Curtiss's reply made an attempt at diplomacy. But he left no doubt that he was breaking free not only from the AEA, but from Bell's obsession with the tetrahedron. "Mr. Herring showed me a great deal, and I would not be at all surprised if his patents, backed by a strong company, would pretty well control the use of the gyroscope in obtaining automatic equlibrium. This seems to be about the only road to success in securing automatic stability in an aeroplane." It must have been painful indeed for the Bells to learn of this tacit but final rejection of the tetrahedral dream by the cleverest of their protégés.

Trying to preserve the warmth between himself and his patrons, Curtiss wrote to Mabel in her role as "little mother" of the AEA, telling her that he and Herring "would like to have Mr. Bell and the boys with us if they care to come." But he would have known perfectly well that Bell had no use for the controversial Herring, whom Samuel Langley had written off years earlier as "a bumbler." Casey Baldwin, who had soured on Curtiss's vaunted skills as a master of engines, openly scorned the deal in the next issue of the AEA *Bulletin*, which he knew Curtiss would see: "That level-headed American business men should back Mr. Herring has created quite a furor in aeronautical circles. It probably means that Mr. Herring has some more convincing arguments than he has ever made public, or—is it really the Curtiss Company with Mr. Herring's patents to flourish in the eyes of bewildered capitalists? So far as we actually know, the Herring patents are only talking points at present."

The three-member remnant of the AEA took consolation in the impressive flights of the *Silver Dart* and the hope of cutting a deal of their own with the Canadian government. The *Cygnet II* was tried again, and again it failed to fly. Bell fired off a stiff telegram to Curtiss, virtually ordering him to appear for the valedictory meeting of the AEA at Beinn Bhreagh on the last day of March: "HAVE YOUR BUSINESS ARRANGED SO AS TO BE HERE 31ST SURE. VERY IMPORTANT AND YOU WILL REGRET IT ALL YOUR LIFE IF NOT." Curtiss apparently meant to comply. He left Hammondsport en route to Nova Scotia. But then, whether from cold feet or genuine need, he allowed the pressure of new business to detain him in New York. So, regret or no, he failed to show.

From the beginning, Bell had led the AEA with the sense that history hung upon its every move—thus the careful minutes of every meeting and the full records of every experiment. So it was at the end, when he timed the final conclave to go right down to the stroke of midnight on the final day. With him in the Great Hall were Baldwin and McCurdy, their young wives, and a couple of friends. The final motion expressed the group's "high appreciation" to Mabel Bell—who was in Washington with her daughter Daisy, who had given birth to her second child two weeks earlier—for "her loving and sympathetic devotion without which the work of the Association would have come to nought."

"The Aerial Experiment Association is now a thing of the past," Bell wrote a few days later. "It has made its mark upon the history of Aviation and *its work will live*."

ON THE EVENING after the AEA's forlorn finale in Nova Scotia, Katharine Wright was across the Atlantic, sitting in an elegant hall in Paris and watching in delighted amusement as diplomats, scientists, and heirs to ancestral fortunes spoke her family's name again and again. The occasion was the monthly banquet of the Aéro-Club de France. She was the first woman ever to attend a club affair. As she listened, French sophisticates were raising goblets of champagne and toasting her teetotaler father for the aviator sons he had given the world. Six months earlier, teaching fifteen-year-old Ohioans to conjugate Latin verbs, she could not have imagined a less likely scene. Yet it summed up the transformation that had taken place in the life of her family. On two continents, people were reading newspaper stories about what she looked like and what she wore, how she spoke. And if she was famous, her brothers were something more than famous.

She was in France because her brothers, again, had needed her. Wilbur, in a fit of homesickness, had asked them all to come over—Kate, Orville, and Milton. He had hoped to be home by Thanksgiving 1908, then by Christmas, but there had been simply too much still to do. New machines had to be built for European exhibitions and races in the flying season of 1909. There was business to conduct with the governments of Germany and Italy, work to supervise with their affiliates. And Will's French contract required him to train other aviators in the use of the machine. Daily flying would be impossible at Camp d'Auvours in the wintertime; he had barely endured the cold in the December flight that took the Michelin Prize. But in the south of France, he could complete his training in balmy conditions and provide his family with a rest.

"Can't you come over for a couple of months?" he pleaded in a letter to Dayton. "I do not see how I can get home, and yet I am crazy to see some of the home folks." At the resort town of Pau, he learned, he could find good grounds for flying and a quiet haven for his family. "I know that you love 'Old Steele' [High School]," he coaxed, "but I think you would love it still better if the briny deep separated it from you for a while. We will be needing a social manager and can pay enough salary to make the proposition attractive, so do not worry about the six per day the school board gives you for peripateting about Old Steele's classic halls."

In fact, she needed no coaxing. All her life she had been reading about the cities and museums of Europe, and she yearned to see them herself. She was more than ready for a break from life at 7 Hawthorn. Her father, just past his eightieth birthday, had come home from a Church trip exhausted and "very dull" in the mind. Orville, still on crutches, was too frail to get around alone. So she went along at his side twice a day as he hobbled to the shop to look in on Charlie Taylor. But it was too cold there for him to sit for long—"he can't stand a bit of cold"—so back they would come to Hawthorn Street, where "we keep the house like a bake-oven the whole time." For an hour every evening Kate massaged Orville's legs—until they imported a powerful Swiss masseur from the YMCA—and because Orville said he was too weak even to dictate correspondence, she spent a good part of each day "thumping off letters for brother till I am black and blue in the face." Yet Orville, submerged as usual in his own concerns, seemed puzzled that she was not returning to work. Six weeks after the crash at Fort Myer, just before Will's invitation, she was at the end of her patience. "I wish you would come home, Jullum. Nobody else takes a particle of responsibility. They leave everything on me. I am about played out but Orv doesn't realize it a bit."

Nor, now, would Orville make up his mind about the trip to Europe. Just in case, Kate checked schedules for trains and ocean liners and ordered a new dress made, all the while managing the house and caring for her brother and checking on her father and maintaining ties at Steele High. "I am so tired I want to weep the whole time!" Then, at last, Orville exchanged his crutches for a cane and said he was well enough to go. Kate rushed to buy tickets, pack trunks, and make arrangements for Carrie Kayler and her husband to stay with Milton. Just after New Year's she and Orville raced to New York to catch the German liner *Kaiser Wilhelm der Grosse* in its first voyage of 1909.

And then her life changed.

AT THE PARIS STATION the two arriving Wrights were greeted by a minor mob scene—French and American reporters; Hart Berg and his stylish wife; and there, among "a lot of others," and dressed, astonishingly, in silk hat and evening clothes, their beloved brother, "old 'Jullum.' " Old, yet new, too. Quiet and cool as ever, the gray eyes focused and shrewd, he was slightly fuller in the face after eight months of French cuisine. But that wasn't the difference. It was the realm he occupied. He was no longer just Kate's to admire and appreciate, an obscure man of hidden talent and character. Now he was recognized and celebrated on the streets of a great foreign capital; pointed out and fawned over; admired and revered—a hero. And she was being drawn into that realm. There were telegrams of welcome, photographers' flashes in her face. Arnold Fordyce, a sleek Frenchman who had visited Dayton on behalf of his employer, the publisher Henri Letellier, presented her with a great bouquet of American Beauty roses surrounding an American flag. "It was to smile," she said.

FOR THREE MONTHS KATE lived like an American heiress on the traditional Grand Tour of Europe. After "a strenuous week of shopping" in Paris, led by

Mrs. Berg, whom Kate thought "pretty as a picture and about the best dressed woman I ever saw," she was whisked to the spa town of Pau in southwestern France and welcomed at the Grand Hôtel Gassion, next door to the chateau where the French king Henry IV had been born four centuries earlier. While her brothers taught three French Army officers to fly, she strolled the town's mile-long promenade and gazed across foothills to the southern horizon, where the snow-crested Pyrenees marked the Spanish frontier. "I never saw anything so lovely." For two hours each morning she took lessons from a French tutor, and soon was fluent enough to take tea with the mayor and attend a soirée for a thousand in the Wrights' honor. She met Arthur Balfour, the British statesman; dined with Lord Northcliffe, the portly young publisher of the London *Daily Mail*, who had founded his fortune on the bicycling craze; and practiced her French each day for a week with the son of Prime Minister Georges Clemenceau, who afterward sent her a giant box of candied fruit. A French count took her up in a hot-air balloon, and on February 15, 1909, not to be outdone, her brother seated her next to him for a seven-minute ride in his aeroplane. She thus became one of the first women ever to fly in a heavier-than-air machine, and in the classic Wright slang for anything deemed very good, she reported to her father, "Them *is* fine."

Five days later, very early in the morning, she was introduced to King Alfonso XIII of Spain. She stood with the king as he watched Wilbur fly, then breakfasted with him. "The King was very enthusiastic and was wild to go with Will but he had to promise that he wouldn't." It was the king's mother, it turned out, who refused to let him fly.

After six weeks of social calls by day and gaiety by night, she forced herself to stay in her room for most of two days. "I have had too much excitement seeing people." Then King Edward VII of England arrived with his retinue to view the spectacle. "The weather was fine and the show was a great success," Kate reported. "Jullam made a flight alone and then took me."

Next, Paris again, to be feted by the Aéro-Club and to see the sights; then Rome, where the mayor—"a Jew!" she told the bishop—personally guided Kate and Orville through the Capitoline Museum. Will assembled the demonstration flyer in an automobile shop on the Flaminian Way, Julius Caesar's route to Gaul, and flew it over a long, sloping plain strewn with stone ruins of ancient aqueducts and villas. As at Le Mans and Pau, crowds came to watch and notables sought introductions to the three Ohioans. Half-laughing at the comedy of it, Kate now found the company of dignitaries quite commonplace. J. Pierpont Morgan and his family were merely "very pleasant people," while Lloyd Carpenter Griscom, the American ambassador, was "a nice sort of man." Duke "What's-His-Name" was forgettable, and when King Victor Emmanuel XX of Italy asked to see flights at 8:00 A.M., she teased her antimonarchist father: "The Kings *are* a nuisance all right. They always come at such unearthly hours." While "His Gracious &c." snapped dozens of photos of Wilbur in flight, Kate kept company at the flying field with a young German captain, who was observing on behalf of Kaiser Wilhelm II. The captain, too, sent her a dozen American Beauties.

Wilbur Wright was suddenly a full-fledged European fad. Imitations of his cap turned up in stores across the continent. His face was caricatured in newspapers, magazines, and picture postcards. Inevitably, songwriters began to envision the opportunities the Wrights had presented to young men intent on impressing their sweethearts. One French lyricist visualized the delicious things that might happen *"Dans mon Aéroplane"*:

Not long ago little old Suzanne
Said to her lover
Oh! How bored I am
I don't like cars any more,
I don't like horses
I want something new.
He answered: Darling,
The other day
I got you a gift.
It's something really swell
That I bought from Monsieur Wright.
Oh! Come, oh! Come,
Come up in my aeroplane
It's just like a bird
It stays up in the air as it should.
Oh! Come, oh! Come,
Come on, little old Suzanne
You'll go crazy, honey
When you've seen my little bird.

EVENTS BROKE in staccato succession.

In the middle of June, Dayton honored the brothers in a two-day "gala."

On June 26, Glenn Curtiss made his first sale—a "Golden Flyer" purchased by the Aeronautic Society of New York, an offshoot of the Aero Club of America. The Wrights sued, charging patent infringement.

Three days later, Orville began a series of spectacular flights at Fort Myer, fulfilling the terms of the Wrights' contract with the Army.

On July 25, Louis Blériot flew his monoplane 23.5 miles across the English Channel, electrifying the world. Blériot had barely landed when he began to prepare for the first great aviation meet, to be held in France in less than a month.

THE *GRANDE SEMAINE DE L'AVIATION DE LA CHAMPAGNE* was to be the major international sporting event of 1909, eclipsing even the great automobile races that had captured sporting attention each summer since the turn of the century. Among its organizers was Ernest Archdeacon, on the lookout as ever for France's next chance to regain primacy in flying. He and his friends persuaded the Aéro-Club to sanction the event and the region's great champagne producers—Mumm, Veuve Cliquot, and Möet et Chandon among them—to pay for it. Near the ancient cathedral town of Rheims, a hundred

"THERE ARE NO ISLANDS ANY MORE."

Louis Blériot crosses the English Channel, July 25, 1909

miles northeast of Paris, thousands of acres on the Plain of Bétheny were cleared and groomed to create a gigantic rectangular course ten kilometers long. Special trains were assigned to carry aviation enthusiasts from Paris to Rheims, and new tracks laid to take them directly to the flying course. A temporary village was constructed to greet the expected crowds, and sponsors' champagne was stocked to quench their summer thirst. The president, the premier, and several Cabinet members made plans to attend. So did statesmen and soldiers from as close as Britain and as far as Japan. While her husband stalked big game in Africa, Edith Kermit Roosevelt, first lady of the United States until only a few months earlier, arrived with several of her children.

The grandstands, bedecked with hundreds of French tricolors, held fifty thousand, but many thousands more gathered to watch from the perimeter of the field and the low hills nearby. In all, some five hundred thousand tickets would be sold during the week. Even in the days before the meet itself, crowds came to watch the preparations, and went home bitterly disappointed whenever conditions forced the officials to cancel practice flights. They had to be told repeatedly that despite all the preparations, successful flying depended entirely on the whims of the weather. "A horse race, an automobile contest, and even a regatta can be held in pelting rain and a tearing wind," the *Times* of London explained, "but the aeroplanes . . . cannot live in a storm. The automatic stability of the various descriptions of flying machines is doubtless much greater than was that of the first aeroplane which astonished the world 18 months ago, and the device of twisting the wings or of using *ailerons* to keep the machine on an even keel has proved most useful; but the stability problem has not yet been satisfactorily solved."

The classic biplane designs of the Wrights and the Voisins had been thought superior in stability to the new monoplane machines. But now, with Louis Blériot's wildly acclaimed flight across the Channel, monoplanes were all the buzz at Rheims, for speed and stability as well as for beauty. In fact, Blériot raised a ruckus when officials changed the rules to allow more contestants to compete in the evening, when the winds tended to abate. His new monoplane could fly in a stiff wind, he claimed, and he did not care to surrender that advantage by allowing his rivals to let hours of daylight pass as they waited for the dead calms of evening. Blériot won the point, and evening flights were restricted. Nonetheless, Blériot stayed on the ground when the wind blew.

Seven aeroplane competitions were planned. The richest prize, and the most coveted, was the *Coupe Internationale d'Aviation Gordon-Bennett*. James Gordon Bennett was not just the publisher of the *New York Herald* and a notorious transcontinental rake; he also published the *Herald*'s European sister, the English-language *Paris Herald*, forerunner of the *International Herald-Tribune*. This prize, to be awarded to the fastest machine at the meet, was said to represent the world's championship in aviation. The publisher set the purse at ten thousand dollars and commissioned an elaborate silver trophy—a winged child bearing a Wright flyer over its head. That was the plum that attracted Glenn Curtiss's eye. Though the Wrights stayed at home, three French pilots were to fly machines manufactured by the Wrights' French company. For Curtiss to defeat those machines and carry home the trophy bearing their likeness would make an especially savory triumph.

Curtiss arrived on August 12 with the parts of a brand-new biplane packed in four small crates. James Gordon Bennett himself eyed this meager load and asked in disbelief, "Those few packages?" But French aviators who saw the assembled Curtiss machine pronounced it "a dangerous competitor." It was a sleeker version of the Golden Flyer, with every unnecessary scrap of weight removed. Curtiss fitted it with a water-cooled eight-cylinder engine capable of fifty horsepower. Yet its capabilities were quite uncertain. The machine had been thrown together in Hammondsport at the last possible minute—"the greatest rush job I have ever undertaken"—with no time for even a single test flight before Curtiss had to race to New York to catch his ship. The rush continued upon his arrival in the port of Cherbourg, where he found out there wasn't even enough time to send the machine to Rheims by normal freight. In a burst of charity toward the cause of international aviation, French railway officials allowed him to haul the crates aboard the train as personal luggage, so long as he shared a compartment with them.

The event was billed as international, but Curtiss and George Cockburn of Great Britain were the only non-French entrants. The key French competitor would be Louis Blériot. The French were intensely curious about Curtiss, especially after learning of the Wrights' lawsuit against him, and they were inclined to favor his position. "For the time, at least, he is quite as popular as the Wrights were, if not more so," an American reporter said, "for they declined the issue when they were invited to take part in the grand tournament, while Curtiss pluckily accepted it. . . . When the Parisians learned that Curtiss had come there practically at his own expense, and that he had been doing a lot of hard work in obscurity while the Wrights got all of America's praise, they warmed up to him more than ever."

When Curtiss learned that Blériot had installed an eighty-horsepower motor in his feather-light monoplane, he concluded that "my chances were very slim indeed, if in fact they had not entirely disappeared." In fact, his speeds in test flights were impressive. But Blériot and his fellow Frenchmen had machines in reserve. If they cracked up in one contest, they could enter new machines in the next. Curtiss had only one machine built for one purpose—speed. If he entered an early contest and anything went wrong, he might forfeit his chances for the entire meet. So he staked everything on the final event, the speed challenge he cared for most—the race for the Gordon Bennett Cup—and kept his name out of all the others. Americans attending the meet wanted to see their flag entered for every competition, and they told Curtiss so with increasing vehemence. He resisted. He had only to point to the field to show them why he had to be cautious. It was littered with wrecks every day. On one practice flight Curtiss counted twelve broken aeroplanes on the ground below him.

HENRI FARMAN won the one-hundred-thousand-franc distance prize with a long, slow drone of nearly 112 miles in just over three hours.

Curtiss was allowed a trial run before the speed contest. He made it on a windless day, and was startled to find himself bumped and jolted as if in an automobile on a dirt road. Curtiss concluded that rising columns of heated air were responsible. His time was the best turned in during the trials and he concluded that this "boiling" or turbulent air actually enhanced his speed.

When his turn came for the official run, he took the machine up to an altitude of five hundred feet. He borrowed the trick from bicycle and motorcycle racing—he would start at the highest point on the track, then take advantage of gravity while shooting downward to cross the start line. He whipped around the course twice in fifteen minutes and fifty seconds, cutting around the pylons at extremely small margins of error, for an average speed of more than forty-six miles per hour.

Blériot flew last. Curtiss thought he was hitting sixty miles per hour. Perhaps on the straightaways Blériot did fly that fast. But when the final times were announced, Curtiss's screaming turns had won him the race by six seconds. Newspapers proclaimed him "CHAMPION AVIATOR OF THE WORLD."

FROM RHEIMS, Curtiss hurried to Brescia, Italy, where he made several demonstration flights, including his first with a passenger. This was the flamboyant Italian poet and novelist Gabriele d'Annunzio, who climbed out of the aeroplane all but overcome: "Until now I have never really lived! Life on earth is a creeping, crawling business. It is in the air that one feels the glory of being a man and of conquering the elements. There is the exquisite smooth-

" 'CHAMPION AVIATOR OF THE WORLD' "

Glenn Curtiss winning the Gordon Bennett Cup at Rheims

ness of motion and the joy of gliding through space—It is wonderful! Can I not express it in poetry? I might try."

RHEIMS INAUGURATED a series of great aviation meets. Attitudes toward aviation on both sides of the Atlantic moved directly from disbelief and astonishment to visions of an aeronautical paradise just around the corner, with aeroplanes replacing automobiles and "aerial buses" to convey commuters. There was a sudden sense of epochs shifting. In postcards, posters, and advertisements, commercial artists seized upon the beauty of the new monoplanes. Thinking back over the year since Wilbur Wright's first flights at Le Mans, a Paris journalist wrote: "Everything that has happened astonishes you, surprises your imagination, leaves you deeply moved and disconcerted, your head a bit dizzy as if you'd had too much to drink."

That was hardly an uncommon sensation at Rheims, where the sponsors made back twice their investment in prizes on champagne sales alone. Yet some kept a clear head and watched more closely.

One such observer was Julien Ripley, a friend of Alexander Graham Bell. Bell had asked Ripley for a full report on the Rheims meet, and he gave it shortly after the meet ended. Standing out in Ripley's memory was a plain fact that others already were forgetting amid the hoopla and excitement—that aviators and spectators alike had had to wait for endless hours for the wind to die down. Flights, when they occurred, were wonderful. But they could not occur at all in the atmospheric conditions that prevailed in most places at most times. "It is evident the serious effect that the slightest breeze has on all the aeroplanes," Ripley told Bell. "When Curtiss flew for the cup there was no apparent stirring of the air. . . . Yet Curtiss said that when he got going the air seemed boiling and he was nearly thrown from his seat."

IN AN AGE OF WORLD'S FAIRS that lasted for months and years, the two-week Hudson-Fulton Celebration, planned for the fall of 1909 in New York, seemed comparatively modest. Its purpose was to mark the progress of American civilization since the explorer Henry Hudson's voyage up the Hudson (then North) River in 1609 and the inventor Robert Fulton's navigation of the river by steam power in 1807. If the anniversaries did not coincide exactly, it was close enough for the New York state legislature, which appointed a commission in 1905 to organize naval exhibitions, "great land parades," pageants, and festivals stretching from Staten Island two hundred miles north to Troy and Cohoes. All events were to be educational—"the most careful pains were taken to avoid anything of a commercial tincture"—and open to the public free of charge. It was not to be a celebration of victory in war, the Hudson-Fulton Celebration Commission declared, but a celebration of peace, "civil concord," and "material prosperity."

In the spring of 1908, the flurry of news stories about flying machines led William Berri, a member of the Commission, to suggest it would be "striking, even dramatic," to mark "the climax of three centuries of progress" with the first aerial navigation of the Hudson. His fellow commissioners agreed, and a Committee on Aeronautics was established. Efforts were made to organize an international aviation meet, but Farman, Blériot and Delagrange demanded "enormous sums" to take part, so the committee entered negotiations with two American exhibitors, Wilbur Wright and Glenn Curtiss. Both agreed to participate, though on different terms. Wright agreed that, conditions permitting, he would "give at New York a full illustration of the possibilities of flight through the air by his aeroplane." Curtiss's agreement was more specific; he would attempt to fly from Governors Island in New York Harbor up the Hudson to the northern tip of Manhattan, a distance of roughly ten miles, where he would turn around at Grant's Tomb and fly back to the island.

*Scientific American*, now among the Wrights' strongest advocates, called its readers' "special attention to this contest, among other reasons because we see in it an opportunity for America to win back some of that prestige which she undoubtedly lost when, by her indifference to the claims of the Wright brothers, she drove them to find more appreciative treatment at the hands of people of an alien tongue and race. For there is no denying that our attitude to the new art . . . of flying has been altogether unworthy of a country which claims to be particularly solicitous of the inventor, and ever ready to encourage the man who can present us with a novel and useful idea embodied in practical mechanical form."

Will arrived in New York with Charlie Taylor early on the morning of Sunday, September 19. Hudson-Fulton officials ferried them out to Governors Island to have a look around. The island had been recently doubled in size with landfill brought over from the excavation of the Lexington Avenue subway. The Army designated the sandy fill, an expanse of about a hundred acres, as a flying field. The view north from the island was one of the most spectacular in the Western hemisphere. From east to west it took in the Brooklyn waterfront, the Brooklyn Bridge, the towers of Lower Manhattan, and the two other islands in the harbor—Ellis Island, the doorstep to the United States for millions of European immigrants, and Bedloe's Island, home of the Statue of Liberty in her twenty-fifth year. To Battery Park at the tip of Manhattan it was half a mile; to the statue, a bit more.

Any flight from here would be the first that anyone besides Louis Blériot had made over water. And this was not the open water of the English Channel, where an aeroplane with a bad engine might glide down gently wherever it needed to and be rescued by a following boat. Scores of ships and boats plied the harbor on an average day, and the number would multiply many times during the celebration. A flight up the Hudson would be a flight through a canyon, with the skyscrapers of Manhattan on one side, the Palisades of New Jersey on the other, and a spiky floor of battleships beneath. But the earth below was the lesser part of the danger, and at least could be surveyed in advance. The utterly unknown danger was the nature of the wind in this urban wilderness. No one had flown through a city before. No one could know how the man-made escarpments would affect the behavior of the air. In short, these flights would be by far the most dangerous ever tried.

Will said nothing of the danger.

"I have not come here to astonish the world," he told reporters. "I don't believe in that kind of thing. . . . I expect to make an average flight and give everybody a chance to see what an aeroplane is like in the air."

When asked about flying over skyscrapers, he was careful not to commit himself. "That would be quite possible, especially in the lower end of the island in its narrowest part. I do not say that I shall attempt that flight, but there is no reason why it should not be made."

Would it be dangerous to fly over ships in the harbor?

"An aeroplane should be able to go anywhere."

TWO IDENTICAL WOODEN SHEDS had been erected by the veranda of the officers' club at the Governors Island Army post, in one corner of the field designated for launchings and landings. Each shed was about thirty-five feet long, twenty feet wide, and twenty feet high. So the Wright and Curtiss aeroplanes were assembled within a few yards of each other.

Reporters hung around the sheds all week. One of them observed closely, and detected distinctly different atmospheres in the two camps. In the Wright shed only two men did all the work. The inventor was "glum" with the reporters, though "interesting in the extreme; you know he is unusual the instant you see him." He was "a man who has evidently forgotten how to talk, a man who seems to have no desire to make friends, a man whose whole soul seems to be wrapped up in the strange freak of the air. . . ." Though his sparse hair was turning gray at the temples, Wright looked younger than in his photographs, "thin, supple, graceful, quick upon his feet and evidently quicker still of thought." He worked in shirt-sleeves, never with overalls or an apron. "You cannot think of him in such togs or with spots of grease or dabs of oil upon him. He is far too neat." His mechanic was "equally taciturn . . . a man who seems to understand the every thought of Wright and to anticipate his every wish as well as the every need of the strange machine." Occasionally the reporters heard "strange noises" from inside the shed, but very little talk. They must have wondered about the canoe that was on hand; Will intended to mount it on the flyer's underside, to keep it buoyant in case of a crash in the water.

In the other shed was "a far different group of men" and "a much smaller and better finished bird." Where Wright was silent, Curtiss bantered at length with the journalists, and "is the friend of every one of them." He didn't put his hands on the machine much. "He has brought men down from his upstate factory to do that for him." In his features and dress he looked "more like the young English university men one sees spending vacations in . . . the continental watering places than a man who has done things." But unlike such idlers he was "usually in a hurry, and . . . thinking about something." The observer guessed Curtiss to be "the forerunner of the type who will take up real aeronautics within the next year, young men who like difficult and dangerous things because they are difficult and dangerous."

Curtiss was again in a rush to prepare a suitable machine. The machine he had flown at Rheims was only a couple of miles away, but it was unavailable for flying. While Curtiss was in Europe, Augustus Herring had promised the people at Wanamaker's department store they could display the aeroplane to customers during the Hudson-Fulton celebration.

CURTISS AND HIS WIFE arrived at Governors Island to find Wilbur Wright already at work, his hands covered in grease.

"How do you do, Mr. Curtiss? You will excuse me from shaking hands, for you see what they are like. How are you feeling since your return?"

"Very good," Curtiss said.

"Really, you ought to."

Apparently the two had spoken at a meeting of the Aero Club before Curtiss's departure for France, for Will now asked if Curtiss had used his "suggestions" for the flight at Rheims.

"Some of them," Curtiss said.

"Yes," Will said. "You must have followed some of my advice."

Anyone who knew Will would have realized he was having some private fun with words.

KNOTS OF ONLOOKERS came as close as they could—mostly Army officers and their families and friends. Among them was a Columbia College boy named Grover Cleveland Loening, first vice president of the new Aero Club of Columbia and a quivering enthusiast for all things aeronautical. His mother had wangled him a pass to the island, and he brought a letter of introduction to Wilbur Wright from his mother's friend, the banker August Belmont. Loening got through the screen of soldiers around the Wright shed and thrust his letter at Wright, who glanced at it, said nothing, and turned away. Loening stood in one spot watching the work for three hours. Then Wright tossed him a rag and told him to get to work. Loening came back every day, and kept working.

In the formal language of the Hudson-Fulton contracts there was no mention of a competition between Wright and Curtiss. The planners wanted only an exhibition—in fact, two separate exhibitions. It was clear they regarded

Wright as the larger draw, at least at the time the contracts were drawn up, before Curtiss's triumph at Rheims. Wright was to make any flight of his choice of not less than ten miles with awards totaling fifteen thousand dollars. Curtiss was asked to fly from Governors Island up the Hudson River to Grant's Tomb, at the northern end of the city, and return. This was a distance of some twenty-three miles, though Curtiss's fee was to be only five thousand dollars. There would be no way of saying which man had won—as long as both men flew. Still, the reporters were acutely aware of the growing rivalry between the two men, and speculation arose even about the practice flights—how long they would be, and how daring. At some point, apparently in the Curtiss camp, rumors began to circulate about a flight around the Statue of Liberty.

"The winds will probably be rather strong." Will wrote Orville, "but it will bother Curtiss more than me unless I am mistaken."

BOTH MACHINES were ready on September 28, but the wind blew at twenty miles per hour or more. The aviators waited until the sun went down, but the wind stayed brisk. Wright said he had to test the flyer with the canoe attached, and he could not do so in any considerable wind. Curtiss quashed the speculation about flying over skyscrapers: "I wouldn't fly over the buildings of the city if they deeded to me everything that I passed over."

That night, as usual, Will returned to Manhattan and slept at the Park Hotel. Curtiss spent the night in his shed on the island. He and his mechanics arose early on Wednesday, September 29, and readied the machine. A mist lay over the island and the harbor. But Curtiss whirled the propeller, got in the machine, and trundled off over the landfill and soon was lost to sight in the fog. Several minutes later, the aeroplane could be seen taxiing back to the shed. When reporters arrived on the island a little later, Curtiss's publicity man told them the aviator had made a short flight, and the afternoon papers reported it.

Grover Loening, the kid from Columbia, was there. He had come to the island early each morning since Wilbur had handed him his rag. Years later, Loening said the breeze that morning was blowing from behind him toward the point where Curtiss rumbled into the mist. To leave the ground, Loening said, Curtiss would have had to come back in Loening's direction. But Loening neither saw nor heard the aeroplane. He concluded that "Curtiss never got off the ground, because the required run into the wind from where he had vanished in the fog would have brought him right by where I stood, and certainly the plane could at least have been heard. Also Curtiss never could, in my opinion in that morning fog, again have located the landing area on the island." Loening was a partisan for Will; no disinterested observer attested that Curtiss actually had made a flight.

When Will arrived on the island shortly before 9:00 A.M., Loening told him what he thought he had seen, or not seen, and "Wilbur was furious at this controversy, openly despised Curtiss, was convinced he was not only faking but doing so with a cheap scheme to hurt the Wrights." Loening told reporters, too, but no paper repeated the charge. When reporters asked Wright to comment on Curtiss's early-morning performance, he only said: "I am glad the conditions are so favorable."

WITHIN MINUTES Will and Charlie and the soldiers in khaki were pulling the machine out of the shed and laying the launch rail straight into the breeze, due west. Across the little strip of water called Buttermilk Channel, spectators waiting on the Brooklyn waterfront saw the stir of activity and pressed close to the water. On the island, soldiers had to warn spectators to clear away to the sides or the aviator would not fly at all.

Will and Charlie took hold of the propellers and yanked. It took them a dozen tries before the engine caught. Then "the blades whizzed round so fast that two blurs of gray were all that the eye could catch at the rear of the machine." Will listened to the engine for a moment, then turned his Scotch-plaid cap backward, climbed into his seat, and nodded, and Charlie began to push. Within a hundred feet the machine outdistanced him. As the reporter from *The New York Press* saw it, the flyer was almost at the end of the rail when Wright "gave his forward planes a sharp tilt. The air, striking beneath them, raised the whole machine twenty feet into the air before a man could take a breath." The machine circled the launching area twice, then "headed straight across in the direction of Buttermilk Channel. The entire garrison of the island was out on the parade ground, and a cheer went up that carried clear over to the Brooklyn shore and was re-echoed by the eager watchers there."

Over the water, the machine banked and headed north toward Manhattan, then continued in a sweeping circle around the northern end of Governors Island and back down its western edge.

Not many could have been watching, for the flight had been quite unexpected. For all those who happened to be there, it was their first sight of a thing they had perhaps believed impossible or somehow exaggerated. But there had been no exaggeration, they now realized. The machine was flying, and not in a sick and shaky imitation of flight, but boldly, under "the complete mastery" of the "lean, clean-shaven" navigator, his features and movements perfectly visible at a height of thirty or forty feet. "He sat rigid, his legs braced firmly against the narrow rail in front, his left hand on the lever that regulates the forward planes, his right on the one that does the double duty of manipulating the rudder and adjusting the flexible tips of the great wings so that the machine balances itself on curves. He had no time to spare for the cheers that greeted him. It hardly seemed he heard them."

The crews of nearby tugs and steamers scrambled to their whistles and let loose with salutes. But the signal that something was happening came too late. By the time people at the foot of Manhattan or in the close-in blocks of Brooklyn got to the water, the machine had disappeared, skimming over Castle Williams and back down onto the sand. Will had been in the air for only seven minutes and ten seconds.

WORD SPREAD NORTH from the Battery that Wilbur Wright had flown. People began rushing into Battery Park, crowding against the long seawall and filling the piers. Soon the roofs and windows of buildings overlooking the harbor were jammed, and a carnival of ferryboats, tugboats, and pleasure craft streamed out to find a good spot to watch the next flight.

At 10:17 they were rewarded.

In the induction center on Ellis Island, immigrants waiting in lines were startled by a rising din outside. Interpreters explained in several languages that a machine was flying over New York Harbor. Reporters who learned that much did not say if the explanation tended to comfort the newcomers or increase their sense of alarm.

Boats cut their engines and rocked in the waves. Many tied their whistles open. Passengers and crew members watched the machine heading toward them across the water. It came on and passed over, moving directly toward Bedloe's Island. As usual, people of farming backgrounds thought it sounded like a threshing machine.

It approached the statue, passed by the head on the Manhattan side, and flew straight on.

On the Jersey shore, people saw the machine bank and sweep into a tight half-circle, then head away, back over the harbor. Now every skipper in the harbor opened his steam whistle. The noise grew deafening. On the ferryboat *Queens*, five hundred passengers grabbed their hats and waved. Just ahead lay a far greater hulk in the harbor. It was the Cunard liner *Lusitania*, outbound for Liverpool. Passengers on deck waved white handkerchiefs. The flying machine came on and flew just overhead, and the liner let loose with a volcanic blast of steam.

"A NEW THING CAME TODAY."

Wilbur circles the Statue of Liberty

A hundred feet up, the roar and the heat enveloped Will. The flyer rocked and flew on.

APPROACHING THE STATUE OF LIBERTY from the west, the aeroplane accelerated to fifty miles per hour. Then Will slowed and went into a banking circle around the waist of the statue.

Reporters heard different responses in different places—or imputed their own responses to the crowds. One at the foot of Manhattan said that while the flight lasted, cheers rose from Battery Park and the rooftops and the windows and merged into a sustained roar. Another said the people around him "stood with mouths open, but silent, as a rule. . . .

"A new thing came today, a thing which New York had never seen before and waiting thousands felt a new sensation, felt their throats tighten, failed to understand why their eyes did not see as well as they have before. . . . Nearly all saw a flying biplane for the first time. It was an absolutely new sensation, one which will not be forgotten for some time by those who experienced it. New sensations are rare in these modern days."

On Governors Island, photographers sprinted out onto the landfill with a crowd of spectators in their wake. Will, approaching, saw an archipelago of upturned faces where there was supposed to be only open sand. He found a slim clear spot and slammed to earth much harder than usual.

*The New York Times* man got close enough to hear a quick exchange between pilot and mechanic:

"Goes pretty well, Charlie."

"Looks all right to me, Will."

As more reporters crowded around, he said, "That is the worst landing I have made in years, and it is a wonder I did not smash the machine to pieces." Apart from that, "he was as self-contained as ever."

JUST BEFORE 1:00 P.M., Will and a Hudson-Fulton official went by launch to the Battery and passed unnoticed through the crowd, a black derby pulled down tight on Will's head. Several blocks north, at the Singer Building, they ate a quick lunch and went up to the roof. Curtiss was supposed to meet them; the idea was for the aviators to see the city from above and get a sense for the wind conditions at high altitude. Curtiss was late, so Will surveyed New York without him.

Extras hit the streets. The word now was that Wright would fly up the Hudson and circle the battleships late that same afternoon. From the lunch hour on, nearly every boat that pushed off from the Battery pier was loaded with hopeful spectators, and "the slopes of the old parade ground on the island looked like the lawns of Central Park on a morning in May."

But the wind came up again. Curtiss's machine stayed in its shed. At about 5:30, soldiers pulled the Wright machine out to the launch rail. This time, with the wind brisk, Will flew only a short circle. But the view of the aeroplane from the island was lovely, the wings catching the setting sun and turning a silvery-pink.

• • •

TALK OF THE CIRCLE FLIGHT at the Statue of Liberty filtered through the city and spread through the nation as the evening papers were bought and delivered.

"The sad part of it was that a comparatively small number saw it," one correspondent wrote. "True, the Battery was crowded, there were a privileged few on the army post and many on passing boats. That was all. There was no time for the thousands to gather, no warning to them, no time to give one." But the official flights were expected the next day. "Then, too, it is up to Curtiss to make good. He has done the talking and the jollying. Wright has merely made history and seemingly forgotten all about it tonight. . . .

"The sight took your breath away. It was all so new, all so totally different, and more thrilling than one thought it was going to be. There is nothing else like it. It is worth going far to see. It must be worth going half around the world to try."

THE NEXT DAY, September 30, outside Berlin, Orville Wright broke his own world record in altitude by flying to a height of nine hundred feet.

In Buffalo, Judge John R. Hazel of the U.S. Circuit Court ordered the Herring-Curtiss Company and Glenn Curtiss to stop making aeroplanes on grounds that they infringed upon patents owned by Wilbur and Orville Wright.

In New York, it began to rain, and the wind stiffened. Neither Will nor Curtiss would fly in the threatening weather.

IN DAYTON, Milton was having trouble sleeping in "these exciting times." He seized the papers every morning and afternoon and wrote letters of support to Will. "You know I have the greatest interest in what you do." The latest papers said the weather in New York was poor, and there had been no attempts to fly since the Statue of Liberty flight. "That flight struck some here as sensational, and the applause attending it as overwhelming. But six or seven miles or minutes are small to you.

"I hope your shop is as usual on Sunday. It gives you immense advantage, whatever is forgone. Trust in God, and be true, and you will have the victory. I have tried it so much, and often. . . .

"I trust you for the greatest precaution, and the greatest courage and achievements."

ONE OF THE WRITERS covering the Hudson-Fulton Celebration was Arthur Ruhl, the *Collier's* man who had tracked the Wrights at Kitty Hawk the year before and pleased them with his play-by-play account of reporters fighting the sand and mosquitoes. He thought most New Yorkers had anticipated the pageant with little enthusiasm. It was hard to cheer for events of one hundred and three hundred years earlier. Life in New York was already an uproarious pageant. Plans for the Hudson-Fulton hoopla sounded "like stopping the battle of Waterloo to shoot off a few skyrockets," and "the 'regular' New Yorker effected to be bored by the whole affair." But the natives' indifference faded when parades began to move up the avenues and replicas of Fulton's *Clermont* and Hudson's *Half Moon* appeared among modern naval squadrons on the river. Watching from a high window, Ruhl thought people "seemed to spring out of the pavement and the walls . . . They were like wheat running into and filling the interstices of a grain elevator . . . No one could see these hordes moving at the bottom of the canyon streets and remain indifferent." Close to the river, hundreds of automobiles crowded together, each seizing a vantage point, with women and children standing on passenger seats to try to see. The autos gave "an impression of something brilliantly strident, the beginning of a new mechanical age that will be as different from ours as ours is from that of Hudson or Fulton."

CURTISS'S CONTRACT was due to expire at midnight on Saturday, October 2. He was supposed to leave for an exhibition in St. Louis on Monday. But the wind blew hard all Saturday, and he would not risk a flight. He decided to stay one more day. He knew Will would not fly on Sunday.

But Sunday morning the wind was still blowing. Curtiss sat at his shed all day, waiting. At 5:30 P.M., as the light was beginning to fade, he had the aeroplane prepared and took off. He turned and headed in the direction of the Statue of Liberty, but then observers saw the aeroplane "swerve and careen as if it were about to fall." Curtiss swung back around and landed. "I didn't like it up there," he said. "I seemed to strike a boiling point in the air. The machine was wobbling and I thought it would feel better on solid ground. I could have kept on flying, but to make the turn would have carried me over the water and I didn't care for that. I am looking out for Number One."

He left the city that night.

THE WIND STILL BLEW the next morning. Sunday or no, winds or no, the skids of the Wright machine left the sand just before 10:00.

The machine scooted over the harbor at about twenty feet above the water, heading toward the gap between the Statue of Liberty and Ellis Island. Approaching a great liner, Will rose and kept rising, gaining speed until he reached an altitude of about two hundred feet. Then he banked a little to the right, taking a northward course up the Hudson River.

This route was not on the Celebration's official schedule of events; it was his to choose. With the turn up the Hudson it was clear he had chosen the route the Hudson-Fulton planners had assigned to Glenn Curtiss.

In the harbor, warships of five nations cruised beneath him with countless ferries and pleasure craft and working boats. Sailors stood and hurrahed, though few voices could be heard over the single, "deafening shriek" of ships' and boats' whistles and horns. The noise caused a chain reaction that moved swiftly up the Hudson, setting more whistles blowing, which in turn drew people rushing to the river all along the western side of Manhattan.

He flew against the wind on his way north. Several times spectators saw the machine veer from side to side, as if struggling to regain its control in gusts that squirted out from the canyons of the streets. The machine swung "upwards and downwards like a ship in a gale." He passed so close to the masts of the British cruiser *Drake* that people thought the machine would be impaled. But after each wobble the wings returned to level, "as if they were being driven on wheels over a sheet of clear ice." By the time he reached Riverside Park there were tens of thousands of people there.

He stayed about two hundred feet above the river, high enough to see meadows beyond the New Jersey Palisades, the stone cliffs that border the river on the west. To the regret of spectators, he stayed close to that side of the river, to avoid the city's billowing wind gusts.

Approaching Grant's Tomb he banked into a great circle, passing over HMS *Inflexible*, where some sort of massed noisemakers made a din like thunder. He turned and headed south, back down the Hudson.

With the wind at his back, he made the return trip in roughly half the time it had taken him to get to Grant's Tomb.

At Governor's Island, "a mild form of hysteria" overtook the crowd as the machine came low and skidded across the sand, thirty-three minutes after it had left.

A circle formed as the aviator got out and circled the aeroplane, touching cables and surfaces. One man said: "He doesn't seem to give a damn, does he?"

"It was an interesting trip," he wrote his father, "and at times rather exciting."

IT WAS ANNOUNCED that he would make a much longer flight that afternoon, flying up the East River (including a pass under the roadway of the Brooklyn Bridge); up the Harlem River; rounding the turn at Spuyten Duyvil at the far northern tip of the island; and returning down the Hudson to Governors Island with perhaps a swing over neighboring Newark, New Jersey.

He and Charlie yanked the propellers to crank the engine. There was a roar, then an explosion. A cylinder shot straight upward, tearing a hole in the upper wing, then crashed to the ground a few feet away. He picked it up. Then he faced the reporters, gave a little shrug and smiled.

"It is all over, gentlemen."

THE NEXT MORNING, in his room at the Park Hotel, Will had only a short time before catching a train for Washington, where he was supposed to begin the training of Army flyers. Among his callers was Cass Gilbert, one of the nation's leading architects, a designer of state capitols, churches, and great Beaux-Arts residences. He would shortly design the Woolworth Building, the Gothic landmark in lower Manhattan that would be the world's tallest skyscraper for twenty years. The pretext for his visit was to invite Wright to the annual banquet of the American Institute of Architects, of which Gilbert was president. But mainly he wanted to get a look at the man who had created a new thing in the world.

Gilbert was the sort of man Will might have become had he gone to college. Nine years older than Will, and a native Ohioan, he had studied architecture at MIT and in Europe, then worked for the great New York firm of McKim, Mead and White for several years in the 1890s before establishing his own practice. With his waxed mustache and elegant dress, he made a sharp contrast to his host.

They chatted for only a few minutes—about their roots in the Midwest, and flying in cities, and what aeroplanes might mean for the architecture of the future. A couple of hours later, at his office, Gilbert dictated a memo to record his impressions while they remained fresh. He thought that "a generation or two hence," it might be interesting to read "such a record precisely as we would now be interested in reading of Robert Fulton, the inventor of the steamboat.

"His personality interested me very much," Gilbert said. His face and clothing were inconspicuous and unremarkable. "There is absolutely nothing romantic or distinguished in his dress, appearance or manner." He noticed that Wright fiddled constantly with a small card he held in one hand, "as though some physical or mental strain through which he had passed had keyed him up to a point where his hand must be in motion, not at all as though he were impatient of his visitor or eager to leave him. . . . He occasionally looked straight at me with a very frank, clear expression but more often looked slightly to the right and downwards."

> Very simple and direct and of few words, modestly spoken. He smiled occasionally with a sort of half smile that did not give me the impression of much exuberance of spirit but rather of a provincial boy who had an underlying sense of humor and a perfect confidence in himself but with a slightly provincial cynicism as to how seriously the other man might regard him or his views. He was . . . probably very keenly sensitive, and on the whole rather the type of high grade, intelligent and well read mechanic whom I occasionally meet in connection with building work. He looked like the student and the shop man rather than the man of affairs.

When Gilbert rose to leave, he put his hand on Will's arm, and, "looking him straight in the face . . . very seriously," said: "Mr. Wright, I want to tell you that in common with all of your countrymen we are proud of you and of what you have accomplished and the way you have gone about it. The serious men of this country appreciate that you are working seriously to accomplish a scientific result and hope that you will keep right on. The real men are glad that you don't make an acrobatic or circus performance out of the machine as some others seem inclined to do."

Will said: "I never was much on circus performances, anyway."

Gilbert said again that the country was proud of him, and "while I was saying this he shook hands with me warmly and looked me straight in the face evidently deeply appreciating the sincerity with which it was said, and I felt that he fully responded to it although he did not express it in words."

# EPILOGUE

“FOR MERITORIOUS INVESTIGATIONS IN CONNECTION WITH THE SCIENCE OF AERODROMICS”

Charles Walcott, Wilbur, Alexander Graham Bell, and Orville after the Wrights were awarded the first Langley Medal on February 10, 1910

BEFORE THE HUDSON-FULTON CELEBRATION, there had been some real danger that the mounting achievements of Glenn Curtiss and the European aviators would obscure the Wrights' claim to be the true founders and pioneers of human flight. But Will's showing in one of the world's great cities, with the New York press watching and writing, settled the issue for good, at least in the court of public opinion. Certainly Will himself believed there was little more to prove.

Perhaps, too, he had sensed that the Wrights' rivals, Curtiss included, soon would pass them in endurance, altitude, and speed—as, very soon, they did. Better, then, to seize this one critical opportunity to show up his chief rival, and leave the arena as the undisputed champion, than to go on and on, like a fighter past his prime. In 1897, after the flights of the unmanned aerodromes, Samuel Langley had said, "I have brought to a close the portion of the work which seemed to be specially mine." Then he had tried to do more. Will intended not to repeat the mistake. Soon after his exhibitions in New York, he said he would not fly again.

He changed his mind only once. On May 25, 1910, he rode as Orville's passenger for six and a half minutes over Huffman Prairie, the only time the brothers flew together. Then Bishop Wright climbed into the passenger's seat and went aloft with his youngest son in his own first flight.

Orville flew often that year, mostly to train pilots for the exhibition team the brothers had established. (Once, finding himself unable to descend from a height of 1,500 feet in a calm sky, he realized he was caught in a broad thermal updraft, one of the invisible elevators ridden by the soaring birds the brothers had sought to imitate—a "most thrilling" sensation.) In the fall of 1911 he returned to Kitty Hawk to test a new glider and set a world's record for soaring—nine minutes and forty-five seconds—that stood for ten years.

But by then gliding seemed little more than an antiquarian pastime. Aviation already had charged ahead into an era of spectacles and stunts that simultaneously enchanted and repelled the public. It had begun at Rheims, where Eugène Lefebvre shocked the crowds by diving toward the ground, then pulling up at the last moment to make a soaring ascent. He died in a Wright flyer two weeks later. "He was entirely too daring," Hart Berg told Will, "and never seemed to look over his machine as I was accustomed to see you do." Ferdinand Ferber said Lefebvre's death could not slow the progress of aviation, as "we are too thirsty for air, for space, and for speed to delay the realization of a discovery that history has been waiting for such a long time! " Ferber himself was dead a month later, crushed under the engine of his Voisin machine. Leon Delagrange was killed just after New Year's 1910 when his Blériot monoplane crashed before thousands of spectators in Bordeaux. Many more pilots died soon, including several members of the Wrights' exhibition team. Flight was thrilling to pilots and spectators alike, and the crowds yearned for feats of daring. It was hard for young men to exert the scrupulous care with which the Wrights had tempered their physical courage.

FRIENDSHIPS DIED, too. On the day that Will circled the Statue of Liberty, George Spratt, brooding at his home in Pennsylvania, wrote the brothers "to ask if you do not think I can claim a share of your success?" He had persuaded himself the Wrights owed him far more credit than he had received for ideas about the simultaneous measurement of lift and drag in their wind tunnel experiments. Will, "a trifle hurt," acknowledged Spratt's contribution but reminded him they had used his basic idea in a device quite different from the one he had envisioned. In return, Will pointed out, the brothers had given Spratt their entire store of wind-tunnel data—a treasure chest that contained the keys to flight. "Has your idea yielded you yourself tables as comprehensive and accurate as those you received through us?" He asked Spratt to "believe me ever your friend," but the relationship was beyond repair. Years later, when Orville requested copies of the brothers' early letters for a book he hoped to write on the invention of the aeroplane, Spratt refused. The Wrights had spurned his appeals for a three-way partnership, or at least for due recognition. Therefore, "I do not see how you can give a correct account of this without bringing yourselves into open censure." Orville let the matter drop, "not wishing to add to the sorrow of an already unhappy life."

The brothers' relations with Octave Chanute, after years of "secretly nursed bitterness" on both sides, collapsed in a terrible exchange of letters in 1910. Again, the chief issue was proper credit for ideas about flight that now, it was clear, had been not only important but historic. When Chanute said, "I am afraid, my friend, that your usually sound judgment has been warped by the desire for great wealth," Will, incensed, unleashed his own litany of grievances against Chanute and said: "As to inordinate desire for wealth, you are the only person acquainted with us who has ever made such an accusation." When, after three months, no response had come from Chicago, Will tried to heal the wound, saying, "My brother and I do not form many intimate friendships and do not lightly give them up." Chanute, in failing health, responded in kind, saying he hoped "to resume our former relations." But he died soon afterward, in November 1910. In a memorial article, Will said "his labors had vast influence in bringing about the era of human flight. . . . In patience and goodness of heart he has rarely been surpassed. Few men were more universally respected and loved."

CREDIT WAS also on the mind of Alexander Graham Bell—not for himself, but for his late friend, Samuel Langley. Soon after the Wrights' epochal flights in 1908, Bell helped to inaugurate a program by which the Smithsonian would award an annual Langley Medal for "specially meritorious investigations in connection with the science of aerodromics and its application to aviation." The Wrights were chosen as the first recipients, and two identical gold medals were struck at the Paris Mint. Yet when Bell was chosen to speak at the awards ceremony at the Smithsonian on February 10, 1910, the inventor barely mentioned the recipients' names. He congratulated the Wrights for bringing "the aerodrome" to "the commercial and practical stage," just as Langley had predicted someone would. But it was Langley himself, Bell declared, who must be recognized as "the great pioneer of aerial flight"; who had divined "Langley's Law," which "opens up enormous possibilities for the aerodrome of the future"; and who had constructed "a perfectly good flying machine" that failed to fly only because of its faulty launcher. "Who can say what a third trial might have demonstrated?"

This view, already well rooted among Langley's friends, spread to others in aeronautical circles, especially those who resented the Wrights' insistence on claiming a financial reward for their work. One carrier of the legend was Charles Manly. When Manly died in 1927 at the age of 51, he was still planning that third and successful trial of the great aerodrome, with his son in the pilot's seat. Another loyalist was the perpetually late Stephen Balzer, who also went to his grave believing in the Langley machine, though not in Charles Manly. In 1931, *The New York Times* reported that Balzer, "strong and active" at age 70, was "still confident that if a man with some knowledge of the air had been chosen to fly the ill-fated Langley machine instead of an engineer it might have been the first to fly." Of course, in the fall of 1903, no one but the Wright brothers had possessed such knowledge.

THE WHOLE aim of Will's efforts from 1905 to 1909 had been to avoid "business cares and vexatious law suits" so that he and his brother could concentrate on science. In fact, "business cares and vexatious law suits" were precisely what he got.

Soon after the Hudson-Fulton flights, the brothers incorporated the Wright Company and began to manufacture aeroplanes for sale. In the end, their plan of selling to governments had not made them independent of business cares, and they concluded they must once again be builders and sellers, not experimenters. They designed a large, lovely home, to be called Hawthorn Hill, in the Dayton suburb of Oakwood. But there was little time to enjoy their new affluence. Business worries pressed in on them, yet business came second to their legal battles, chiefly against Glenn Curtiss. Will assumed the leading role in the boardroom and the courtroom alike. It was harsh, exhausting, nerve-racking work. He pursued Curtiss with a sense that he had no alternative, but with deep regret for years wasted. "We had hoped in 1906 to sell our invention to governments for enough money to satisfy our needs and then devote our time to science," he wrote to a friend in France in January 1912, "but the jealousy of certain persons blocked this plan, and compelled us to rely on our patents and commercial exploitation. . . . When we think what we might have accomplished if we had been able to devote this time to experiments, we feel very sad, but it is always easier to deal with things than with men, and no one can direct his life entirely as he would choose."

Four months later, after many weeks away from home, Will became ill in Boston. When he returned to Dayton, doctors diagnosed malaria, then typhoid fever. He died early on the morning of May 30, 1912.

"A short life, full of consequences," his shocked father wrote that night in his diary. "An unfailing intellect, imperturbable temper, great self-reliance and as great modesty, seeing the right clearly, pursuing it steadfastly, he lived and died."

ORVILLE AND KATE blamed Curtiss for Will's death. They believed the legal battles, which they attributed chiefly to Curtiss's avarice, had weakened their brother's constitution and made him susceptible to the fever that took his life. Pressing forward in court, they won a final victory early in 1914, when the U.S. Circuit Court of Appeals declared the Wrights' patent to be not only valid but a pioneer patent, and thus entitled to the broadest possible protection, covering ailerons as well as wing-warping. The Wright Company announced it would charge a 20 percent royalty on every aeroplane manufactured in the United States, though Orville said he would follow a "policy of leniency" with every company but one—Curtiss's.

Curtiss responded with an extraordinary maneuver. He asked and received permission from Charles Walcott, Langley's successor as secretary of the Smithsonian, to haul the great aerodrome out of storage and fly it. This, he hoped, would prove the machine had been capable of flight and thus undermine the pioneer status of the Wrights' patent. In the summers of 1914 and 1915, Curtiss succeeded in making several short, straight hops over the lake at Hammondsport. He and his collaborator, Albert Zahm, a friend-turned-foe of the Wrights, claimed the trials had been made "without modification" of the original design except for the addition of pontoons. Photographs alone—not to mention the detailed report of Lorin Wright, whom Orville sent as an observer—showed that Curtiss and Zahm in fact had made many modifications, chiefly to improve the shape and strength of the wings. Yet Charles Walcott accepted the trials as conclusive. It became an article of faith in Smithsonian literature—and on the label of the aerodrome on display in the Smithsonian's Arts and Industries Building—that Langley had built "the first man-carrying aeroplane in the history of the world capable of sustained free flight." The Smithsonian granted that the Wrights had been first to fly, but not the inventors of the aeroplane. And it was the inventing, not the flying, that really mattered to Orville. "There were thousands of men who could have taken our 1903 machine into the air for the first flight," he told one correspondent, "but I believe there was no one else in the world at that time beside Wilbur and myself who possessed the scientific data for building a machine that would fly."

For Curtiss, the "capable of flight" controversy became irrelevant soon after his hops in the souped-up aerodrome. His chief interest was money, and he got all he could have wanted with the coming of World War I. From 1914 to 1917, he built planes without ailerons to avoid the Wrights' patent, then shipped them to Canada and England, where ailerons were attached. When the United States entered the war, the government ordered the pooling of all U.S. aeroplane patents, freeing Curtiss of the Wrights' shadow at last. Thousands of Curtiss planes were sold to the Allies, and his JN model, fondly nicknamed the "Jenny," became the darling of aviation. At the age of 40, Curtiss found himself a multimillionaire and a captain of industry. He lost interest in aviation after the war and shifted to real estate, becoming a major player in the south Florida land boom of the 1920s. He died in 1930.

FOR ORVILLE and Katharine, the Smithsonian's distortion of history became the consuming focus of their lives. Orville, utterly uninterested in corporate leadership, had sold his stake in the Wright Company in 1915. What mattered to him now was not the future but the past—to defend his and his brothers' claim to history. For years he lobbied to induce the Institution's leaders to change their position, or at least to submit the dispute to an impartial board of inquiry. When they refused, he took the extreme step of sending the 1903 flyer abroad for permanent display in the Science Museum of London. "I had thought that truth eventually must prevail," he told the press, "but I have found that silent truth cannot withstand error aided by continued propaganda. . . . I regret more than anyone else that this course was necessary."

He made that decision alone. His father had died in 1917, and Kate no longer lived at Hawthorn Hill.

In 1925, at the age of 50, she fell in love with an old friend from Oberlin College, Henry Haskell, who had become editor of the *Kansas City Star*. She had corresponded for years with Haskell and his wife. But two years after his wife died, Haskell told Kate he had loved her since college, setting off the great crisis of her life. Through weeks of daily correspondence that alternated between exhilaration and anguish, she accepted his approach—though she had come to believe "that men were not the least interested in me, except as a friend"—and concluded that she loved "dear Harry" as well. For a time she could bearly bring herself to consider a new life. "I don't see how Orv could get along without me. And it would break my heart not to stay by him after all he has been to me." She kept the affair secret, with her "terrible conscience" telling her that to leave Orville would be to betray him. She told Haskell:

> It hasn't been the usual sister and brother relation. I am sure we have been more to each other than many husbands and wives and that our obligation to each other is really different from the usual one. I feel sure Orv would feel the same way if the same thing came up with him. Orv built this house with the idea of my being here with him just as much as any husband builds for his wife. Everything has been planned for the future with the idea that we would be together always.

She worried about Orville's health, never robust since the accident at Fort Myer; about the strain of the Smithsonian problem; about the effect on Orville if she married Haskell; about the effect on Haskell if she remained with Orville. Yet she kept her composure. "Altogether, Harry, my world is in a state of great disquiet and uneasiness. But I am not discouraged or unhappy. In the background there are all the things that are 'true and lovely and of good report'—the real things that Will and Orv did and what they were and Orv is. Nothing can touch that and I am going to get along with this awful trouble as well as I can."

Still, her worst forebodings about Orville came true. When she decided to marry Haskell and broke the news to her brother, Orville refused to give her his blessing or to attend the wedding. She made a new home with Haskell but apparently never escaped a deep despair over the break. When Kate's closest friend in Dayton pleaded with her to visit, she replied: "I can't go to Dayton yet. . . . I don't see how I ever can. . . . In my imagination I walk through that house, looking for Little Brother, and at all the dear familiar things that made my home. But I never find Little Brother and I have lost my old home forever, I fear."

In 1929, on the eve of a trip to Europe, she became ill with pneumonia. Haskell sent a pleading telegram to Orville, who reached her bedside before she died.

ORVILLE LIVED nearly twenty more years, dividing his time between Dayton and his summer home in Canada, on Georgian Bay. He enjoyed the company of close friends and of his nieces and nephews and their families. He jousted with the Smithsonian. With bored distaste, he handled the massive correspondence that came to him as the sole surviving "father of flight." When in Dayton, he drove each day to the laboratory he had built in the old neighborhood, but he did little more than tinker with household gadgets. If any more proof was needed of his brother's dominant drive during their years together, it became obvious in the quiet, uneventful days of Orville's life after 1912.

He watched the proliferation of airplanes—he called them "aeroplanes" long after the public had discarded the early usage—with astonishment. During World War I he had told a reporter that "Wilbur and I always thought that the principal use of the aeroplane for a number of years would be military, yet had we been told . . . that hundreds of thousands of aeroplanes would be used in a single war, we would have been as skeptical of that statement as some of the rest of you were of the aeroplane itself." Nearly thirty years later, days after the bombing of Hiroshima and Nagasaki, he told a friend, "I once thought the aeroplane would end wars. I now wonder whether the aeroplane and the atomic bomb can do it."

For years, Kate had begged him to write a history of the brothers' achievements. After her death, others did, too, including Charles Lindbergh, who became Orville's friend after his solo flight of the Atlantic in 1927. They served together on the National Advisory Committee on Aeronautics, the precursor of National Aeronautics and Space Administration. When Lindbergh learned that Orville had suppressed the manuscript of one biography and tried to suppress another because they were "too personal," he urged the older man to write his own version. Orville said he still might do it someday. Writers' attempts to capture the early days of flight in words were "never quite accurate," he said, as "no one else quite understands the spirit and conditions of those times." Finally Orville agreed to a biography by Fred Kelly, a journalist and friend. But he never wrote his own book.

In 1942, Smithsonian Secretary Charles Abbot—no longer the young astronomer who trembled whenever Samuel Langley summoned him—agreed to publish a clear, official statement that the Wrights, not Langley, had invented the airplane. Six years later, on January 30, 1948, Orville died of heart failure. Only then was the 1903 flyer brought back to the United States. Lindbergh's *Spirit of St. Louis* was moved from the central position of honor in the North Hall of the Smithsonian's Arts and Industries Building, and the flyer was hung in its place. In 1976 the machine was moved to the new National Air and Space Museum, where it hangs in the center of the main hall, overlooking the grassy Mall where Samuel Langley, coattails flying, once chased his flying toys.

At the original ceremony to dedicate the flyer on December 17, 1948, the brothers' oldest nephew, Milton Wright, was invited to describe what his uncles had done.

"The aeroplane means many things to many people," he said. "To some it may be a vehicle for romantic adventure or simply quick transportation. To others it may be a military weapon or a means of relieving suffering. To me it represents the fabric, the glue, the spruce, the sheet metal, and the wire which, put together under commonplace circumstances but with knowledge and skill, gave substance to dreams and fulfillment to hopes."

# ACKNOWLEDGMENTS

IF THERE'S A PLACE where overwriting is common and entirely called for, it's at the end, where acts of assistance and grace are noted. Restraint and subtlety don't seem right when an author needs the words to say what he feels for the people who helped him write his book. To all those who helped me, take my word for it: I've thrown in all the adjectives, adverbs, and hyperbole I can muster, but they don't do justice to the way I feel.

To the Lukas Prize Project, a joint program of Harvard University's Nieman Foundation for Journalism and the Columbia University Graduate School of Journalism, I offer thanks for its J. Anthony Lukas Work-in-Progress Award. All who care about good nonfiction owe a debt to Linda Healey and the Project itself for working to preserve the memory of that peerless writer, journalist, and historian.

Books bring new friends. Nick Engler of the Wright Brothers Aeroplane Company, West Milton, Ohio, was unfailingly generous in sharing his encyclopedic knowledge and deep insights about the Wright brothers, gained through years of extraordinary work in replicating and flying the Wrights' machines. The greatest fun in all my research came when I watched Engler's replicas fly, and those moments were enormously helpful to my effort to bring the Wrights to life. Several of his ideas found their way into the narrative, and his critique of the manuscript saved this neophyte of flight from committing numerous errors. Whatever merits the book may have owe much to this one-of-a-kind historian, craftsman, author, and educator.

I also had the good luck during this time to become friends with the great Wright authority Fred Howard, author of *Wilbur and Orville*, who sent good tips and lent good books, then applied an eagle eye to the manuscript, much to my benefit.

I was equally lucky in an old friendship. At the beginning, Jonathan Marwil talked me through the planning. At the end he offered insights that led to many improvements, small and large.

For assistance with research, thanks go to Adam Pasick, Ceceile Kay Richter, the uncompensated but effective Emily Wilson-Tobin, and especially to Devon Thomas, an ever-ready and expert librarian-researcher who delivered results at express speed. For tracking down difficult questions in Dayton, and catching errors, thanks go to Louis Chmiel, a tough and tireless detective on the trail of the Wrights.

The aviation historian Charles Gibbs-Smith once said that "bothering librarians is perhaps a legitimate and inevitable proclivity of writers the world over; but I think we all tend to take their long-suffering and ever-willing co-operation too much for granted." In that spirit of appreciation for unsung heroes, I sing wholehearted thanks and praise to these archivists and librarians: Dawne Dewey and John Sanford of Wright State University; Leonard Bruno of the Manuscripts Division of the Library of Congress; Tracy Elizabeth Robinson, William Cox, and Alan Bain of the Smithsonian Institution Archives; Kristine Kaske and Patricia Williams of the National Air and Space Museum Archives; William Baxter of the National Air and Space Museum Library; Randy Neuman of the United Brethren Historical Center at Huntington College; Bryan Skib, Donna Bradshaw, Anne Beaubien, Deborah Heiden, Barbara Beaton, Judy Avery, and the staff of the Buhr Shelving Facility, all of the University of Michigan Libraries; John White of the Southern Historical Collection, University of North Carolina at Chapel Hill; Terry Hoover of the Research Center of the Henry Ford Museum & Greenfield Village, in Dearborn, Michigan; Nancy R. Horlacher of Carillon Historical Park in Dayton; and Tom Hilberg of the Medina County (Ohio) Historical Society.

Among all these, special appreciation is due to John Sanford at Wright State for his expertise in the Wright's archives, his patience with my questions, his speed in responding to my requests, and his generosity in sharing his own research on the Hudson-Fulton Celebration.

Ellis Yochelson of the National Museum of Natural History, biographer of Charles Walcott and paleontologist, gave freely of his vast knowledge of the history of the Smithsonian Institution. Other scholars, researchers, and authorities who answered questions and suggested leads include Patricia Whitesell, director and curator of the University of Michigan's Detroit Observatory; Dr. John Shy and Dr. Arlene Shy of the University of Michigan; Dr. Tom D. Crouch and Dr. Peter L. Jakab of the National Air and Space Museum; Darrell Collins, historian of the Wright Brothers National Memorial at Kill Devil Hills, North Carolina; Dr. Robert V. Bruce of Boston University; Dr. Ross Petty of Babson College; Susan Bushouse Foran; Ann Honious and Bob Peterson of the Dayton Aviation Heritage National Historical Park; Dr. Joe W. McDaniel, of Dayton; Betty Darst, of Dayton; Mary Ann Johnson, of Dayton; and Lucy Johnson of the British Association for the Advancement of Science.

Special thanks to Kim Holien, historian of the U.S. Army's Fort Myer and

Fort McNair, for sharing many items from his files of Wright memorabilia; to John Root, Kim Flottum, and Jim Thompson, of the A.I. Root Company in Medina, Ohio, for helping me learn about the remarkable life of Amos Root; and to the pilot and hang-gliding champion Dudley Mead, for telling me how it feels to glide in the Wrights' machines.

For permission to examine the minutes of meetings of the Regents of the Smithsonian Institution, thanks to the Office of the Secretary and Kathy Boi.

Friends and family gave essential help at many moments. In Arlington, Virginia, Jim, Lisa, Paris, and Ben Mitzelfeld shared their home and their enthusiasm. Across the Potomac, Laurie Leitch and Elizabeth McDonnell did the same. David Garrigus, who trod many of the same trails as I for his first-rate film documentary, "Kitty Hawk: The Wright Brothers' Journey of Invention," gave tips and leads at many points. Kim Burton helped in so many spare moments that they added up to some very large total. Dr. Thomas Segall and Ann Segall, M.S.W., shared their thoughts about the dynamics of the Wright family. Thanks to all those who participated in the Great but Ultimately Irrelevant Wright Title Survey, and congratulations to Penny Schreiber, the only contestant who nailed it.

I'm deeply grateful to Jamie Evans of Hass/MS&L for volunteering his artistic eye and his scarce time to help conceive and shepherd the illustrative drawings, and to the artists David Messing and David MacArthur for the fine drawings themselves. Thanks, too, to Terry Mahar, for technical wizardry.

One person types the words, but most books grow out of a far more complex process of labor and sacrifice on the part of entire families, and the families need help. Dave and Peggy Farrell offered my family a combination of inspiration, enthusiasm, and concrete aid that adds up to a gift of extraordinary generosity. Liz, Jon, and Peter Jacobs were a constant source of encouragement, ideas, and fun, as were Cathy Angelocci, Karen Holzhausen, Randy and Patti Milgrom, Bob and Sarah Dawson, Doug Evett, John Bebow, Karl Leif Bates, the other members of the Unnamed Group—Richard Campbell, Rob Pasick, John Bacon, Rick Ratliff, David Stringer, and Jerry Burton—Floyd and Elizabeth Erickson; and James and Dorothy Tobin, to whom the book is dedicated.

Claire and Lizzie Tobin tolerated "all Wright brothers, all the time"—even enjoyed it, I think.

I have been very lucky to fall in with a creative, professional, and hardworking production team at the Free Press. Thanks especially to Casey Reivich for her gracious patience and close attention to all-important details; to Tom Lau for a lovely jacket illustration; and to that Free Press associate, Will Nichols, for keeping me on time, or pretty close.

The largest debts come last:

No more hand-wringing, please, about the passing of the great editors. The breed survives in Bruce Nichols. He conceived this book. Then he coached, coaxed, comforted, and cheered. He counseled wisely, waited patiently, insisted politely, edited skillfully, remembered everything important, dispensed with trivia, and never got mad, though he might have.

Carol Mann, likewise, is the sort of agent every author wants. She did the deal with consummate professionalism, then stayed with the process down to the wire, providing good advice, emotional sustenance, and friendship. (And thanks to Christy Fletcher, for good work early on.)

Angels, too, help authors. Janice Harayda was mine.

All of which makes it awfully hard to say anything sensible about the contributions of Leesa Erickson Tobin, who gave all of the above (from skilled research to shrewd advice to angelic gifts) and a great deal more, with a strange and wonderful combination of steely fortitude, hard labor, infinite patience, and tender understanding. Freud said we need love and work. Because of her, I have both.

JAMES TOBIN

*Ann Arbor, Michigan*
*September 20, 2002*

# NOTES

## *Abbreviations*

The following initials, acronyms, and abbreviations are used throughout the notes:

*Persons*:

| | | | |
|---|---|---|---|
| WW | Wilbur Wright | SPL | Samuel Pierpont Langley |
| OW | Orville Wright | OC | Octave Chanute |
| MW | Milton Wright | AGB | Alexander Graham Bell |
| KW | Katharine Wright | | |

*Sources*:

| | |
|---|---|
| FC, WBP, LC | Family Correspondence, Wright Brothers papers, Library of Congress |
| GC, WBP, LC | General Correspondence, Wright Brothers papers, Library of Congress |
| SI | Smithsonian Institution |
| SIA | Smithsonian Institution Archives |
| NASM | National Air and Space Museum, Smithsonian Institution |
| WBC, WSU | Wright Brothers Collection, Special Collections and Archives, Paul Laurence Dunbar Library, Wright State University |
| Jakab and Young, eds., *Published Writings* | Peter L. Jakab and Rick Young, eds., *The Published Writings of Wilbur and Orville Wright* (Washington, D.C.: Smithsonian Institution Press, 2000.) |
| McFarland, ed., *Papers* | Marvin W. McFarland, editor, *The Papers of Wilbur and Orville Wright*, vols. 1 and 2 (New York: McGraw-Hill, 1953.) |

## Prologue

PAGE

1 *He could take care:* Woodland Cemetery visit on 5/30/1899, MW diary entry, 5/30/1899, Bishop Milton Wright, *Diaries, 1857–1917* (Dayton, Ohio: Wright State University Libraries, 1999), 506.

2 *"This man is strange and cold":* *New York Times*, 6/1/1912.

2 *"my imagination pictures things":* WW to KW, 6/08/1907, FC, WBP, LC.

2 *The neighbors included:* Heads of households on Wrights' block of Hawthorn Street, U.S. Sanborn map #42, U.S. Census records, 1900, Dayton Aviation Heritage National Historical Park.

2 *"When I saw this house":* John McMahon, *The Wright Brothers: Fathers of Flight* (Boston: Little Brown, 1930) 19–20.

2 *At the rear:* Remodeling, MW diary entry, 5/22/1899, *Diaries, 1857–1917*, 505.

3 *Along the front:* Design and furnishings of the Wrights' home, undated typewritten reports, "Wright House" and "The Wright House"; Ivonette Wright Miller to "dear sirs," 6/25/1977; "Layout of First & Second Floors in the Wright Bros. Home," Lois Gorrell memo, 7/20/1981, "Notes on back of Registrar's copy of photo 188:16081"; E. J. Cutler oral history interview, 1/19/56; all in E.I. #186, Greenfield Village Buildings, Wright Home, Collections of Henry Ford Museum & Greenfield Village Research Center.

3 *"always the danger":* WW to Lou Wright, 6/18/1901, FC, WBP, LC.

3 *He was "an enthusiast":* WW's request for information on flight, WW to Smithsonian Institution, 5/30/1899, McFarland, ed., *Papers*, vol. 1, 4–5.

4 *A man once wrote:* Request for "all Smithsonian publications," Webster Prentiss True, "The Smithsonian Institution," *Smithsonian Scientific Series*, vol. 1 (Washington, D.C.: Smithsonian Institution, 1929), 156.

5 *"I shall count this day":* quoted in Robert V. Bruce, *Bell: Alexander Graham Bell and the Conquest of Solitude* (Ithaca: Cornell University Press, 1973), 361.

5 "A *'flying-machine,' so long a type for ridicule":* SPL, "The 'Flying-Machine,'" *McClure's Magazine*, June 1897, 660.

5 *"the only things of human construction":* Charles M. Manly, *Langley Memoir on Mechanical Flight*, pt. 2, (Washington, D.C.: Government Printing Office, 1911), 128.

6 *"a list of works relating to aerial navigation":* Richard Rathbun to WW, 6/2/1899, GC, WBP, LC.

## *Chapter One:* "The Edge of Wonder"

7 *"reality most like to dreams":* quoted in Laurence Goldstein, *The Flying Machine and Modern Literature* (Bloomington, Ind.: Indiana University Press, 1986), 1–2. Goldstein suggests that poets of flight and inventors of flying machines have been motivated by the same "primordial envy" of the birds. He also quotes the anthropologist Mircea Eliade: "The longing to break the ties that hold [man] in bondage to the earth is not a result of cosmic pressures or of economic insecurity—it is constitutive of man, in that he is a being who enjoys a mode of existence unique in the world. Such a desire to free himself from his limitations, which he feels to be a kind of degradation, and to regain spontaneity and freedom . . . must be ranked among the specific marks of man."

8 *"Like a living thing":* SPL's account of the flight of No. 5, "The 'Flying-Machine,'" *McClure's Magazine*, June 1897, 658–60.

8 *he remembered a day:* SPL's earliest memories, "Uncompleted Memoirs of S.P. Langley Written at Aiken, South Carolina, From February 8 to 20, 1906," box 22b, RU 7003, SIA.

8 *"I cannot remember":* George Brown Goode, "The Three Secretaries," in Goode, ed., *The Smithsonian Institution, 1846–1896: The History of Its First Half Century* (Washington, D.C.: Smithsonian Institution, 1897), 203–04.

9 *But when Langley was finishing:* Goode, "The Three Secretaries," in Goode, ed., *The Smithsonian Institution, 1846–1896*, 201.

10 *"My brother's . . . perseverance":* Goode, "The Three Secretaries," in Goode, ed., *The Smithsonian Institution, 1846–1896*, 205.

10 *Langley revisited the philosopher's:* Cyrus Adler, *I Have Considered the Days* (Philadelphia: Jewish Publication Society of America, 1941), 183–84. Adler said Langley "summer after summer went to England and had the privilege of sitting in a corner of Carlyle's library and hearing the great man talk." After the two shared a four-hour carriage ride, Carlyle told a friend that Langley was "the most sensible American he had ever met." When Langley heard this he said, "Yes, I was

with him for four hours yesterday and never opened my mouth." For the possible influence of Carlyle's advice to "know thy work" on Langley, see Ernest Samuels's introduction to *The Education of Henry Adams* (Boston: Houghton Mifflin, 1974), viii. Samuels said this advice of Carlyle's had registered on Adams, Langley's friend and fellow native Bostonian.

10 *"He had not as yet published anything"*: Andrew Dickson White, "Samuel Pierpont Langley," in "Samuel Pierpont Langley," *Smithsonian Miscellaneous Collections*, Vol. XLIX (Washington: Smithsonian Institution, 1907), 9.

10 *The college had taken it*: Allegheny Observatory in 1867, Donald Leroy Obendorf, "Samuel P. Langley: Solar Scientist, 1867–1891 (Ph.D. dissertation, University of California, Berkeley, 1969), 2–7; White, "Samuel P. Langley," 10.

11 *"his strength burdened"*: Obendorf dissertation, "Samuel P. Langley . . . ," 91.

11 *The time service boosted*: Obendorf dissertation, "Samuel P. Langley . . . ," 9–29. For the role of midwestern observatories in public service, see Patricia S. Whitesell, "Nineteenth-Century Longitude Determinations in the Great Lakes Region: Government-University Collaborations," *Journal of Astronomical History and Heritage*, December 2000, 131–57.

11 *They complained that Langley*: Fight over SPL's pay and privileges, Obendorf dissertation, "Samuel P. Langley . . . ," 48–55.

12 *"the sun's disk is seen"*: SPL, "The Sun: A 'Total' Eclipse," *Scientific American*, July 27, 1878.

12 *And the sun had practical advantages*: Why SPL studied sun, Obendorf dissertation, "Samuel P. Langley . . . ," 76–80.

12 *"perhaps the only celestial object"*: White, "Samuel Pierpont Langley," Cyrus Adler, "Samuel Pierpont Langley," *Annual Report of the Smithsonian for 1906* (Washington, D.C.: Government Printing Office, 1907), 515–33.

13 *"whose actual vastness"*: SPL, *The New Astronomy* (Boston: Houghton Mifflin, 1887), 19.

13 *"One who has sat at a powerful telescope"*: SPL, *New Astronomy*, 17.

13 *Of his hundreds of sunspot*: Obendorf dissertation, "Samuel P. Langley . . . ," 110–16.

14 *He took pride in his attention to detail*: Obendorf dissertation, "Samuel P. Langley . . . ," 210, 157.

14 *"that heat and light were not two different things"*: Goode, "The Three Secretaries," in Goode, ed., *The Smithsonian Institution, 1846–1896*, 218.

14 *"The thoroughness, ingenuity, and beauty"*: An unnamed Smithsonian regent quoted in Paul H. Oehser, *Sons of Science: The Story of the Smithsonian Institution and Its Leaders* (New York: Henry Schuman, 1949), 117–18.

14 *He pushed aides to do*: Obendorf dissertation, "Samuel P. Langley . . . ," 226–30.

14 *He tended to exaggerate*: Obendorf dissertation, "Samuel P. Langley . . . ," 161–68, 193–99, 220–33.

15 *"A certain part of Langley"*: Obendorf dissertation, "Samuel P. Langley . . . ," 227.

15 *"I particularly remember"*: White, "Samuel Pierpont Langley," 18.

15 *"loved to talk with men"*: White, "Samuel Pierpont Langley," 19.

15 *"There was something reaching"*: John A. Brashear., "A Biographical Sketch of S. P. Langley," in *Miscellaneous Scientific Papers of the Allegheny Observatory*, No. 19; reprinted from *Popular Astronomy*, Vol XIV, 1906.

15 *"a strong craving for real society"*: Adler, "Samuel Pierpont Langley."

16 *He attended meetings*: Meetings with medical society and drugstore group, John A. Brashear, "A Biographical Sketch of S. P. Langley," in *Miscellaneous Scientific Papers of the Allegheny Observatory*, No. 19; reprinted from *Popular Astronomy*, Vol XIV, 1906.

16 *"You have seen for yourself"*: Quoted in Obendorf dissertation, "Samuel P. Langley . . . ," 57–58.

16 *"almost every thing outside"*: SPL, "History of the Allegheny Observatory," in John E. Parke, ed., *Recollections of Seventy Years and Historical Gleanings of Allegheny, Pennsylvania* (Boston: Rand, Avery & Co.), 1886.

16 *"In the astronomical circles of the Old World"*: Quoted in Obendorf dissertation, "Samuel P. Langley . . . ," 69.

17 *"a great addition to the university"*: Obendorf dissertation, "Samuel P. Langley . . . ," 57–58.

17 *"in a decidedly roundabout way"*: William Hallock, "What Science Owes to Professor Langley," *New York Times*, March 4, 1906.

17 *"I know nothing about chemistry"*: Quoted in Obendorf dissertation, "Samuel P. Langley . . . ," 81.

17 *"Professor Langley . . . shares the views"*: Richard A. Proctor, *Old and New Astronomy* (London: Longmans, Green, 1892), 333n.

18 *Among the amateurs on the program*: Israel Lancaster at 1886 meeting of AAAS, "Birds that Did Not Soar," *New York Times*, August 24, 1886; Tom D. Crouch, *A Dream of Wings: Americans and the Airplane, 1875–1905* (Washington, D.C.: Smithsonian Institution Press, 1981), 38–41.

18 *"I was brought to think"*: SPL, "The 'Flying-Machine,'" *McClure's Magazine*, June 1897, 648.
18 *"the whole subject"*: *Langley Memoir on Mechanical Flight*, v. 1, 2.
18 *"Not to build a flying-machine,"* SPL, *"The 'Flying Machine,'"* 648.
19 *In one key test:* Experiment with brass plate on whirling arm, SPL, "The 'Flying Machine,'" 649.
19 *scenes from childhood:* SPL, "The 'Flying Machine,'" 650.
20 *"I have no wish or ambition"*: Quoted in Cyrus Adler, "Samuel Pierpont Langley," 515–33.
21 *"to meet the requirements of flight"*: *Langley Memoir*, pt. 1, 7.
21 *"a light wooden frame"*: *Langley Memoir*, v. I, p. 9
21 *"It was a very amusing sight"*: Webster Prentiss True, *The Smithsonian Institution*, Smithsonian Scientific Series, vol. I (Washington, D.C.: Smithsonian Institution Series, Inc., 1929), 290.
22 *"It is enough," he wrote:* SPL, "The 'Flying-Machine,'" 654.
23 *"Isn't that maddening!"*: Robert V. Bruce, *Bell: Alexander Graham Bell and the Conquest of Solitude* (Ithaca, N.Y.: Cornell University Press, 1973), 362.
23 *"Whether from natural disposition"*: Abbot quoted in Obendorf dissertation, "Samuel P. Langley . . . ," 207.
23 *In engineering, the approach:* Tom D. Crouch, *A Dream of Wings: Americans and the Airplane, 1875–1905* (Washington, D.C.: Smithsonian Institution Press, 1981), 140.
23 *"In designing this first aerodrome"*: *Langley Memoir on Mechanical Flight*, pt. 1, 30.
23 *"the lines which Nature"*: *Langley Memoir on Mechanical Flight*, pt. 1, 32.
23 *then alter "the form of construction"*: SPL, "The 'Flying Machine,'" 654.
24 *The term "aerodrome"*: Basil L. Gildersleeve to AGB, 3/15/1913, box 143, AGB papers, LC. See also Crouch, *A Dream of Wings*, 133.
24 *"backward steps"*: SPL, "The History of a Doctrine," *American Journal of Science*, January 1889, 2.
24 "It appeared": All from *Langley Memoir*, pt. 1, pp. 38–66.
25 *"the point was reached"*: SPL, "The Flying-Machine," 654.
25 *"pitch up or down"*: SPL to AGB, box 131, AGB papers, LC.
25 *On May 6, 1896:* SPL, "The 'Flying Machine,' 658–59. See also *Langley Memoir on Mechanical Flight*, pt. 1, 106–08. Langley's article in *McClure's* includes Bell's description of the flight, which appeared as a letter in *Science*, 5/22/1896.
26 *"a man who cannot be replaced"*: Minutes of the SI Regents, 1/27/1897, SIA.
27 *"the severe strain of his scientific labors"*: Cyrus Adler, "Samuel Pierpont Langley," *Annual Report of the Smithsonian Institution for 1906*, (Washington, D.C.: U.S. Government Printing Office, 1907, 527.
27 *He asked the regents:* Minutes of the SI Regents, 1/27/1897, SIA.
27 *"You have done what no one"*: Gardiner Hubbard to SPL, 4/27/1897, Box 36, RU 7003, SIA.
27 *"I have passed sixty years"*: SPL, draft of letter to Gardiner Hubbard, 4/28/1897, "Private Wastebook," 1888–1894 [sic], NASM/SI archives, 476–78. Several days later, Langley consented to meet with a patent attorney sent by Hubbard. "I told him I was not expecting to take out patents, yet was enquiring at the urging of friends whether the thing was patentable. That I knew that almost anything could be patented and did not doubt that no end of patents could be secured here. What I wanted to know was whether they would probably be worth anything." The upshot of the exchange is unknown, but it appears that Langley never followed through. SPL, entry of 5/4/1897, "Private Wastebook," 1888–1894 [sic], 482, NASM archives.
28 *"If anyone were to put"*: SPL to OC, 6/8/1897, Chanute papers, LC.
28 *Chanute offered encouragement:* OC to SPL, 6/11/1897, letterbooks, Chanute papers, LC.
28 *"a collection of church steeples"*: quoted in Cynthia R. Field, Richard E. Stamm, and Heather P. Ewing, *The Castle: An Illustrated History of the Smithsonian Building* (Washington, D.C.: Smithsonian Institution Press, 1993), 122.
29 *"Speed, then, is indispensable here"*: SPL, "The 'Flying-Machine,'" 650.
29 *"Swift Camilla"*: SPL, "The 'Flying-Machine,'" 650.
29 *"anyone who is disposed"*: SPL to OC, 12/[?]/1897, Chanute papers, LC.
29 *"I know of nobody"*: OC to SPL, 12/11/1897, letterbooks, Chanute papers, LC.
29 *"an athletic, breezy type of man"*: Charles G. Abbot, *Adventures in the World of Science* (Washington, D.C.: Public Affairs Press, 1958), 97.
30 *That morning, after discussion:* SPL-Walcott meeting of 3/21/1898, "Memorandum of conversation with Mr. Walcott," 3/21/1898, Langley scrapbook, 1897–98, RU 7003, SIA.
30 *Walcott moved with astonishing speed:* Memoranda, 3/25/1898, 3/28/1898, 3/30/1898, 3/31/1898, 4/2/1898, Langley scrapbook, 1897–98, RU 7003, SIA.

30 *The President was much pleased:* Memorandum, 3/28/1898, Langley scrapbook, 1897–98, RU 7003, SIA.
30 *"The machine has worked":* Quoted in Archibald D. Turnbull and Clifford L. Lord, *History of United States Naval Aviation* (New Haven: Yale University Press, 1949), 1.
31 *He "was by no means eager":* SPL's meeting with aerodrome advisory committee, "Memorandum," 4/6/1898, Langley scrapbook, 1897–98, RU 7003, SIA.
31 *no financial interest:* SPL to Stimson Brown, 4/16/1898, Langley scrapbook, 1897–98, RU 7003, SIA. Langley's letter suggests he made this assertion to members of the advisory committee during their visit to the Smithsonian on 4/6/1898.
32 *For the committee's report said:* Committee's report on feasibility of Langley's manned aerodrome project, successive drafts of the report, including SPL's recommended changes, appear in Langley scrapbook, 1897–98, RU 7003, SIA.
32 *"the gentlemen . . . are honorable":* Paul Beckwith to SPL, 6/10/1898, Langley scrapbook, 1897–98, RU 7003, SIA.
32 *"Have you any young man":* SPL to R.H. Thurston, in Meyer, ed., *Langley's Aero Engine of 1903*, 31.
32 *Manly came from an old Virginia family:* C. B. Veal, "Manly, The Engineer," *S.A.E. Journal*, April 1939.
33 *The secretary began by emphasizing:* SPL's meeting with Board of Ordnance and Fortification, 11/9/1898, SPL, "Memorandum of verbal statements made to the Board of Ordnance & Fortification," 11/9/1898, Langley scrapbook, 1897–98, RU 7003, SIA. These remarks are quoted from Langley's memorandum, written on the day of the meeting, about his presentation to the Board of Ordnance and Fortification. I have altered verbs to the tenses Langley would have used in his spoken remarks before the board.
34 *That afternoon, the board agreed:* BOF's approval of SPL's funding request, "Extract from Proceedings of Board of Ordnance and Fortification," 11/9/1898, Langley scrapbook, 1897–98, RU 7003, SIA.
34 *Someone at the War Department:* Press stories of BOF's allotment to SPL, "Flying Machines in War," *Washington Post*, 11/11/1898; "Flying Machines in War," *New York Times*, 11/11/1899.
34 *"I could not undertake":* SPL to I. N. Lewis, 11/12/1898, Langley scrapbook, 1897–98, RU 7003, SIA.
34 *"fullest discretion":* Telegram, I. N. Lewis to SPL, 12/16/1898, Langley scrapbook, 1897–98, RU 7003, SIA.
34 *"the flying-machine has always been":* "Wait for the Flying-machine," *Washington Post*, 11/14/1898.
35 *Langley, confident of his purpose:* Robert B. Meyer Jr., ed., *Langley's Aero Engine of 1903* (Washington, D.C.: Smithsonian Institution Press, 1971), 7, 29–35.

## *Chapter Two:* "A Slight Possibility"

37 *"because those he otherwise harmonized with":* H. A. Thompson, *Our Bishops* (Dayton, Ohio: United Brethren Publishing House, 1904), 527.
37 *As a boy in the 1840s:* For MW's background, education, and early ministry, see Daryl Melvin Elliott, "Bishop Milton Wright and the Quest for a Christian America," Ph.D. dissertation, Drew University, 1992, 12–107; Tom D. Crouch, *The Bishop's Boys: A Life of Wilbur and Orville Wright* (New York: Norton, 1989), 23–43.
37 *Milton sharpened all his:* MW's views on slavery, Elliot dissertation, "Bishop Milton Wright . . . ," 86–94.
38 *"steady, continued and systematic investigation":* Thompson, *Our Bishops*, 535.
38 *"a man of strong convictions":* Thompson, *Our Bishops*, 527.
38 *Worst of all about the secret:* Antisecrecy views of MW and United Brethren, Elliott dissertation, "Bishop Milton Wright . . . ," 167–81.
39 *As Eddie and Orville listened:* WW's faked phonograph speech, Fred C. Kelly, "Traits of the Wright Brothers," *Technology Review*, June 1949, 505.
39 *Among other sports:* John R. MacMahon, *The Wright Brothers: Fathers of Flight* (Boston: Little, Brown & Co., 1930), 41.
40 *They made plans for him:* MW, "Wilbur Wright Born in Henry County," *New Castle (Ind.) Daily Times*, 6/11/1909, scrapbooks, WBP, LC.
40 *classes in trigonometry and Greek:* Fred Howard, *Wilbur and Orville: A Biography of the Wright Brothers* (New York: Knopf, 1987), 5.

40 *But an accident intruded:* There is no contemporary account of the injury except MW's brief diary reference (see next note), though even that entry was recorded months after the fact, and in any case, MW was away from home at the time of the injury. Much later, MW described the injury and the "nervous palpitations of the heart" briefly in a newspaper article, "Wilbur Wright Born in Henry County," *New Castle (Ind.) Daily Times*, 6/11/1909, scrapbooks, WBP, LC. Tom D. Crouch notes the uncertainty of the timing, in *The Bishop's Boys: A Life of Wilbur and Orville Wright* (New York: Norton, 1989), 74–75. John R. McMahon, whose account was based on interviews with family members, not including WW, describes the accident and mentions "a long period of delicate health if not semi-invalidism, with a diet confined to liquids, eggs and toast. It seemed to every one that the boy was handicapped for life," *The Wright Brothers: Fathers of Flight* (Boston: Little, Brown & Co., 1930), 49–50. Fred C. Kelly, who interviewed OW extensively in later life, refers to "a heart disorder from which he did not completely recover for several years," *The Wright Brothers* (New York: Harcourt, Brace, 1943), 26.

40 *"Wilbur's health was restored":* Diary entry of 12/31/1886, Bishop Milton Wright, *Diaries: 1857–1917* (Dayton, Ohio: Wright State University Libraries, 1999), 261.

40 *"might be time and money wasted":* WW to MW, 9/12/1894, FC, WBP, LC.

40 *"a declining, not a suffering invalid":* MW, "Wilbur Wright Born in Henry County," *New Castle (Ind.) Daily Times*, 6/11/1909, scrapbooks, WBP, LC.

40 *catch her breath:* Lorin Wright to MW, 11/23/1885, FC, WBP, LC.

41 *She was very bright:* Susan Wright's intelligence, skill at math, McMahon, *Wright Brothers: Fathers of Flight*, 15–16, 22–23.

41 *"adapting household tools":* Kelly, *Wright Brothers*, 26.

41 *"with a faithfulness":* MW, "Wilbur Wright Born in Henry County," *New Castle (Ind.) Daily Times*, 6/11/1909, scrapbooks, WBP, LC.

41 *"What does Will do?"* LW to KW, 11/12/1888, FC, WBP, LC.

41 *helped his father with church business:* Howard, *Wilbur and Orville*, 5.

41 *"perhaps nearly equaled":* MW, "Wilbur Wright Born in Henry County," *New Castle (Ind.) Daily Times*, 6/11/1909, scrapbooks, WBP, LC.

41 *Traveling in Europe:* Ivonette Wright Miller, comp., *Wright Reminiscences*.

42 *After eight years:* Battle between Brethren "Radicals" and "Liberals," Crouch, *Bishop's Boys*, 51–52, 60–69, 78–82; Elliott dissertation, "Bishop Milton Wright . . . ," 186–252.

43 *"They try to smile":* WW to MW, 10/10/1892, 10/10/1892, FC, WBP, LC. For an example of WW's early argumentative style, see his pamphlet, "Scenes in the Church Commission," 1888, file 4, box 8, WBC, WSU. A characteristic passage expresses Wilbur's view of the Church Commission's refusal to meet in public. "We were, of course, greatly disappointed, but we could offer no reasonable objection, for it seemed entirely proper, and indeed fitting, that a body meeting to legislate secrecy in, should also legislate in secret."

44 *When he was seventeen:* OW's printing press, Crouch, *The Bishop's Boys*, 96.

44 *It was a big operation:* Charlotte K. and August E. Brunsman, "Wright & Wright, Printers: The 'Other' Career of Wilbur and Orville," *Printing History*, vol. X, no. 1, 1988, 2.

44 *Soon he started a small:* Wright brothers as printers, Crouch, *The Bishop's Boys*, 93–103.

44 *Dayton had become:* Dayton as a center of manufacturing and invention, Mark Bernstein, *Grand Eccentrics: Turning the Century: Dayton and the Inventing of America* (Wilmington, Ohio: Orange Frazer Press, 1996), 15–18.

45 *But it was a great craze:* Robert A. Smith, *A Social History of the Bicycle* (New York: American Heritage Press, 1972), 17–47.

46 *"I have been thinking":* Renewed college plan, WW to MW, 9/12/1894, FC, WBP, LC.

46 *"Yes," he replied:* MW to WW, 9/15/1894, FC, WBP, LC.

47 *"Learn all you can about housework":* MW to KW, 10/15/1887.

47 *"You have a good mind":* MW to KW, 5/30/1888, FC, WBP, LC.

48 *"But for you," Milton told:* MW to KW, 8/9/1899, FC, WBP, LC.

48 *She referred to it in a letter:* KW to Henry J. Haskell, 6/16/1925, Katharine Wright Haskell papers, Western Historical Manuscript Collection, University of Missouri-Kansas City (hereafter KWH papers.)

48 *She persuaded herself:* KW to Henry J. Haskell, 6/16/1925, KWH papers.

48 *"was briefly attracted by one of the girls":* John R. McMahon, *The Wright Brothers: Fathers of Flight* (Boston: Little, Brown and Co., 1930), 40–41.

48 *"Orv used to say it was up to Will":* Charles E. Taylor, as told to Robert S. Ball, "My Story of the

Wright Brothers," *Collier's Weekly*, 12/25/1948, reprinted in Jakab and Young, eds., *Published Writings*, 289–90.

48 *"there was a reason"*: KW to Henry J. Haskell, 6/22/1925, KWH papers.

49 *"putting up the swings"*: Thomas R. Coles, "The 'Wright Boys' as a School-Mate Knew Them," *Out West*, January 1910, 36–38 [typescript copy in E.I. #186, "Greenfield Village Buildings—Wright Cycle Shop," collections of Henry Ford Museum & Greenfield Village Research Center.

49 *Raised in the suburbs*: Lilienthal's background and early experiments, introductory essay by Gustav Lilienthal in Otto Lilienthal, *Birdflight as the Basis of Aviation* (London: Longmans, Green, and Co., 1911), xi–xxiv.

49 *the act was photographed*: Lilienthal photographs as innovation, John D. Anderson, Jr., *Introduction to Flight: Its Engineering and History* (New York: McGraw-Hill, 1978), 10.

50 *"No one can realize"*: Article about Lilienthal, "Vernon," "The Flying Man: Otto Lilienthal's Flying Machine," *McClure's Magazine*, September 1894, 323–31.

51 *"It is a difficult task"*: Otto Lilienthal, "Practical Experiments in Soaring," *Annual Report of the Smithsonian Institution for the Year Ending June 30, 1893* (Washington, D.C.: Government Printing Office, 1894), 197.

52 *"like a prophet crying"*: WW, "What Mouillard Did," in Published Writings, 172.

52 *"O! blind humanity!"*: Mouillard, "The Empire of the Air: An Ornithological Essay on the Flight of Birds," *Annual Report of the Smithsonian Institution, 1893*, 397.

52 *"must thoroughly know"*: Mouillard, "Empire of the Air," 413.

52 *"The evolution of the flying machine"*: Lilienthal, "Practical Experiments in Soaring."

52 *"I feel well convinced"*: Mouillard, "Empire of the Air," 398.

52 *"continual practice"*: Lilienthal, "Practical Experiments in Soaring."

53 *Octave Chanute, too*: Octave Chanute, *Progress in Flying Machines* (New York: *The American Engineer and Railroad Journal*, 1894), 257.

54 *"rocked its body"*: WW testimony, appearing as Appendix G, *Aeronautical Journal*, July-September 1916, 118.

54 *"it tilted so that"*: WW testimony, appearing as Appendix G, *Aeronautical Journal*, July–September 1916, 118.

54 *Gear idea for shifting angles of wings*: OW deposition, 1/13/20, McFarland, ed., *Papers*, vol. 1, 8.

55 *cardboard box*: Deposition of OW, 1/13/1920, McFarland, ed., *Papers*, vol. 1, 8.

55 *Will built a kite*: OW deposition, McFarland, ed., vol. 1, *Papers*, vol. 1, 10–11.

## *Chapter Three*: "Some Practical Experiments"

57 *Samuel Langley enjoyed*: J. Gordon Vaeth, *Langley: Man of Science and Flight* (New York: Ronald Press Co., 1966), 61.

57 *" 'dim religious light' "*: Abbot, *Adventures in the World of Science*, 31; SPL to Richard Rathbun, 3/30/1900, box 56, RU 31, SIA.

58 *"it is expected to be"*: SPL, "Preliminary Report of Progress of Work on Account of the Board of Ordnance & Fortification, from January 1, 1899 to October 14, 1899," box 96, RU 31, SIA. Langley recorded further details of work completed to this date in his entry of 10/8/1899, Aerodromics 26, Private Wastebook 1898–1903, NASM archive.

58 *Then Langley led his guests*: BOF's inspection of 11/21/1899, Charles Manly's waste book entry, 11/21/1899, in Meyer, ed., *Langley's Aero Engine of 1903*, 55.

58 *With his original allotment*: SPL learns of actual 1898 allotment, SPL, entry of 10/8/1899, Aerodromics 26, Private Wastebook 1898–1903, NASM archive, SI.

58 *By January, the original*: BOF allotment runs out and is replenished, Richard Rathbun to SPL, 1/12/1900, box 45; I. N. Lewis to SPL, 1/22/1900, box 43, RU 31, SIA.

59 *"practically no means"*: Memorandum, Manly to SPL, 1/15/1900, in Meyer, ed., *Langley's Aero Engine of 1903*, 59.

59 *"integrity of purpose"*: Manly to SPL, 2/24/1900, box 45, RU 31, SIA.

59 *"I have had to practically guarantee"*: Manly to Balzer, 1/10/1900, in Meyer, ed., *Langley's Aero Engine of 1903*, 57.

59 *"was often unfairly impatient"*: C. G. Abbot, "Samuel Pierpont Langley," *Smithsonian Miscellaneous Collections*, vol. 92, no. 8 (Washington, D.C.: Smithsonian Institution, 1934), 2.

59 *"Same old nervous driving energy"*: Abbot, *Adventures in the World of Science*, 14.

59 *"Do you know why the Secretary never married?"*: Cyrus Adler, *I Have Considered the Days* (Philadelphia: Jewish Publication Society of America, 1941), 250.
60 *The secretary wanted work done*: SPL's high standards of craftsmanship, Tom D. Crouch, *A Dream of Wings: Americans and the Airplane, 1875–1905* (Washington, D.C.: Smithsonian Institution Press, 1981), 147–48.
60 *"William, my hat!"*: C. G. Abbot, "Samuel Pierpont Langley," *Smithsonian Miscellaneous Collections*, vol. 92, no. 8 (Washington D.C.: Smithsonian Institution, 1934), 3.
60 *"I used to have cold shivers"*: Abbot, *Adventures in the World of Science*, 27.
60 *"It's correct already"*: Abbot, *Adventures in the World of Science*, 27.
60 *Virtually no one but Langley*: SPL in his own home, Andrew Dickson White, "Samuel Pierpont Langley," in "Samuel Pierpont Langley," *Smithsonian Miscellaneous Collections*, part of Vol. XLIX (Washington: Smithsonian Institution, 1907), 23–24. Little evidence survives to shed light on Langley's home in northwest Washington. In this memorial after Langley's death, White said: "[H]e lived in remarkable seclusion, not accessible in his own home save to very few."
61 *"our friends the reporters"*: SPL to W. Hallett Phillips, box 17, RU 7003, SIA.
61 *By his order only three*: SPL's keys to the Castle, Richard Rathbun to SPL, 12/1/1903, box 56, RU 31, SIA.
61 *"a sort of holy place"*: Abbot, *Adventures in the World of Science*, 32.
61 *he often displayed the awkwardness*: Abbot, *Adventures in the World of Science*, 14, 32.
61 *"not only revered, but loved"*: White, Andrew Dickson, "Samuel Pierpont Langley," in "Samuel Pierpont Langley," *Smithsonian Miscellaneous Collections*, Vol. XLIX (Washington, D.C.: Smithsonian Institution, 1907), 23–24.
61 *"shell of hauteur"*: C. G. Abbot, "Samuel Pierpont Langley," *Smithsonian Miscellaneous Collections*, vol. 92, no. 8 (Washington D.C.: Smithsonian Institution, 1934), 2.
61 *"his longing for friendship"*: Helen Waldo Burnside to Cyrus Adler, 2/12/1907, box 22b, RU 7003, SIA.
61 *Bell's closest friend*: Robert V. Bruce, *Bell: Alexander Graham Bell and the Conquest of Solitude* (Ithaca, N.Y.: Cornell University Press, 1973), 313.
61 *"a staunch friend," "a great man"*: Abbot, *Adventures in the World of Science*, 15–16.
61 *Langley offered to house*: Sarah F. I. Goode to SPL [undated copy], box 22b, RU 7003, SIA.
61 *the secretary quietly cut up*: Abbot, *Adventures in the World of Science*, 16.
61 *"his habit, when he once trusted a man"*: remarks of Charles Walcott in untitled manuscript report of remarks made at Smithsonian Institution upon SPL's death, 3/1/1906, box 14, RU 7003, SIA.
62 *His aides made Langley's quest*: Aides' commitment to SPL's aerodrome enterprise; concern for SPL's and Smithsonian's reputation, Adler, *I Have Considered the Days*, 187–88.
62 *"with the utmost speed"*: SPL, "Experiments With the Langley Aerodrome," *Annual Report of the Smithsonian Institution, 1905* (Washington, D.C.: Government Printing Office, 1905), 114.
62 *"strain every point"*: Balzer to Manly, in Manly to SPL, 7/9/1899, in Meyer, *Langley's Aero Engine of 1903*, 41.
62 *Manly spoke up for Balzer*: Manly to SPL, 2/21/1900, box 45, RU 31, SIA.
63 *"I desire to say"*: Stephen Balzer to SPL, 3/1/1900, box 11, RU 31, SIA.
63 *"very uneasy"*: Manly to Balzer, 3/20/1900, in Meyer, ed., *Langley's Aero Engine of 1903*, 61.
63 *"almost as much superior in skill"*: SPL to Robert Ridgway, 3/29/1900, reprinted in *Langley Memoir on Mechanical Flight*, pt. 2, 285–89.
63 *Langley watched the great carrion-eaters*: SPL's study of the Jamaican turkey vulture, SPL's "Notes on the soaring of the 'John Crow' made principally during a journey to Jamaica, in March, 1900," box 40, RU 31, SIA.
64 *Specimens of "John Crow"*: SPL to Richard Rathbun, 3/30/1900, box 56, RU 31, SIA; R. Ridgway to SPL, 5/26/1900, box 40, RU 31, SIA; Abbot, *Adventures in the World of Science*, 21.
64 *"I have been noting this ability to guide"*: SPL to Manly, 4/16/1900, reprinted in *Langley Memoir on Mechanical Flight*, pt. 2, appendix, 293.
64 *He was the son of learned*: "Octave Chanute: A Biographical Sketch," *Cassier's Magazine*, March 1898; Tom D. Crouch, *A Dream of Wings: Americans and the Airplane, 1875–1905* (Washington, D.C.: Smithsonian Institution Press, 1981), 22–26.
65 *Chanute had begun to pursue an interest*: Chanute's entrance into aeronautical studies, Chanute, "Experiments in Flying," *McClure's Magazine*, June 1900, 127; Crouch, *A Dream of Wings*, 22–27; WW to MW, 10/24/1901, FC, WBP, LC.
66 *Yet anyone who skimmed*: Samples from Chanute's *Progress in Flying Machines* (New York: The American Engineer and Railroad Journal, 1894), 33, 35–37, 64.

67 *"question can be answered in the affirmative"*: Chanute, *Progress in Flying Machines*, preface.
67 *"only after Lilienthal"*: Chanute, "Experiments in Flying," *McClure's Magazine*, June 1900, 127.
68 *"With every gust of the wind"*: "Men Fly in Midair," *Chicago Tribune*, 6/24/1896.
68 A *second machine*: For Chanute's glider experiments of 1896–97, see Chanute's own "Recent Experiments in Gliding Flight," James Means, ed., *The Aeronautical Annual, 1897* (Boston: W. B. Clarke & Co., 1897); and "Experiments in Flying," *McClure's Magazine*, June 1900, 127–33; and Tom D. Crouch, A *Dream of Wings: Americans and the Airplane, 1875–1905* (Washington, D.C.: Smithsonian Institution Press, 1981), 175–202. Quotations from Chanute's writings in this section are taken from both articles above.
69 *But a letter that came to him*: WW's first letter to OC, 5/13/1900, McFarland, ed., *Papers*, vol. I, 15–19.
71 *Chanute immediately wrote*: OC's first response to WW, 5/17/1900, McFarland, ed., *Papers*, vol. I, 19–21.
71 *"experiments upon experiments"*: SPL to Manly, 4/6/1900, in Meyer, ed., *Langley's Aero Engine of 1903*, 62–63. See also *Langley Memoir on Mechanical Flight*, pt. 2, 218.
72 *Finally, Balzer got the engine*: George Wells diary, entry of 5/15/1900, in Meyer, ed., *Langley's Aero Engine of 1903*, 64.
72 *Balzer's machinists*: George Wells diary, entry of 5/21/1900, in Meyer, ed., *Langley's Aero Engine of 1903*, 65–66.
72 *Manly advanced money*: George Wells diary, entry of 5/30/1900, in Meyer, ed., *Langley's Aero Engine of 1903*, 66.
72 *"very discouraging and exasperating delays"*: George Wells diary, entry of 6/3/1900, in Meyer, ed., *Langley's Aero Engine of 1903*, 66.
72 *"not to over-tax your strength"*: SPL to Manly, 6/15/1900, in Meyer, ed., *Langley's Aero Engine of 1903*, 69; George Wells diary, entry of 6/3/1900, *ibid.*, 66.
72 *"very encouraging" performance*: Manly to SPL, 6/13/1900, in Meyer, ed., *Langley's Aero Engine of 1903*, 67–68.
73 *"was in itself such a conclusion"*: SPL to Manly, 6/15/1900, in Meyer, ed., *Langley's Aero Engine of 1903*, 69.
73 *"that the engine will be successful"*: Manly to SPL, 6/19/1900, Meyer, ed., *Langley's Aero Engine of 1903*, 73.
73 *"The chief object of your visit"*: SPL to Manly, 7/11/1900, Meyer, ed., *Langley's Aero Engine of 1903*, 78.
73 *he finally dashed off notes*: For a description of WW's lost letter to the Kitty Hawk weather station, see Fred C. Kelly, ed., *Miracle at Kitty Hawk: The Letters of Wilbur & Orville Wright* (New York: Farrar, Straus and Young, 1951), 25n. The report of a Wright letter to Myrtle Beach was given by William Tate to his son, E. W. Tate, who told it to a newspaper reporter much later. See "Tar Heel's Courtesy Appealed to Wrights," *Durham Morning Herald*, 5/10/1970, North Carolina Collection Clipping File through 1975, UNC Library, Chapel Hill. See also Stephen Kirk, *First in Flight: The Wright Brothers in North Carolina* (Winston-Salem, N.C.: John F. Blair, 1995), 25–27.
73 *"You would find here"*: William J. Tate to WW, 8/18/1900, in Kelly, *Miracle at Kitty Hawk*, 25–26.
74 *"My imagination pictures things"*: WW to KW, 6/8/1907, FC, WBP, LC. This capacity "to think pictorially and spatially," to have "a vision of what the object or structure should look like and how it will work," is at the center of Peter L. Jakab's analysis of the Wrights' creative process in *Visions of a Flying Machine: The Wright Brothers and the Process of Invention* (Washington, D.C.: Smithsonian Institution Press, 1990), 4–5.
74 *"inability to balance and steer"*: WW, "Some Aeronautical Experiments," McFarland, ed., *Papers*, vol. 1, 99–100. This account of the theory and calculations that entered into the design of the 1900 glider is drawn from WW to OC, 11/16/1900, McFarland, ed., *Papers*, vol. 1, 40–44; Nick Engler, "Lift and Drift: The Story of the Wright Brothers' Wind Tunnel" (Wright Brothers Aeroplane Company, 2002; also available at www.wright-brothers.org); Tom D. Crouch, *The Bishop's Boys: A Life of Wilbur and Orville Wright* (New York: Norton, 1989), 175–180; and Jakab, *Visions of a Flying Machine*, 63–82. The author is grateful to Nick Engler for assistance and clarification.
76 *"We are getting even"*: KW to MW, 8/22/1900, FC, LC.
76 *"added one more year"*: KW to MW, 9/5/1900, FC, LC.
76 *"I am intending to start"*: WW to MW, 9/3/1900, Kelly, *Miracle at Kitty Hawk*, 27.

77 *Will promised he would be careful:* KW in 1900, WW to KW, 9/20/1908, McFarland, ed., *Papers*, vol. 2, 925–27.
77 *"It is neither necessary nor practical"*: quoted in Valerie Modman, *The Road to Kitty Hawk* (Alexandria, Va.: Time-Life Books, 1980), 68.
78 *his application was denied:* Hiram S. Maxim, letter to *The Times* of London, "Flying Machines," 6/16/1896.
78 *Hearing Langley tell:* SPL's meeting with Hiram Maxim, SPL, memorandum, 7/10/1900, in Meyer, ed., *Langley's Aero Engine of 1903*, 77.
78 *But in Paris, de Dion:* SPL's meeting with Comte Albert de Dion, SPL, memorandum of 7/30/1900, in Meyer, ed., *Langley's Aero Engine of 1903*, 79.
78 *Large reputations weighed:* Abbot, *Adventures in the World of Science*, 19.
78 *"In view of the advice"*: SPL to Manly, 7/30/1900, in Meyer, ed., *Langley's Aero Engine of 1903*, 80.
78 *and tested the engine:* Manly to SPL, 8/21/1900, in Meyer, ed., *Langley's Aero Engine of 1903*, 80–83.
79 *"I should be disposed to begin afresh"*: SPL to Richard Rathbun, 9/3/1900, box 40, RU 31, SIA.
79 *"I regret that you have:* Richard Rathbun to SPL, 8/14/1900, box 56, RU 31, SIA.
79 *As Will arrived:* The great hurricane of 1900 is recounted in Erik Larson, *Isaac's Storm: A Man, A Time, and the Deadliest Hurricane in History* (New York: Crown, 1999).
79 *"I supposed you knew"*: WW to MW, 9/9/1900, FC, LC.
80 *Finally Will found:* WW's 1900 boat trip to Kitty Hawk, "Fragmentary Memorandum by Wilbur Wright," circa 9/13/1900, *Papers*, vol. 1, 23–25.
80 *The Tates lived:* WW to MW, 9/23/1900, McFarland, ed., *Papers*, vol. 1, 25–27.
80 *"about as desolate a region"*: OC, "The Wright Brothers' Flights," *The Independent*, 6/4/1908, 1287–88.
80 *Now, three weeks later:* William J. Tate, "With the Wrights at Kitty Hawk: Anniversary of First Flight Twenty-Five Years Ago," in Jakab and Young, eds., *Published Writings*, 280.
81 *On Wednesday, October 3:* This account relies on John R. McMahon, *The Wright Brothers: Fathers of Flight* (Boston: Little, Brown and Company, 1930), 83–85. After conversations with Orville Wright some twenty years after the fact, McMahon reported that Orville also boarded the glider on the first day of trials. But Tom Crouch and Peter L. Jakab suspect this claim. In their Kitty Hawk trials of 1901, when the Wrights began to make exact records of every flight, including the operator, Wilbur was the operator in every recorded gliding experiment except one flight on July 29, 1901. It seems likely—though not certain—that Orville began to share the operator's work fully only in 1902. Thus it seems unlikely that he made a flight in 1900. This account also draws on the author's experience in assisting with tests of Nick Engler's authentic replica of the 1900 glider, with Dudley Mead as operator, at Jockey's Ridge State Park near Kitty Hawk in September 2000.
82 *The kite had lessons to teach:* Nick Engler to the author, 10/1/2000. The brothers said little about giving up the tower. In 1903, WW received an inquiry from an experimenter named Merrill, who planned to experiment with a "soaring tower." Apparently referring to the Kitty Hawk experiments of 1900, WW told OC: "Our plans for experimenting without forward motion all failed on account of lack of suitable winds. Possibly Mr. Merrill intends to use lighter load per sq. ft. of wing area. If so he may succeed." WW to OC, 9/9/1903, McFarland, ed., *Papers*, vol. 2, 353.
83 *"the strip of beach"*: Arthur Ruhl, "History at Kill Devil Hill," *Collier's Weekly*, 5/30/1908.
83 *"that flight represented only a small part"*: H. H. Brimley, "Market Gunning," reprinted in David Stick, ed., *An Outer Banks Reader* (Chapel Hill: University of North Carolina Press, 1998), 29–32.
83 *"the sand is the greatest thing"*: OW to KW, 10/14/1900, FC, LC.
84 *"No bird soars in a calm"*: WW's Notebook A[1900–] 1901, McFarland, ed., *Papers*, vol. 1, 34–37.
85 *Otto Lilienthal and Octave:* WW's defiance of Lilienthal's and Chanute's warnings about large gliders, Nick Engler to the author, 9/12/2000. Engler said he was especially struck by WW's courage when he, Engler, made glides in his authentic replica of the Wrights' 1900 machine. He noted the experiments of 1900 as characteristic of Wilbur's supreme self-confidence, once he had convinced himself of a truth. For Wilbur, Engler said, "truth trumped authority."
85 *Will had hoped for hours:* WW's glides of 1900, WW, "Some Aeronautical Experiments," *Papers*, vol. 1, 105–07.
86 *"with the experience I have acquired"*: SPL to Rathbun, 9/21/1900, box 40, RU 31, SIA. For SPL's visit to the French Military Balloon Park, see also SPL, memorandum, 9/20/1900, box 40, RU 31, SIA.

86 *now felt entirely convinced:* Manly to Stephen Balzer, 9/28/1900, in Meyer, ed., *Langley's Aero Engine of 1903*, 88–89. See Meyer's footnote; he believed Manly "had devised an erroneous theory to explain the operation of the rotary engine."
86 *Immediately he got:* Manly to SPL, 9/11/1900, in Meyer, ed., *Langley's Aero Engine of 1903*, 86–87.
87 *"either . . . that he will build something":* SPL to Rathbun, 9/29/1900, in Meyer, ed., *Langley's Aero Engine of 1903*, 90–91.
87 *"Manly can build here":* Rathbun to SPL, 10/9/1900, in Meyer, ed., *Langley's Aero Engine of 1903*, 91. Meyer identifies this as a cable from Manly to SPL, but the author is more likely Rathbun.

## *Chapter Four:* "Truth and Error Intimately Mixed"

88 *"my . . . ideas":* "I" versus "we" in WW's correspondence—there is a notable exception to Wilbur's use of the first person singular in the early months of his project. On November 27, 1899, writing to the Instrument Division of the U.S. Weather Bureau to request data on wind velocities, Wilbur wrote, "We have been doing some experimenting with kites, with a view to constructing one capable of sustaining a man. . . . We expect to carry the experiment further next year. . . . We would like to learn if possible . . ." Wilbur wrote the letter on stationery headed "Wright Cycle Co.," whereas his other early correspondence was usually written on plain paper. Possibly it struck him as more credible to imply to the Weather Bureau that this unusual undertaking was a corporate matter rather than an individual quest. WW to Instrument Division, U.S. Weather Bureau, 11/27/1899, McFarland, ed., *Papers*, vol. 1, 12.
89 *"often chided Wilbur":* Fred C. Kelly, ed., *Miracle at Kitty Hawk: The Letters of Wilbur & Orville Wright* (New York: Farrar, Straus and Young, 1951), 8. The statement is Kelly's. He interviewed OW extensively in preparing the only biography that OW authorized, *The Wright Brothers* (New York: Harcourt, Brace & Co., 1943). In *One Day at Kitty Hawk: The Untold Story of the Wright Brothers* (New York: Crowell, 1975), John Evangelist Walsh made a pointed case for WW's primacy, and noted that " 'I' becomes 'we' and 'my' becomes 'our' " in WW's letters after the trip to Kitty Hawk in 1900. See especially Walsh's long footnote, "The Kelly Biography," 251–52. Walsh argued that OW, after WW's death, tried, through his influence on Kelly's book and in other ways, to encourage the inaccurate view that "it was not Wilbur but Orville himself who had supplied all the initial theorizing, the original scientific insight, by means of which the elusive principles of flight had been uncovered." Walsh's case, if not conclusive, is carefully documented and worth considering. He appears to go too far when he says: "Judging from the brothers' subsequent extreme care in maintaining a united image, some sort of pact or agreement may have been made at Kitty Hawk—the advantage to Wilbur of having the close assistance of his brother's considerable mechanical skills and the alert sympathy growing from long association is obvious."
89 *"From the time we were little children":* McFarland, ed., *Papers*, vol. 1, v.
89 *had the habit of solidarity:* For a discussion of forces encouraging unity within the Wright family, see Adrian J. Kinnane, "A House United: Morality and Invention in the Wright Brothers' Home," *Psychohistory Review*, Spring 1988, 367–97.
90 *"I love to scrap with Orv":* "Wright Cycle Shop," Oral history interview with Harold S. and Ivonette Wright Miller, 9/17/1989, 57–58, E.I. #186, Henry Ford Museum & Greenfield Village Research Center.
90 *"No truth is without some mixture of error":* WW to George Spratt, 4/27/1903, GC, WBP, LC.
90 *Their niece, Ivonette:* "Wright Cycle Shop," Oral history interview with Harold S. and Ivonette Wright Miller, 9/17/1989, 58, E.I. #186, box 4, Henry Ford Museum & Greenfield Village Research Center.
91 *Every day at lunch:* Carrie Kayler's memory of WW and cap, Kelly, ed., *Miracle at Kitty Hawk*, 19.
91 *he had somehow mastered the harmonica:* "At the Wright Home . . . ," Oral history with Harold S. and Ivonette Wright Miller, 9/19/1989, 16–17, E.I. #186, Henry Ford Museum & Greenfield Village Research Center.
91 *"placing his hands on the seat of a chair":* MW, "Wilbur Wright Born in Henry County," *New Castle* (Ind.) *Daily Times*, 6/1/1909, scrapbooks, WBP, LC.
91 *Orville's "peculiar spells":* WW to MW, 7/20/1907, FC, WBP, LC.
92 *"Orville never seemed to tire":* Ivonette Wright Miller, comp., *Wright Reminiscences* (Air Ford Museum Foundation, 1978), 2–3.
92 *Orville was perhaps the better mathematician:* "Wright Cycle Shop," Oral history interview with

Harold S. and Ivonette Wright Miller, 9/17/1989, 63, E.I. #186, box 4, Henry Ford Museum & Greenfield Village Research Center.

92 *"Wilbur was good at starting things"*: "Wright Cycle Shop," Oral history interview with Harold S. and Ivonette Wright Miller, 9/17/1989, 63, E.I. #186, box 4, Henry Ford Museum & Greenfield Village Research Center.

92 *"Orv seems to be"*: KW to WW, 6/1/1907, FC, WBP, LC.

93 *"When I learned that you intended"*: WW's letter to Lou Wright regarding Herbert Wright, 6/18/1901, FC, WBP, LC.

94 *"a sort of tournament"*: OC to Edward Huffaker, 6/1/01, Chanute papers, LC.

95 *"extract instruction from its failure"*: OC to WW, 6/29/1901, McFarland, ed., *Papers*, vol. 1, 57–58.

95 *"the leading authority"*: WW to Reuchlin Wright, 7/3/1901, FC, WBP, LC.

95 *"We are of course delighted,"*: WW to OC, 7/6/01, McFarland, ed., *Papers*, vol. 1, 67.

95 *"As the Wrights approached"*: Timing of SPL's trip and talk with Chanute, OC to James Means, 6/23/1901, letterbook, 9/25/1900–1/20/1902, Chanute papers, LC.

96 *a monthly payroll of more than*: Costs of aerodrome staff, Tom D. Crouch, *A Dream of Wings: Americans and the Airplane, 1875–1909* (Washington, D.C.: Smithsonian Institution Press, 1981), 273.

96 *For weeks that spring Manly raced*: Manly to SPL, 6/4/1901, box 45, RU 31, SIA.

96 *"showed conclusively that the balancing"*: Charles M. Manly, *Langley Memoir on Mechanical Flight*, Pt. II, 232.

96 *"engineering knowledge . . . is of less consequence"*: Quoted in Tom D. Crouch, *A Dream of Wings: Americans and the Airplane, 1875–1905* (Washington, D.C.: Smithsonian Institution Press, 1981), 281–82.

97 *"I am now, as I have been"*: Manly to SPL, 6/1/1901, box 45, RU 31, SIA.

97 *"I could not doubt"*: SPL's account, "The Fire Walk Ceremony in Tahiti," appeared first as a letter in *Nature*, 8/22/1901, 397–99, then as an article, with photographs, in *Annual Report of the Smithsonian Institution*, 1901 (Washington, D.C.: Government Printing Office, 1902), 539–44.

99 *"nourished a scientific passion for doubt"*: Henry Adams, *The Education of Henry Adams* (Boston: Houghton Mifflin, 1974), 377.

99 *he wrote a short essay*: SPL's essay, "The Laws of Nature," was read to the Philosophical Society of Washington on May 10, 1902, and appeared in the *Annual Report of the Smithsonian Institution, 1901* (Washington, D.C.: Government Printing Office, 1902), 545–52.

101 *Huffaker was a farm boy*: Crouch, *A Dream of Wings*, 85–86.

101 *"Well, Huffaker is as God made him"*: Cyrus Adler, *I Have Considered the Days*, 251.

102 *"one of the most acute observers"*: SPL introduction to E. C. Huffaker, "On Soaring Flight," *Smithsonian Report*, 1898.

102 *"Their work is done"*: Entries of July 19, 20, 22 and 23, Chanute-Huffaker diary, LC.

102 *"in a mighty cloud"*: War with mosquitos, OW to KW, 7/28/01, FC, WBP, LC.

103 *"The mosquitos have been almost unbearable"*: WW to MW, 7/26/1901, FC, WBP, LC.

103 *Huffaker did not*: WW to OC, 10/18/1901, McFarland, ed., *Papers*, vol. 1, 137–38.

103 *"He is intelligent"*: WW to MW, 7/26/1901, FC, WBP, LC.

104 *"not good for a day's work"*: George Spratt to OC, 10/4/1899.

104 *"nothing . . . but food and clothing"*: Spratt to OC, 10/17/1899, box 45, Chanute papers, LC.

105 *He considered a future*: Spratt's background and ambitions, Spratt to OC, 12/28/1898, box 45; Spratt to OC, 1/28/1902, box 43, Chanute papers, LC; Spratt to WW, 4/23/1903, GC, WBP, LC.

105 *"saying if the Lord intended"*: Spratt to OC, 10/17/1899, box 45, Chanute papers, LC.

105 *"something that is a remarkably close second"*: Spratt to OC, 1/28/1902, box 43, Chanute papers, LC.

107 *"To the person who had never attempted"*: Testimony of WW, 2/15/1912, reprinted as Appendix G, *The Aeronautical Journal*, July–September 1916, 119. My reports of the sensations associated with launching and piloting Wright gliders is based on my experiences while assisting the flight experimenter and historian Nick Engler on the Outer Banks in September 2000 and October 2001. Engler's authentic reproductions of the Wrights' aircraft, painstakingly created on the basis of years of research, allow one to experience what the Wrights went through down to minute details.

108 *"Mr. Huffaker remarked"*: WW to MW, 7/26/1901, FC, WBP, LC.

108 *"the very fix"*: OW to KW, 7/28/01, McFarland, ed., *Papers*, vol. 1, 75.

108 *Orville took a turn*: WW's diary A, 7/29/01, in McFarland, ed., *Papers*, vol. 1, 76. Wright biogra-

phers have said that Orville Wright made no flights at Kitty Hawk until 1902. But the entry for July 29, 1901, in the Chanute-Huffaker diary notes "Orville Wright in the machine" in the fourth glide attempt of the day. Chanute-Huffaker diary, KW's copy, FC-LC.

109 *"The balancing of a gliding"*: WW, "Some Aeronautical Experiments," in McFarland, *Papers*, vol. 1, 101.

109 *their plan had been simple and effective*: Wrights' understanding of aerodynamic forces and approach to glider design, Peter L. Jakab, *Visions of a Flying Machine: The Wright Brothers and the Process of Invention* (Washington, D.C.: Smithsonian Institution Press, 1990), 63–82.

111 *The list of negatives was longer*: WW's Diary A, 7/30/1901, McFarland, *Papers 1*, 77–78.

111 *"The Wrights have no high regard"*: Chanute-Huffaker diary, 7/29/1901.

111 *"I took it as a joke"*: WW to George A. Spratt, McFarland, *Papers*, vol. 1, 118–19.

111 *On August 8 the wind*: Chanute-Huffaker diary, entry of 8/11/1901.

112 *"The control of the machine"*: WW, "Some Aeronautical Experiments," McFarland, ed., *Papers*, vol. 1, 111–12.

112 *A similar landing sent him tumbling*: Huffaker recorded that sharp turns occurred in flights of August 9, though WW did not mention them in his own diary.

112 *"Upturned wing seems to fall behind"*: McFarland, ed., *Papers*, vol. 1, 82.

112 *"a very unlooked for result"*: WW to OC, 8/22/01, *Papers*, vol. 1, 84.

113 *"Truth and error were everywhere so intimately mixed"*: Testimony of Wilbur Wright, 2/15/1912, reprinted as Appendix G, *The Aeronautical Journal*, July–September 1916, 103.

113 *they watched a buzzard sail*: Chanute-Huffaker diary, 8/15/1901.

113 *"He looked rather sheepish."*: WW to George A. Spratt, 9/21/1901, McFarland, ed., *Papers*, vol. 1, 118.

113 *"We doubted that we would ever resume"*: Testimony of Wilbur Wright, 2/15/1912, reprinted as Appendix G, *The Aeronautical Journal*, July–September 1916, 120. Wilbur also recalled his pessimistic prediction of 1901 in a speech to the Aero Club of France in 1908. See McFarland, *Papers*, vol. 2, 934–35. For Will's cold, see KW to MW, 8/26/01, *Papers*, vol. 1, 84.

113 *"the present status of the work"*: I. N. Lewis to SPL, 9/20/1901, box 96, RU 31, SIA.

113 *"personnel . . . had almost entirely changed"*: Charles Manly to SPL, 9/27/1901, box 96, RU 31, SIA.

114 *The slow schedule was*: SPL's reports on aerodrome to Board of Ordnance and Fortification in fall 1901, draft, SPL to I. N. Lewis, 10/1/1901; "Report of Progress of Aerodromic Work on Account of the Board of Ordnance and Fortification from January 1st, 1899 to October 1, 1901," box 96, RU 31, SIA.

114 *He had other sources*: Additional funding for aerodrome work after BOF money runs out, AGB to SPL, June 1898, box 131, AGB papers, LC; Robert B. Meyer, Jr., ed., *Langley's Aero Engine of 1903* (Washington, D.C.: Smithsonian Institution Press, 1971), 107–08.

114 *Manly tested a new set*: Manly's failed test of water jackets, Meyer, ed., *Langley's Aero Engine of 1903*, 107.

## *Chapter Five*: "The Possibility of Exactness"

115 *"I think you have performed"*: WW to OC, 8/22/1901, OC to WW, 8/23/1901, McFarland, ed., *Papers*, vol. 1, 83–84.

115 *In fact, the old engineer*: For Chanute's failure to understand wing-warping, see Tom D. Crouch, *The Bishop's Boys: A Life of Wilbur and Orville Wright* (New York: Norton, 1989), 200–201.

116 *Chanute conveyed the invitation*: OC invites WW to speak, OC to WW, 8/29/01, McFarland, ed., *Papers*, vol. 1, 91.

116 *"Will was about to refuse"*: KW to MW, 9/3/01, McFarland, ed., *Papers*, vol. 1, 92.

116 *"After your kindness"*: WW to OC, 9/2/01, McFarland, ed., *Papers*, vol. 1, 92.

117 *"May they make it"*: OC to WW, 9/5/01, WW to OC, 9/6/01, McFarland, ed., *Papers*, vol. 1, 93.

117 *"We don't hear anything"*: KW to MW, 9/3/01, McFarland, ed., *Papers*, vol. 1, 92. KW's full remark was, "We don't hear anything but flying machine and engine from morning til night." The engine she referred to was an engine being installed in the back room of the bicycle shop, not an engine for a flying machine.

118 *"The basis of Professor Newcomb's character"*: William Alvord, "Address of the Retiring President of the Society, in Awarding the Bruce Medal to Professor Simon Newcomb," *Publications of the Astronomical Society of the Pacific*, 4/2/1898.

119 *Newcomb posited two*: See Bruce, *Bell*, 431, on the flaw in Newcomb's reasoning.

119 *"The first successful flyer"*: Simon Newcomb, "Is the Airship Coming?" *McClure's Magazine*, September 1901, 432–35.
120 *"Those who ventured"*: Chanute's introduction, "Some Aeronautical Experiments," *Journal of the Western Society of Engineers*, December 1901, reprinted in Jakab and Young, eds., *Published Writings*, 114.
120 *He had taken the train:* WW before Western Society of Engineers presentation, WW to MW, 10/24/1901; KW to MW, 9/25/1901, FC, WBP, LC.
120 *"clothes do make the man"*: KW to MW, 9/11/1901, FC, WBP, LC.
121 *"When this one feature has been worked out"*: Marvin McFarland said the brothers would have toned down their optimism had the speech been made a few months later, when they realized that important problems—particularly that of propellers—remained to be explored and solved. McFarland, ed., *Papers*, vol. 1, 100n.
121 *He led the engineers:* WW's September 1901 address to Western Society of Engineers, no text of the spoken address survives. WW edited the speech before it appeared as an article in the *Journal of the Western Society of Engineers*, December 1901, 489–510, and edited it again for its appearance in the *Annual Report of the Board of Regents of the Smithsonian Institution, 1902* (Washington, D.C.: Government Printing Office, 1903), 133–48. Jakab and Young, eds., *Published Writings*, includes the Smithsonian version of "Some Aeronautical Experiments," along with Chanute's introductory remarks. McFarland, ed., *Papers*, vol. 1, 99–118, presents the version that appeared in the *Journal of the Western Society of Engineers*.
123 *"His faith in Lilienthal's tables"*: WW to George A. Spratt, 9/21/1901, McFarland, ed., *Papers*, vol. 1, 119.
123 *"I am amused with your apology"*: OC to WW, 11/27/1901, McFarland, ed., *Papers*, vol. 1, 164.
123 *"a devilish good paper"*: OC to WW, 11/27/1901, McFarland, ed., *Papers*, vol. 1, 168.
123 *Chanute urged Will:* OC offers encouragement and financial aid, OC to WW, 10/20/1901, McFarland, ed., *Papers*, vol. 1, 139.
123 *"If we did not feel that the time"*: WW to OC, 10/24/1901, McFarland ed., *Papers*, vol. 1, 142.
123 *Here Orville voiced:* OW's caution about publishing claims, Kelly, *The Wright Brothers*, 73–74.
125 *"We have been experimenting"*: WW to MW, 10/24/1901, FC, WBP, LC.
125 *Kate found the house:* KW in fall 1901, KW to MW, 10/2/1901, FC, WBP, LC.
125 *"The boys are working every night"*: KW to MW, 10/2/1901, 10/12/1901, FC, WBP, LC.
126 *Above the door:* Wrights' bicycle shop. The description is based largely on documents at the Henry Ford Museum & Greenfield Village Research Center, including "The Wright Cycle Company Report: Research Guides"; oral history interviews with Ivonette Wright Miller and Harold Miller, 7/6/1984 and 9/17/1989; "Interview with Mae Voris Regarding the Wright Cycle Shop," 7/19/1984; Susan [Bushouse] to John Wright, 5/8/1984; untitled draft press release (c. 1937); and several typescript reports by museum staff, including "Business at 1127 West Third," "Equipment History of the First Wright Shop," and an untitled memorandum, 5/3/1984, by researcher Susan Bushouse. See also nephew Milton Wright's memories of the shop in Ivonette Wright Miller, *Wright Reminiscences*, 68. For a detailed history of the building that housed the bicycle shop, see Allan Fletcher, "The Wright Brothers' Home and Cycle Shop In Greenfield Village," master's thesis in museum practice, University of Michigan, 1972. Wright researchers have disagreed over the location of the wind tunnel. Charles Taylor, who worked most closely with them, recalled in 1948 that the Wrights did most of their experimenting in the upstairs rooms; see Taylor's "as-told-to" article written with Robert S. Ball, "My Story of the Wright Brothers," *Collier's Weekly*, 12/25/1948, reprinted in Jakab and Young, *Published Writings*, 286. Susan Bushouse Foran, who did thorough research on the Wrights' shop equipment for a restoration at Greenfield Village in the 1980s, believed the wind tunnel experiments probably were conducted in an upstairs room.
127 *"We had taken up aeronautics"*: WW and OW, "The Wright Brothers Aeroplane," *Century Magazine*, September 1908, reprinted in Jakab and Young, eds., *Published Writings*, 28.
127 *"a multitude of variations"*: WW and OW, "The Wright Brothers Aeroplane," *Century Magazine*, September 1908, reprinted in Jakab and Young, *Published Writings*, 28. The names of both Wrights appeared on this article, but Orville was the actual author.
127 *It appears that Will:* Construction of model wing surfaces for wind tunnel, WW to George A. Spratt, 2/14/1902, McFarland, ed., *Papers*, vol. 1, 216; Fred Howard's Appendix II.C., "Wright Model Aerofoils," McFarland, ed., *Papers*, vol. 1, 556–72.
127 *They constructed the balances:* For the wind tunnel and balances, see Nick Engler, "Lift and Drift:

The Story of the Wright Brothers' Wind Tunnel" (Wright Brothers Aeroplane Company, 2002), also available at www.wright-brothers.org; the author is indebted to Engler for sharing his extensive knowledge of the wind tunnel tests, based on his reproduction of the tunnel and the lift and drift balances. On the difficulty of keeping the balances assembled, see Crouch, *Bishop's Boys*, 224–25. Crouch, who attempted the feat with a replica, said "the slightest jar would dislodge the many pins on which the various parts rested, reducing the device to an assortment of bits and pieces on the tunnel floor. Reassembly was exasperating, something akin to building a house of cards; one slip and the entire edifice collapsed." Engler reports that the balances become easier to handle with repetition.

128 *"It is perfectly marvelous"*: OC to WW, 11/18/1901, McFarland, ed., *Papers*, vol. 1, 156.
129 *"little better than guess-work"*: WW and OW, "The Wright Brothers Aeroplane," *Century Magazine*, September 1908, reprinted in Jakab and Young, *Published Writings*, 28.
129 *"The 'monumental work' of a great American"*: WW to OC, 11/2/1901, McFarland, ed., *Papers*, vol. 1, 148.
129 *"#12 has the highest dynamic efficiency"*: WW to OC, 12/15/1901, McFarland, ed., *Papers*, vol. 1, 181.
129 *With that, the brothers*: Ending wind tunnel experiments, WW to George Spratt, 12/15/01, 1/1/02, McFarland, ed., *Papers*, vol. 1, 181–82.
130 *"Of course nothing would give me greater pleasure"*: WW to OC, 12/23/1902, McFarland, ed., *Papers*, vol. I, 187.
131 *The skids occurred*: For explanations of the skidding problem and the original design of the 1902 vertical tail, see Peter L. Jakab, *Visions of a Flying Machine: The Wright Brothers and the Process of Invention* (Washington, D.C.: Smithsonian Institution Press, 1990), 157–59, and Kelly, *Wright Brothers*, 79–80.
131 *"The matter of lateral stability"*: WW to OC, 2/7/1902, McFarland, ed., *Papers*, vol. 1, 212.

## *Chapter Six*: "A Thousand Glides"

132 *"It is conceded"*: *Dayton Daily News* coverage, "Modern Flying Machine a Dayton Product," 1/25/1902.
134 *"Pigeon Flies"*: Alberto Santos-Dumont, *My Airships* (London: Grant Richards, 1904), 19–20.
135 *Alberto, then eighteen*: Santos-Dumont's background, education, and early airship flights, Peter Wykeham, *Santos-Dumont: A Study in Obsession* (London: Putnam, 1962). For a good summary, see Douglas Botting, *The Giant Airships* (Alexandria, Va.: Time-Life Books, 1980), 22–29.
135 *" 'Man flies!' old fellow!"*: Santos-Dumont, *My Airships*, 20–21.
135 *Santos-Dumont was an honored guest*: Roosevelt and Santos-Dumont, "President Would Like to Try Flying Machine," *Washington Times*, 4/17/1902.
136 *"The salvation of one's self"*: MW, "The Spiritual Warfare," *Religious Telescope*, 6/18/1873, quoted in Daryl M. Elliott, "Bishop Milton Wright and the Quest for a Christian America," Ph.D. dissertation, Drew University, 1992, 170.
136 *"There is something rotten"*: WW to MW, 6/2/1897, quoted in Elliot, "Bishop Milton Wright . . . ," 285.
136 *The bishop expected*: MW and Millard Keiter controversy, Elliot, "Bishop Milton Wright . . . ," 282–96.
137 *"inconceivable, incomprehensible and incredible"*: WW to Reuchlin Wright, 5/20/1902, FC, WBP, LC.
137 *"Some members must have felt"*: WW to MW, 2/15/1902, FC, WBP, LC.
137 *"Feel loss of sleep"*: Milton Wright, *Diaries, 1857–1917* (Dayton, Ohio: Wright State University Libraries, 1999), entry of 5/30/1902, 569. See also entries of 4/4/1902 and 4/5/1902, and MW to KW, 9/3/1902, FC, WBP, LC.
137 *"imperious"*: Corydon L. Wood, "Bishop Milton Wright," *The Christian Conservator*, 5/16/1917.
137 *"esteems itself"*: Halleck Floyd, "The Ego," *The Christian Conservator*, 5/24/1904, quoted in Elliot, "Bishop Milton Wright . . ." 306. For Milton's awareness of personal criticisms leveled against him, see also MW to WW, 8/28/1902, FC, WBP, LC.
137 *"Will and Orv seem to think"*: KW to MW, 8/20/1902.
137 *"Not one of that crowd"*: KW to MW, 9/3/1902, FC, WBP, LC.
138 *"my chief regret"*: WW to MW, 2/15/1902, FC, WBP, LC.
138 *"attending to some matters"*: WW to OC, 5/16/1902, McFarland, ed., *Papers*, vol. 1, 233.
138 *"If we go to Kitty Hawk"*: WW to OC, 5/29/1902, McFarland, ed., *Papers*, vol. 1, 233.

138 *He came to Dayton:* Will's instructions to Chanute, WW to OC, 7/9/1902, McFarland, ed., *Papers*, vol. 1, 237.
138 *"When my father and myself"*: WW, "The Church Trial at Huntington," pamphlet, United Brethren Historical Center, Huntington College.
139 *"Will spins the sewing machine"*: KW to MW, 8/20/1902, FC, WBP, LC.
139 *They scissored the fabric:* Parts and process in building the 1902 Wright glider, author's interview with Wright reconstructionist Nick Engler, January 2002; Nick Engler, "Building the 1902 Wright Glider," website of the Wright Brothers Aeroplane Company, www.first-to-fly.com.
140 *"They really ought to get away"*: KW to MW, 8/20/1902, FC, WBP, LC.
140 *"I am sorry that Will"*: KW to MW, 8/31/1902, FC, WBP, LC.
140 *"I always feel"*: KW to MW, 8/26/1902, 9/2/1902, FC, WBP, LC.
140 *"Dad seems worried"*: KW to WW and OW, 9/4/02, FC, WBP, LC.
140 *"the success or failure of the first flight"*: Manly to SPL, 1/1/1901, SPL wastebook entry, NASM archive, SI.
140 *"gave an appearance of grace and strength"*: *Langley Memoir on Mechanical Flight*, pt. 2, 186.
140 *"If one is building bridges"*: *Langley Memoir on Mechanical Flight*, pt. 2, 186.
141 *"Everything in the work has got to be so light"*: SPL, "The 'Flying-Machine,' " *McClure's Magazine*, June 1897, 654.
141 "To determine more accurately": *Langley Memoir on Mechanical Flight*, pt. 2, 197.
141 "Mr. Manly adds that the entire vibration": SPL wastebook entry, 5/12/1901, NASM archive, SI.
141 "Although this wing": *Langley Memoir on Mechanical Flight*, pt. 2, 200.
141 "Although an engine may develop sufficient power": SPL, "Experiments With the Langley Aerodrome," *Annual Report of the Smithsonian Institution, 1904* (Washington, D.C.: Government Printing Office, 1905), 117.
141 "Great difficulties . . . were experienced": *Langley Memoir on Mechanical Flight*, pt. 2, 254.
141 "These engines were to be of nearly double the power": SPL, "Experiments With the Langley Aerodrome," *Annual Report of the Smithsonian Institution, 1904* (Washington, D.C.: Government Printing Office, 1905), 117.
142 *"the work lengthened out"*: Manly wastebook entry, 1/1/1901, NASM archive, SI.
142 *Once that was achieved: Langley Memoir on Mechanical Flight*, pt. 2, 217.
143 *"Considering the experience of ten years"*: SPL's "Memoranda of Captain Lewis, given on his visit of April 26, when he went through the shops," SPL wastebook entry, 4/26/1902, NASM archive, SI.
143 *"first one part and then another"*: Charles Manly to SPL, 8/15/1902, Meyer, ed., *Langley's Aero Engine of 1903*, 120.
143 *"The injury to the propeller"*: SPL to Manly, 8/18/1902, Meyer, ed., *Langley's Aero Engine of 1903*, 120–21.
143 *He had often thought:* SPL's ideas about the possibility of rocket engines for aircraft, Russell J. Parkinson, "Doctor Langley's Paradox: Two Letters Suggesting the Development of Rockets," *Smithsonian Miscellaneous Collections*, vol. 140, no. 3, 8/31/1960.
144 *"I do not expect"*: OC to SPL, 4/26/1902, box 18, RU 31, SIA.
144 *"my brother and myself"*: WW to OC, 7/9/1902, *Papers*, vol. 1, 237–38.
144 *"jealous disposition"*: WW to OC, 9/5/1902, *Papers*, vol. 1, 247–48.
144 *"let him rebuild"*: OC to SPL, 10/21/1902, quoted in McFarland, ed., *Papers*, vol. 1, p. 274n.
144 *the stolid Chanute:* For Chanute's strained relations with Herring, see *Papers*, vol. 1, Appendix IV, 649–52n. For WW's efforts to dissuade Chanute from visiting Kitty Hawk in 1902, see especially WW to OC, 6/2/1902, *Papers*, vol. 1, 235–36.
145 *the work of rebuilding the camp:* WW to KW, 8/31/1902, WW to MW, 8/31/1902, FC, WBP, LC; OW diary, 8/29–9/6/1902, LC; WW to OC, 9/21/1902, *Papers*, vol. 1, 257.
145 *"after a bullet"*: WW to KW, 8/31/1902, FC, WBP, LC.
145 *Their shelves were soon:* OW diary, 8/30/1902, FC, WBP, LC.
145 *"We fitted up"*: WW to Spratt, 9/16/1902, McFarland, ed., *Papers*, vol. 1, 253.
145 *"I never saw men so wrapped up"*: Remarks of John T. Daniels quoted by W. O. Sanders, "Then We Quit Laughing," *Collier's Weekly*, 9/17/1927, reprinted in Peter L. Jakab and Rick Young, *The Published Writings of Wilbur and Orville Wright* (Washington, D.C.: Smithsonian Institution Press, 2000), 275.
146 *"uniform courtesy to everyone"*: W. J. Tate, "I Was Host to Wright Brothers at Kitty Hawk," *U.S. Air Services*, December 1943, reprinted in Jakab and Young, *Published Writings*, 283–85.
146 *"a pair of crazy fools"*: Wrights imitating gulls; remarks of John T. Daniels quoted by W. O.

Sanders, "Then We Quit Laughing," *Collier's Weekly*, 9/17/1927, reprinted in Jakab and Young, *Published Writings*, 275.

146 *"gliding downwards through a rising current"*: WW, "Experiments and Observations in Soaring Flight," reprinted in Jakab and Young, eds., *Published Writings*, 133–34.

146 *the Wrights observed:* Bird-watching, WW, "Experiments and Observations in Soaring Flight," in Jakab and Young, *Published Writings*, 140; OW diary, 9/4/1902, 9/6/1902, LC. The Wrights were observing the effects of the rising columns of heated air later called "thermals."

147 *"There is no question"*: WW, "Experiments and Observations in Soaring Flight," in Jakab and Young, *Published Writings*, 141–42.

147 *"When once it becomes possible"*: WW, "Experiments and Observations in Soaring Flight," in Jakab and Young, *Published Writings*, 137.

148 *Yet the glider weighed:* OW diary, 9/25/1902, WBP, LC.

148 *"It was built to withstand"*: WW, "Experiments and Observations in Soaring Flight," in *Papers*, vol. 1, 319. Peter L. Jakab points out the importance of the gliders' sheer durability to the Wrights' success: It allowed them to develop the art of flying through hundreds of glides, with a minimum of breakage—not to mention injury—and repair. *Visions of a Flying Machine*, 171.

148 *the brothers tested:* OW diary, 9/10/1902; 9/12/1902, LC; OW, "How We Invented the Aeroplane," 18.

148 *"It's a wonder"*: KW to WW and OW, 9/4/1902, FC ,WBP, LC.

149 *"You will have to get used"*: OW to KW, 9/11/02, FC, WBP, LC.

149 *More on their minds:* MW's trial, Elliott, "Bishop Milton Wright . . . ," 296.

149 *the brothers' reactions differed:* OW to KW, 9/7/1902; WW to MW, 9/7/1902, FC, WBP, LC.

149 *"We are not going to let this thing"*: KW to MW, 9/9/1902, FC, WBP, LC.

149 *"We will bear in mind"*: WW to MW, 9/12/1902, FC, WBP, LC.

150 *Will found himself aloft:* WW, "Experiments and Observations in Soaring Flight," in Jakab, *Published Writings*, 134–35.

150 *"The machine flew beautifully"*: OW diary, 9/22/02, LC.

150 *It simply looked more stable:* Author's observation of Engler reproductions.

150 *He began the morning:* OW's early 1902 glides and crash, OW diary, 9/23/1902, LC; WW to OC, 9/23/1902, *Papers*, vol. 1, 261; WW, "Experiments and Observations in Soaring Flight," in Jakab and Young, eds., *Published Writings*, 135–36; OW to KW, 9/29/1902, FC-LC.

151 *in the path of an S:* OW diary, 9/29/1902, LC.

151 *"just as a sledge"*: WW testimony, 2/15/1912, reprinted as Appendix G, *The Aeronautical Journal*, July–September 1916, 121. The Wrights' "well-digging" episodes were the first tailspins. For an especially helpful explanation of the phenomenon, see Harry Combs, *Kill Devil Hill: Discovering the Secret of the Wright Brothers* (Boston: Houghton Mifflin, 1979), 167–68.

151 *look of a sportman's camp:* OW diary, 9/30/1902, *Papers*, vol. 1 266; OW to KW, 10/23/1902, *Papers*, vol. 1, 280.

152 *a perception dawned:* WW testimony, 2/15/1912, reprinted as Appendix G, *The Aeronautical Journal*, July–September 1916, 121; OW, *How We Invented the Airplane: An Illustrated History* (New York: Dover, 1953, 1958), 19.

152 *the tail should be movable:* OW's diary B, 10/3/1902, McFarland, ed., *Papers*, vol. 1, 269. On this page also see note 3. McFarland gives the story of Orville's proposal over breakfast without citing a source, saying only that it was "an amusing story told by Orville Wright in later years." Fred G. Kelly, who interviewed Orville extensively, and Harry Combs, who attributes the tale to Lorin Wright, used variations of McFarland's version, also without citing a written source. Fred Howard, in *Wilbur and Orville*, 87–89, explores the confusing differences in the brothers' later versions of what happened, and rightly notes that the most reliable source is Orville's diary entry of October 3, 1902, which says simply, "While lying awake last night, I studied out a new vertical rudder." Unquestionably he would have presented his idea the next morning. See Kelly, *The Wright Brothers* (New York: Harcourt, Brace, 1943), 81–83; Combs, *Kill Devil Hill*, 168–69.

153 *an effort to overtake Wilbur:* OC to SPL, 10/21/1902, quoted in McFarland, ed., *Papers*, vol. 1, 274n.

153 *"Langley tried the cast iron way"*: OC to James Means, 5/31/1897, box 24, Chanute papers, LC. For background on Herring, see Crouch, *A Dream of Wings*, 172–74, 179–222.

153 *But on only his second try:* Failure of Chanute glider at Kitty Hawk, 1902, OW, Diary B, 10/5–10/11/1902, McFarland, ed., *Papers*, vol. 1, 270–73; OC to SPL, 10/21/1902, box 18, RU 31, SIA.

153 *"no trouble in the control"*: OW, Diary B, 10/10/1902, McFarland, ed., *Papers*, vol. I, 272.

154 *"You may hear some interesting news"*: OC to WW, 10/16/1902, McFarland, ed., *Papers*, vol. 1, 276.
154 *"doubtless . . . got some new and valuable ideas"*: OC to SPL, 10/21/1902, box 18, RU 31, SIA.
154 *Chanute said the Wright brothers*: Chanute's letters to SPL about Wrights, OC to SPL, 10/23/1902, 10/27/1902, 11/21/1902, box 18, RU 31, SIA.
154 *"as it would be exceedingly doubtful"*: WW to SPL, 10/20/1902, box 69, RU 31, SIA.
155 *"All she needs is a coat of feathers"*: Quoted in Kelly, *Wright Brothers*, 80.
155 *"We now hold all the records!"*: OW to KW, 10/23/1902, McFarland, ed., *Papers*, vol. 1, 279–80.
155 *The sensations of the pilot:* Author's interview with Dudley Mead, based on Mead's piloting of Nick Engler's authentic Wright replicas at Jockey's Ridge State Park, North Carolina.
155 *"Unfortunately, unlike many who use"*: OW to Arthur W. Page, 1/30/1923, McFarland, ed., *Papers*, vol. 2, 1135–36.
156 *"the management of a flying machine"*: WW, "Experiments and Observations in Soaring Flight," in Jakab and Young, eds., *Published Writings*, 136.
156 *"A thousand glides is equivalent"*: WW, "Experiments and Observations in Soaring Flight," in Jakab and Young, eds., *Published Writings*, 137.
156 *To imitate the conditions:* Manly's use of "artificially produced air drafts acting on the carburetor," Manly to SPL, 8/29/1902, Meyer, ed., *Langley's Aero Engine of 1903*, 121.
156 *"goodwill and sympathy"*: SPL to Manly, 9/9/1902, Meyer, ed., *Langley's Aero Engine of 1903*, 121.
157 *Then came a "totally unexpected"*: Manly's propeller problems of September 1902, Manly to SPL, 9/17/1902, Manly to SPL 9/26/1902, Meyer, ed., *Langley's Aero Engine of 1903*, 122.
157 *"Experiments with the Langley aerodrome"*: "Flying Machine Cost $50,000," *New York Daily Tribune*, 11/6/1902.
157 *"Mr. Chanute feels that the Wright Brothers"*: SPL memorandum, 12/7/1902, RU 31, box 18, SIA.
158 *"especially of their means of control"*: SPL to OC, 12/7/1902, quoted in McFarland, ed., *Papers*, vol. 1, 290, note 4.
158 *"It is not at all probable"*: Quoted by OC to SPL, 12/21/1902, RU 31, box 18, SIA; WW to OC, 12/11/1902, McFarland, ed., *Papers*, vol. 1, 290.
158 *So they sketched a design:* Design and construction of 1903 engine, Orville Wright, "How We Made the First Flight," *Flying*, December 1913, reprinted in Jakab and Young, eds., *Published Writings*, 40–49; Charles E. Taylor, as told to Robert S. Ball, "My Story of the Wright Brothers," *Collier's Weekly*, 12/25/1948, reprinted in Jakab and Young, eds., *Published Writings*, 287–88; OW to George Spratt, 6/7/1903, McFarland, ed., *Papers*, vol. 1, 313.
159 *"there is a very considerable analogy"*: quoted in Fred Howard, *Wilbur and Orville* (New York: Knopf, 1987), 107.
159 *"We have recently done a little experimenting"*: WW to George Spratt, 12/29/1902, McFarland, *Papers*, vol. 1, 292–93.
159 *"they would get into terrific"*: Propeller problem-solving and design, Charles E. Taylor, as told to Robert S. Ball, "My Story of the Wright Brothers," *Collier's Weekly*, 12/25/1948, reprinted in Jakab and Young, eds., *Published Writings*, 290; OW and WW, "The Wright Brothers' Aeroplane," *Century Magazine*, September 1908, reprinted in Jakab and Young, eds., *Published Writings*, 30; OW, "How We Made the First Flight," *Flying*, December 1913, reprinted in Jakab and Young, eds., *Published Writings*, 41. The authoritative documentary study of the Wrights' work on propellers was prepared by Fred Howard, later a distinguished Wright biographer, and appears as Appendix III, "The Wright Propellers," in McFarland, ed., *Papers*, vol. 1, 594–640. For a thorough analysis, see Harry Combs, *Kill Devil Hill: Discovering the Secret of the Wright Brothers* (Boston: Houghton Mifflin, 1979), 179–86. In the late 1970s, Combs asserted that "there is no way that one can overdramatize what the Wrights accomplished in their propeller design. The idea of developing a formula to incorporate all the information they needed into an analytical form was something that would challenge a modern computer programmer."
160 *"We worked out a theory"*: OW to George Spratt, 6/7/1903, McFarland, ed., *Papers*, vol. 1, 313.
160 *"Permit me to introduce"*: OC to WW, 12/19/1902, McFarland, ed., *Papers*, vol. 1, 291.
160 *"the wonderful progress"*: quoted in Alfred Gollin, *No Longer an Island: Britain and the Wright Brothers, 1902–1909 (Heinemann: London, 1984)*, 26.
161 *"We both liked him"*: WW to OC, 9/19/1903, McFarland, ed., *Papers*, vol. 1, 355.
162 *"the aviator turns"*: *"Locomotion aerienne en amerique," Le Monde Illustre*, 3/28/1903, scrapbooks, WBP, LC.
162 *"very kindly received,"* OC to WW, 4/11/1903, McFarland, ed., *Papers*, vol. 1, 304.
162 *"the great German scientist"*: OC's remarks about the Wrights as covered in French press, Charles

Gibbs-Smith, *The Rebirth of European Aviation, 1902–1908* (London: Her Majesty's Stationery Office, 1974), 56–61.

163 *"with all due charity"*: Gibbs-Smith, *Rebirth of European Aviation*, 60.

163 *"For most of the listeners"*: quoted in Tom D. Crouch, *The Bishop's Boys: A Life of Wilbur and Orville Wright* (New York: Norton, 1989), 252.

164 *"With our experience"*: Ferber's letter to Ernest Archdeacon is quoted in Archdeacon's own article, "Mr. Chanute in Paris," *La Locomotion*, 4/11/1903, reprinted as appendix 4, part B, in McFarland, ed., *Papers*, vol. 1, 654–59. For more on Ferber's reaction, see Gibbs-Smith, *Rebirth of European Aviation*, 64.

164 *"entirely solved"*: quoted in Gibbs-Smith, *Rebirth of European Aviation*, 65–66.

165 *"It is very bad policy"*: WW's 1903 address to the Western Society of Engineers and remarks afterward, "Experiments and Observations in Soaring Flight," *Journal of the Western Society of Engineers*, December, 1903, reprinted in McFarland, ed., *Papers*, vol. 1, 318–335.

166 *His detailed plan of work for early December 1902*: Handwritten memo, "The Upper Shop—South Shed," box 45, RU 7003, SIA.

166 *"innumerable other details"*: SPL, "Experiments with the Langley Aerodrome," *Annual Report of the Smithsonian Institution, 1904* (Washington, D.C.: Government Printing Office, 1905), 118.

168 *"The appearance of the machine"*: M. M. Macomb, "Report," 1/6/1904, in *Langley Memoir on Mechanical Flight*, pt. 2, 277.

168 *He instructed Manly*: SPL's instruction to Manly "that the initial test shall be for a voyage not to exceed something 10 or 12 minutes in length," Manly to SPL, 6/[13]/1903, box 45, RU 31, SIA.

## *Chapter Seven:* "Our Turn to Throw"

169 *Langley began to complain*: SPL's dread of Washington heat, SPL to Richard Rathbun, 7/22/1903, box 22b, RU 7003, SIA.

169 *"The weather is warm, moist and heavy"*: Rathbun to Cyrus Adler, 7/24/1903, box 22b, RU 7003, SIA.

170 *"Although plausible attempts have been made"*: "Its First Flight Soon," *New York Daily Tribune*, 7/16/03.

170 *the spectacular narrative lie*: False report of aerodrome dragging houseboat, "Aeroplane Flies at Last," *New York Sun*, 1/31/1903.

170 *Langley, in Boston*: SPL's newspaper habits in 1903, Cyrus Adler, *I Have Considered the Days*, 257.

171 *"considerable nervous excitement"*: John Manly quoted in *Langley Memoir on Mechanical Flight*, pt. 2, 260.

171 *Beale must have had*: Beale/Manly controversy at Chopawamsic Island, "Langley Outstaying Welcome," *New York Daily Tribune*, 8/5/03; "Professor Langley Ordered Off?" *New York Daily Tribune*, 8/8/1903; Charles Manly to Richard Rathbun, 8/7/1903, box 45, RU 31, SIA.

171 *"the danger of misinterpreting"*: "Privacy for Experiment," *New York Daily Tribune*, 8/11/1903.

171 *"believed to be the first"*: "The Langley Airship," *New York Daily Tribune*, 8/20/1903.

172 *"this was . . . not the great test"*: John Manly quoted in *Langley Memoir on Mechanical Flight*, pt. 2, 261.

172 *"instantly there rushed towards us"*: John Manly quoted in *Langley Memoir on Mechanical Flight*, pt. 2, 261.

172 *This "entirely successful" flight*: Manly's report of successful model flight and handling reporters, Manly to Rathbun, 8/8/1903, box 45, RU 31, SIA.

173 *"Langley's Ship Flies"*: *New York Daily Tribune*, 8/9/1903.

173 *"no adequate idea of the wonder and beauty"*: *Langley Memoir on Mechanical Flight*, pt. 2, 260.

173 *"The situation at the club"*: Manly to Rathbun, 8/12/1903, box 45, RU 31, SIA.

173 *"I must . . . become thoroughly familiar"*: Manly to Rathbun, 8/8/1903, box 45, RU 31, SIA.

174 *"Prof. Langley seems to be having"*: WW to OC, 7/22/1903, *Papers*, vol. 1, 344–46.

174 *"bright idea"*: George Spratt to WW and OW, 11/7/1902, GC, WBP, LC.

174 *"I am anxious to have good grounds"*: Spratt to WW and OW, 2/19/1903, GC, WBP, LC. For an excellent summary of Spratt's experimental work, see William F. Trimble, *High Frontier: A History of Aeronautics in Pennsylvania* (Pittsburgh: University of Pittsburgh Press, 1982), 46–55.

174 *"till you get everything clear"*: WW to Spratt, 12/29/1902, GC, WBP, LC.

174 *"I know that an ounce of fact"*: WW to Spratt, 4/12/1903, GC, WBP, LC.

174 *"solved the mystery of the lift"*: Spratt to WW, 4/15/1903, GC, WBP, LC.

174 *"the keys to the heavens"*: Spratt to WW, 4/23/1903, GC, WBP, LC.

174 *"I cannot get the time"*: Spratt to WW, 4/15/1903, GC, WBP, LC.
175 *"How I envy your ability"*: Spratt to WW, 4/15/1903, GC, WBP, LC.
175 *"The application of the principal"*: Spratt to WW, 4/15/1903, GC, WBP, LC.
175 *"I must confess I am at a loss"*: WW to Spratt, 4/20/1903, GC, WBP, LC.
176 *"If one of you could be spared"*: Spratt to WW, 4/23/1903, GC, WBP, LC.
176 *"Orville and I expect to go"*: WW to Spratt, 4/27/1903, GC, WBP, LC.
177 *"Hay and wheat"*: Spratt to OW and WW, 7/12/1903.
177 *"My machine I nearly completed"*: Spratt to WW and OW, 9/13/1903, GC, WBP, LC.
177 *the conference's elders voted:* MW's expulsion from United Brethren, Daryl Elliot, "Bishop Milton Wright and the Quest for a Christian America," Ph.D. dissertation, Drew University, 1992, 296–303.
177 *" 'hand-car,' a corruption"*: WW to KW, 10/18/1903, *Papers*, vol. 1, 366.
177 *In mid-October a gale:* Nature and storms at Kitty Hawk in 1903, McFarland, ed., *Papers*, vol. 1, 365–67; OW, "How We Made the First Flight," *Flying*, December 1913, reprinted in Jakab and Young, eds., *Publishing Writings*, 40–49; OW to KW, 9/26/1903, OW to KW, 10/12/1903, FC, LC. This letter is published in *Papers*, vol. 1, 356–57, but without the sentence, "Every year adds to our comprehension of the wonders of this place."
178 *They made scores of glides:* Glides with 1902 glider in 1903 season, OW to Charles Taylor, 9/28/1903 and 10/15/1903, General Correspondence, LC; OW diary, 10/3/1903, *Papers*, vol. 1, 359–60; OW to KW, 10/12/1903, FC, LC; OW to MW, 10/15/1903, FC, LC; WW to OC, 10/1/1903, *Papers*, vol. 1, 359; WW to MW, 10/4/1903, FC, LC; OW's diary, 10/21/03, *Papers*, vol. 1, 370–71.
179 *This year their guest:* OW to KW, 10/4/1903, FC, WBP, LC.
179 *he would need more horsepower:* Manly to SPL, 8/12/1903, box 45, RU 31, SIA.
179 *"deeply troubled"*: Copy, SPL to Manly, 8/15/1903, box 45, RU 31, SIA.
179 *Considerable planning and coordination:* Plans for delegation to attend aerodrome trial, Manly to Rathbun, 8/24/1903, box 45, RU 31, SIA.
179 *the elaborate assembly procedure:* Aerodrome assembly drill; Manly's reassurance to SPL, Manly to SPL, 8/22/1903, box 45, RU 31, SIA.
180 *"Seeing things on the spot here"*: SPL to Rathbun, 8/27/1903, box 45, RU 31, SIA.
180 *"in spite of every impediment"*: SPL to Mabel Bell, 9/1/[1903], box 131, AGB papers, LC.
180 *But the engine would not start: Langley Memoir on Mechanical Flight*, pt. 2, 262.
181 *"without any warning"*: Propeller smash-up, copy, Manly to SPL, 9/9/1903, box 45, RU 31, SIA.
181 *"The Usual Accident"*: *New York Daily Tribune*, 9/13/1903.
181 *But the tidewater fogs:* Fog ruins wings; repairs, *Langley Memoir on Mechanical Flight*, pt. 2, 262–64.
181 *"Every storm which came anywhere"*: SPL, "Experiments With the Langley Aerodrome," *Annual Report of the Smithsonian Institution, 1904* (Washington, D.C.: Government Printing Office, 1905), 121.
181 *At 10:00 A.M. on October 7:* "Buzzard a Wreck," *Washington Post*, 10/8/1903.
181 *Manly felt "a sudden shock"*: *Langley Memoir on Mechanical Flight*, 265–66.
182 *George Feight, the Wrights':* Feight's letter to the Wrights about Langley's October failure is mentioned in OW to KW, 11/1/1903, FC, WBP, LC.
182 *"I see that Langley"*: WW to OC, 10/16/1903, *Papers*, vol. 1, 364.
183 *After the Englishman learned:* Alexander's missed opportunity, Gollin, *No Longer an Island*, 36–40.
183 *Dan Tate had left:* Dan Tate's work dispute, OW to KW, 11/1/1903, FC, WBP, LC.
183 *"so cold that we could scarcely work"*: OW, "How We Made the First Flight," *Flying*, December 1913, reprinted in Jakab, *Publishing Writings*, 44.
183 *Will began to take:* OW to MW and KW, 11/19/1903, FC, WBP, LC.
183 *"The new machine will be ready"*: OW to KW, 11/1/1903, FC, WBP, LC.
183 *The next day they began:* Details of 1903 flyer's assembly, OW to MW, 10/15/1903, FC, LC; OW's diary, 10/9–11/6/1903, *Papers*, vol. 1, 362–76.
183 *The propeller shafts tore:* OW diary, 11/5/1903, *Papers*, vol. 1, 376–77; OW, "How We Made the First Flight," *Flying*, December 1913, reprinted in Jakab, *Publishing Writings*, 42–44; OW to MW and KW, 11/15/1903, FC, WBP, LC.
183 *"apprehensions of disaster"*: OC to Spratt, 12/19/1903, Chanute papers, LC.
183 *"I do not believe their machine"*: Spratt to OC, 12/[10?]/1903,
184 *Engineers usually allowed:* Worries about margin of safety in engine design, OW to Charles Taylor, 11/23/1903, *Papers*, vol. 1, 385–87.
184 *"We are now quite in doubt"*: OW to MW and KW, 11/15/1903, FC, WBP, LC.

184 *"just the reverse opinion"*: OW to MW and KW, 11/15/1903, *Papers*, vol. 1, 380–81.
184 *They estimated their odds*: OW to MW and KW, 11/19/1903, FC, WBP, LC.
184 *"he nevertheless had more hope"*: OW to MW and KW, 11/15/1903, *Papers*, vol. 1, 380–381.
184 *"too heavy in front"*: Manly to SPL, 10/7/1903, box 45, RU 31, SIA.
185 *"Dismal if not altogether unexpected failure"*: "Langley Airship Fails," *New York Daily Tribune*, 10/8/1903.
185 *"a crushing blow to his theory"*: "Buzzard a Wreck," *Washington Post*, 10/8/1903.
185 *"any stout boy"*: "Prof. Langley's Bird," *Washington Post*, 10/8/1903.
185 *"Notwithstanding the outcome"*: "The Flying Machine Wreck," *Chicago Tribune*, 10/9/1903.
185 *"the front portion of the machine"*: SPL, "Experiments With the Langley Aerodrome," *Annual Report of the Smithsonian Institution, 1904* (Washington, D.C.: Government Printing Office, 1905), 122–23.
185 *"is in no way affected"*: "Langley Explains Fiasco," *Washington Post*, 10/9/1903.
185 *a major storm wrecked*: Costs of tugboat and watercraft wrecked in storm, SPL, "Experiments With the Langley Aerodrome," *Annual Report of the Smithsonian Institution, 1904* (Washington, D.C.: Government Printing Office, 1905), 121–23; *Langley Memoir on Mechanical Flight*, pt. 2 (Washington, D.C.: Smithsonian Institution, 1911), 269.
185 *Yet there had to be*: Decision to make a second test of the great aerodrome, *Langley Memoir on Mechanical Flight*, pt. 2 (Washington, D.C.: Smithsonian Institution, 1911), 270.
186 *removed one small lug*: *Langley Memoir on Mechanical Flight*, pt. 2 (Washington, D.C.: Smithsonian Institution, 1911), 270.
186 *"Day closes in deep gloom"*: OW diary, 11/20/1903, McFarland, ed., *Papers*, vol. 1, 383–84.
186 *"fix anything"*: OW to Charles Taylor, 11/23, 1903, McFarland, ed., *Papers*, vol. 1,385–87.
186 *"We heated the shafts"*: "How We Made the First Flight," *Flying*, December 1913, reprinted in Jakab and Young, eds., *Publishing Writings*, 44.
186 *On the evening of December 6*: Weather on December 6–7, "Cold Day for Airship," *Washington Post*, 12/8/1903.
187 *Langley participated in the annual meeting*: Minutes of the Smithsonian Institution Regents, 12/8/1903, RU 1, SIA.
187 *"it seemed almost disastrous"*: *Langley Memoir on Mechanical Flight*, pt. 2, 271.
187 *"black bottle without a label"*: "Flying Machine Tried and Failed," *Chicago Tribune*, 12/9/1903.
187 *Manly felt "an extreme"*: Manly's report of aerodrome's failure on 12/8/1903, including reports of Reed, McDonald, and other witnesses, *Langley Memoir on Mechanical Flight*, pt. 2, 271–75. SPL's handwritten account appears in his wastebook entry of 12/8/1903, 376–78, NASM archive, Smithsonian Institution. The exact cause of the aerodrome's collapse has remained a matter of debate ever since. The aeronautical historian Tom D. Crouch offers an authoritative view: "[I]n addition to structural shortcomings and leapfrogged problems, the Langley aerodrome seems to have harbored unrecognized aerodynamic defects. Raymond Bisplinghoff, a leading aeronautical engineer, has singled out the craft as a classic case of aeroelastic or wing-torsional divergence. That is, the flexible wings of the aerodrome were given a corkscrew twist as the center of lifting pressure moved rapidly toward the trailing edge at the moment of launch. The flexibility and enormous wing area of the machine magnified this twisting action until the inadequate wing supports failed. Those who have argued that the great aerodrome could have flown if only Langley had abandoned the catapult are wrong. The machine was structurally and aerodynamically unsound." *A Dream of Wings: Americans and the Airplane, 1875–1905* (Washington, D.C.: Smithsonian Institution Press, 1981), 290–91.
189 *Orville Wright, carrying*: OW tells WW of Langley's 12/8/1903 failure, Kelly, *The Wright Brothers*, 94.
190 *"The power is ample"*: WW to MW, 12/14/1903, McFarland, ed., *Papers*, vol. 1, 392–93.
190 *"Repairs took a day"*: Repairs after 12/17/1903 trial and flights of 12/17/1903. OW's diary entries, 12/15/1903, 12/16/1903, 12/17/1903, McFarland, ed., *Papers*, vol. 1, 394–97; WW to OC, 12/28/1903, McFarland, ed., *Papers*, vol. 1, 401–03; "Statement by the Wright Brothers to the Associated Press," 1/5/1904, McFarland, ed., *Papers*, vol. 1, 409–11; OW to James Calvert Smith, 2/11/1933, McFarland, ed., *Papers*, vol. 2, 1161; OW, "How We Made the First Flight," *Flying*, December 1913, in Jakab and Young, eds., *Published Writings*, 40–49; OW as told to Leslie Quirk, "How I Learned to Fly," *Boys' Life*, September 1914, in Jakab and Young, eds., *Published Writings*, 51–57.
192 *"a flight very modest compared with that of birds"*: OW and WW, "The Wright Brothers' Aeroplane," *Century Magazine*, reprinted in Jakab and Young, *Published Writings*, 30.

192 *"it was nevertheless the first in the history of the world"*: OW, "How We Made the First Flight," *Flying*, December 1913, reprinted in Jakab and Young, *Published Writings*, 47.
192 *The Wrights figured up*: Wrights' total cost for experiments 1900–1903 less than one thousand dollars, Fred C. Kelly, *The Wright Brothers* (New York: Harcourt Brace, 1943), 112.

## Interlude

193 *"SUCCESS FOUR FLIGHTS"*: OW to MW, McFarland, ed., *Papers*, vol. I, 397. For a facsimile of the telegram, see Plate 71 in McFarland, ed., *Papers*, vol. 1.
194 *"FLYING MACHINE SOARS 3 MILES . . ."*: *Virginian-Pilot*, 12/18/1903, scrapbooks, WBP, LC.
194 *At 5:30 that afternoon*: Fred C. Kelly, ed., *Miracle at Kitty Hawk: The Letters of Wilbur and Orville Wright* (New York: Farrar, Straus and Young, 1951), 118. Kelly's account of the telegram arriving at 7 Hawthorn is based on the "vivid" recollection of Carrie Kayler Grumbach in 1948. See also entry of 12/17/1903, Bishop Milton Wright, *Diaries, 1857–1917* (Dayton: Wright State University Libraries, 1999), 599–600.
194 *"FIFTY-SEVEN SECONDS"*: Frank Tunison's rejection of 12/17/03 press release, Fred C. Kelly, ed., *Miracle at Kitty Hawk* (New York: Farrar, Straus and Young, 1951), 118; OW to Sam H. Acheson, 1/22/1937, in Kelly, ed., *Miracle at Kitty Hawk*, 429–31. In his letter to Acheson, Orville Wright recounted the story of the exchange between Lorin Wright and Tunison as he, Orville, had heard it, presumably from Lorin, and Orville used the figure "fifty-nine" seconds. In this narrative, I have used the figure fifty-seven, because that was the number that appeared by mistake in the Wrights' telegram of 12/17/1903, and therefore was the number that Tunison would have read and repeated.
195 *"BOYS REPORT FOUR SUCCESSFUL FLIGHTS"*: KW to OC, 12/17/1903, McFarland, ed., *Papers*, vol. 1, 397.
195 *"a propeller working on a perpendicular shaft"*: "Airship After Buyer," *New York Times*, 12/26/1903.
195 *The phanton propeller*: Errors in press accounts of 12/17/1903 flights, "The Latest Flying Machine," *Boston Transcript*, 12/19/1903; "Flying Machine Soars," *Chicago Tribune*, 12/19/1903; "Besieged," undated, unidentified newspaper article; "Airship Is a Success," undated, unidentified newspaper article; "New Air Ship Flies Against Heavy Wind," undated, unidentified Chicago newspaper article.
195 *They tried to set matters*: Wrights' statement about flights of 12/17/1903 to Associated Press, 1/5/1904, McFarland, *Papers*, vol. 1, 409–11.
196 *a letter from Augustus Herring*: Herring's proposal to "join forces" with Wrights, copy, August Herring to WW and OW, 12/26/1903, GC, WBP, LC.
196 *"felt certain that he [Herring]"*: WW to OC, 1/8/1904, McFarland, ed., *Papers*, vol. 1, 412–13.
196 *He took the interurban trolley*: OW's diary, 1/22/1904, McFarland, ed., *Papers*, vol. 1, 417.
196 *He invited his friend*: SPL's dinner with Cyrus Adler on 12/8/1903, Cyrus Adler, *I Have Considered the Days*, 257.
197 *"certain it is that . . . success"*: *Langley Memoir on Mechanical Flight*, pt. 2, 184.
197 *Manly jotted a soothing note*: Manly's reassuring letter to SPL, quoting Chanute's letter regarding the Wrights' flights of 12/17/03, Charles Manly to SPL, 12/25/1903, box 45, RU 31, SIA.
198 *"demonstrated to the world"*: AGB, address to the National Convention of the Navy League of the United States, 1916, reprinted as "Preparedness for Aerial Defense," *Air Power Historian*, October 1955.
199 *"Langley's aerodrome will do nearly everything except"*: "Prof. Langley's Ill-Luck," *Chicago Tribune*, 12/10/1903.
200 *"You can tell Langley"*: quoted in Tom D. Crouch, *The Bishop's Boys: A Life of Wilbur and Orville Wright* (New York: Norton, 1989), 293.
200 *But Langley refused*: Adler, *I Have Considered the Days*, 259.
200 *"his patience and rare philosophy"*: H. H. A. Beach to Cyrus Adler, box 22b, RU 7003, SIA.
200 *"Brashear, I'm ruined"*: David Fairchild, *The World Was My Garden* (New York: Charles Scribner's Sons, 1938), 332–33.
201 *a number of scientists undertook*: Conversation with Daniel Coit Gilman, Adler, *I have Considered the Days*, 258.
201 *"Samuel," she said*: SPL's conversation with his aunt, Julia Goodrich, Adler, *I Have Considered the Days*, 259.
201 *"immensely pleased at your success"*: OC to WW, 12/18/1903, McFarland, ed., *Papers*, vol. 1, 398.
201 *"As you surmise"*: OC to SPL, 1/1/1904, copy in OC letter books, Chanute papers, L.C.

201 *"I think that this success"*: OC to [Lawrence] Hargrave, 1/9/1904, Chanute papers, LC.

202 *"the uses will be limited"*: OC to F. H. Wenham, 1/9/1904, quoted in Charles H. Gibbs-Smith, *The Rebirth of European Aviation, 1902–1908: A Study of the Wright Brothers' Influence* (London: Her Majesty's Stationery Office, 1974), 99.

202 *"I was much distressed"*: OC to SPL, 1/1/1904, copy in OC letter books, Chanute papers, LC.

202 *"From the beginning"*: Wrights' statement about flights of 12/17/1903 to Associated Press, 1/5/1904, McFarland, *Papers*, vol. 1, 409–11.

202 *"paid the freight"*: WW to OC, 1/18/1904, McFarland, ed., *Papers*, 415–16.

202 *"In the clipping which you sent me"*: OC to WW, 1/14/1904, McFarland, ed., *Papers*, vol. 1, 414–15.

202 *"a somewhat general impression"*: WW to OC, 1/18/1904, McFarland, ed., *Papers*, vol. 1, 415.

202 *"for it's very possible"*: quoted in Robert Wohl, *A Passion for Wings: Aviation and the Western Imagination, 1908–1918* (New Haven, Conn.: Yale University Press, 1994), 291, note 24. Wohl gives his source as Andrée Ferber & Robert Ferber, *Les Débuts véritables de l'aviation française* (Paris: Fayard, 1970), an "important collection from Ferber's personal archive."

203 *"The problem cannot be considered"*: Quoted in Charles Gibbs-Smith, *The Rebirth of European Aviation, 1902–1908: A Study of the Wright Brothers' Influence* (London: Her Majesty's Statonery Office, 1974), 106–07.

203 *They planned a trial*: Flights at Huffman Prairie in May 1904, WW and OW, "The Wright Brothers' Aeroplane," *Century Magazine*, September 1908, reprinted in Jakab and Young, *Published Writings*, 31; Fred C. Kelly, *The Wright Brothers*, 123–26; entries of 5/23/1904, 5/25/1904, 5/26/1904, Milton Wright, *Diaries, 1857–1917*, 608; WW to OC, 5/27/1904, McFarland, ed., *Papers*, vol. 1, 437–38. See 436, note 7, for differing reports about the details of these flights.

## *Chapter Eight*: "What Hath God Wrought?"

204 *Hour after hour, they waited*: Wrights waiting for good wind at Huffman Prairie, Tom D. Crouch, *The Bishop's Boys: A Life of Wilbur and Orville Wright* (New York: Norton, 1989), 283.

205 *A return to Kitty Hawk*: Reasons for experimenting elsewhere than Kitty Hawk, WW to George Spratt, 4/12/1905, in Fred C. Kelly, ed., *Miracle at Kitty Hawk: The Letters of Wilbur and Orville Wright* (New York: Farrar, Straus and Young, 1951), 139.

205 *"a prairie-dog town"*: WW to OC, 6/21/1904, McFarland, ed., *Papers*, vol. 1, 441. This letter contains Wilbur's most detailed description of Huffman Prairie.

206 *"we must learn to accommodate"*: WW to OC, 6/21/1904, McFarland, ed., *Papers*, vol. 1, 441.

206 *thought the Wrights were "fools"*: Crouch, *Bishop's Boys*, 279.

206 *In size it was*: Dimensions of 1904 flyer, McFarland, ed., *Papers*, vol. 2, Appendix V, "Aeroplanes and Motors," 1189–90.

206 *Milton's battles within*: MW's church activities in 1904, Daryl Melvin Elliot, "Bishop Milton Wright and the Quest for a Christian America" (Ph.D. dissertation, Drew University, 1992), 302–03.

207 *"like taffy under a hammer blow"*: Fred C. Kelly, *The Wright Brothers* (New York: Harcourt, Brace and Co., 1943), 122.

207 *"We certainly have been 'Jonahed'"*: WW to OC, 6/14/1904, McFarland, ed., *Papers*, vol. 1, 440.

207 *But then came a reason*: Vandal's attack on Santos-Dumont's airship, McFarland, ed., *Papers*, vol. 1, 444, note 3.

207 *"a rather strange affair"*: WW to OC, 7/1/1904, McFarland, ed., *Papers*, vol. 1, 444–45.

207 *"the prospect of a race"*: WW to OC, 7/17/1904, McFarland, ed., *Papers*, vol. 1, 445.

207 His California Arrow *dirigible*: Crouch, *The Bishop's Boys*, 290.

207 *On August 6, both*: Flights of 8/6/1904, WW's Diary E, 1904–05, 8/6/1904, in McFarland, ed., *Papers*, vol. 1, 448.

208 *"It is a pity," he said*: And speeds needed for flight, WW to OC, 8/8/1904, McFarland, ed., *Papers*, vol. 1, 448–49.

208 *"sore all over"*: And OW's crash of 8/24/1904, WW's Diary E, 1904–05, 8/24/1904, in McFarland, ed., *Papers*, vol. 1, 452; OW escapes worse injury, Crouch, *Bishop's Boys*, 283.

208 *French aero enthusiasts read*: Chanute's article, "Aviation in America," and its impact in France, Charles H. Gibbs-Smith, *The Rebirth of European Aviation, 1902–1908: A Study of the Wright Brothers' Influence* (London: Her Majesty's Stationery Office, 1974), 76–81. Gibbs-Smith thoroughly treats both what is included and what is omitted in Chanute's article, and the resulting confusion among French aero enthusiasts.

209 *The early misreporting:* First accounts in France of Wrights' first powered flights, Gibbs-Smith, *Rebirth of European Aviation*, 91–93, 97–99.
209 *Men unwilling to risk:* Responses to Wright flights of Archdeacon, Tatin, and Deutsch de la Meurthe, Gibbs-Smith, *Rebirth of European Aviation*, 105–16.
209 *Ferdinand Ferber, the Army officer:* Ferber's response to 1903 flyer, Gibbs-Smith, *Rebirth of European Aviation*, 100–101.
210 *Ferber gave a lecture:* Voisin's entry into aviation, Gibbs-Smith, *Rebirth of European Aviation*, 126–27.
210 *The children of Dayton:* Description of 1904 catapult launching device, Crouch, *Bishop's Boys*, 284.
210 *the brothers tried the catapult:* First catapult attempts and half-circle flight of 9/7/1904, WW's Diary E, 1904–05, 8/24/1904, in McFarland, ed., *Papers*, vol. 1, 454–55.
211 *"It is difficult nowadays":* James M. Cox, *Journey Through My Years* (New York: Simon & Schuster, 1946), 83.
211 *"I guess the truth":* Kumler quoted in Fred C. Kelly, *The Wright Brothers* (New York: Harcourt, Brace and Co., 1943), 135.
211 *"Frankly," Cox said:* Copy, James M. Cox to Fred C. Kelly, 1/30/1940, GC, WBP, LC.
211 *"I used to chat with them":* Kelly, *Wright Brothers*, 139–40.
212 *"remarkable not in one way":* "Amos I. Root," *Medina (Ohio) Gazette*, 5/1/1923.
215 *Possibly they saw him:* That the Wrights considered Root as a possible investor at least later on, in 1908, is evident in a letter, WW to OC, 1/17/1908, McFarland, *Papers*, vol. 2, 854–55.
216 *"The answer involves a strange point":* Root's chief account of his experience with the Wrights at Huffman Prairie is A. I. Root, "Our Homes: What Hath God Wrought?" *Gleanings in Bee Culture*, 1/1/1905.
219 *"We went out to celebrate Roosevelt's election":* WW to OC, 11/15/1904, McFarland, ed., *Papers*, vol. 1, 464.
219 *His father was watching:* MW and railroad men present at 11/9/1904 flight, MW's diary, 11/9/1904, McFarland, ed., *Papers*, vol. 1, 464; Arthur G. Renstrom, *Wilbur & Orville Wright: A Chronology* (Washington, D.C.: Library of Congress, 1975), 145.
220 *Just before Christmas:* WW notifies Root of approval to publish account of Huffman Prairie flights, "My Flying-Machine Story," *Gleanings in Bee Culture*, 1/1/1905.
220 *He did so in his issue:* Root's accounts of the Wrights' 1904 flights appeared in his "Our Homes" column in *Gleanings in Bee Culture*, 1/1/1905, and a follow-up article, "The Wright Brothers' Flying Machine," in the issue of 1/15/1905. Root kept in touch with the Wrights and wrote several more short articles about their efforts in the form of updates based on press accounts. Most dealt with the Wrights' world-renowned flights in the fall of 1908. Among these are articles in the *Gleanings* issues of 4/1/1905, 11/1/1908, 11/15/1908, 12/1/1908, 12/15/1908, 1/1/1909, and 4/15/1909. Collections of *Gleanings in Bee Culture* are held in several university libraries. I am grateful to John Root, Kim Flottum, and Jim Thompson, of the A. I. Root Company, for making available to me the Root Company's complete run of *Gleanings in Bee Culture*.
220 *This one they rejected: Scientific American*'s rejection of Root article, Kelly, *The Wright Brothers*, 143.

## *Chapter Nine:* "The Clean Air of the Heavens"

222 *"[Will] and I could hardly wait":* Ivonette Wright Miller, "Character Study," in Miller, comp., *Wright Reminiscences* (Air Force Museum Foundation, 1978); 60.
222 *"Perhaps we could go into the manufacture":* "Successful in Flying, the Wrights Guard Their Secret Well," *New York Herald*, 11/25/1906, LC scrapbooks.
223 *"would make further wars":* OW to C. M. Hitchcock, 6/21/1917, McFarland, ed., *Papers*, vol. 2, 1104.
223 *To the sons of:* For Milton Wright's postmillennial view of American history, see Daryl Melvin Elliott, "Bishop Milton Wright and the Quest for a Christian America," Ph.D. dissertation, Drew University, 1992, 125–27.
223 *"A fast flying machine will render":* OC, "The Conquest of the Air," *The Car*, 2/21/1906, scrapbooks, Wright papers, LC.
223 *"We stand ready":* WW to OC, 5/28/05, McFarland, ed., *Papers*, vol. 1, 493–94.
223 *The first military man:* Capper as "scientific soldier," other attributes, Alfred Gollin, *No Longer an Island: Britain and the Wright Brothers, 1902–1909* (London: Heinemann, 1984), 87–88.

223 *"It is of no use whatever"*: Capper quoted in Percy B. Walker, *Early Aviation at Farnborough*, vol. 2 (London: Macdonald, 1974), 12.
224 *"grave doubts concerning the veracity"*: "The Wright Flyer, 1905," *The Aeronautical News*, 5/1/1906, scrapbooks, LC.
224 *"We call our invention 'the Flyer'"*: "Flying Machine Inventors Here," *Norfolk Landmark*, 12/23/1903, scrapbooks, Wright papers, LC.
224 *"a face more of a poet"*: H. M. Weaver to Frank S. Lahm, quoted in "Says Aeroplane Problem is Solved," *New York Herald*, 1/1/1906, scrapbooks, Wright papers, LC.
224 *"One could not help being impressed"*: Remarks of Alexander Ogilvie, Ivonette Wright Miller, comp., *Wright Reminiscences* (Air Force Museum Foundation, Inc., 1978), 191.
224 *"have at least made far greater strides"*: Capper's report quoted in Gollin, *No Longer an Island*, 69–71.
225 *"to ascertain whether this is a subject of interest"*: WW and OW to Robert M. Nevin, in Fred C. Kelly, ed., *Miracle at Kitty Hawk: The Letters of Wilbur and Orville Wright* (New York: Farrar, Straus and Young, 1941), 135–36.
225 *another letter to Colonel Capper*: WW's letter to J. E. Capper, Alfred Gollin, *No Longer an Island: Britain and the Wright Brothers, 1902–1909* (London: Heineman, 1984), 91–92.
225 *"the increasing difficulty"*: WW to J. E. Capper, 1/10/1905, quoted in Gollin, *No Longer an Island*, 91–92.
226 *"I have been [so] skeptical"*: Robert M. Nevin to the secretary of war, 1/21/1905, quoted in Russell J. Parkinson, "Politics, Patents and Planes: Military Aeronautics in the United States, 1863–1907" (Ph.D. dissertation, Duke University, 1963), 210.
226 *"As many requests have been made"*: G. L. Gillespie to Robert M. Nevin, [1/24/1905], Kelly, ed., *Miracle at Kitty Hawk*, 136–37.
227 *"Our consciences are clear"*: WW to OC, 6/1/1905, McFarland, ed., *Papers*, vol. 1, 494–95.
227 *"this may fairly be said"*: Copy, Godfrey L. Cabot to Henry Cabot Lodge, 12/31/1903, GC, WBP, LC. For a good account of the Cabot-Lodge exchange, see Gollin, *No Longer an Island*, 50–52.
227 *In London, Colonel Capper's*: British dealings with Wrights in 1905, Gollin, *No Longer an Island*, 98–107.
228 *So, early in 1904*: Archdeacon and Esnault-Pelterie glider efforts and comments in 1904 and 1905, Gibbs-Smith, *Rebirth of European Aviation*, 103–05, 122–25, 128, 147–48, 152–56.
229 *"risk a voyage to America"*: Ernest Archdeacon to WW and OW, 3/10/1905 (Orville Wright's translation, June 1945), GC, WBP, LC.
229 *"Mr. Archdeacon will find more compliments"*: WW to OC, 4/20/1904, McFarland, ed., *Papers*, vol. 1, 488–89.
229 *"They are evidently learning"*: WW to OC, 4/12/1905, McFarland, ed., *Papers*, vol. 1, 485–87.
230 *"little and contemptible"*: WW to OC, 5/6/1905, McFarland, ed., *Papers*, vol. 1, 490–91.
230 *"It was the decisive battle"*: WW to OC, 5/28/1905, McFarland, ed., *Papers*, vol. 1, 493.
231 *As for Millard Fillmore Keiter*: MW's vindication and reinstatement in 1905, Daryl Melvin Elliott, "Bishop Milton Wright and the Quest for a Christian America" (Ph.D. dissertaton, Drew University, 1992), 310–17.
231 *"the best dividends on the labor"*: WW and OW to Augustus Post, 3/02/06, McFarland, ed., *Papers*, vol. 2, 699–702.
231 *He considered alterations*: Manly to SPL, 3/28/1905, box 45, RU 31, SIA.
231 *"I take this occasion"*: Manly stores equipment, affirms importance of new trials, Manly to SPL, 6/17/1904, Manly to SPL, 11/4/1904, box 45, Record Unit 31, SIA.
231 *"to do otherwise would be to confess"*: Octave Chanute to SPL, 5/25/1905, box 18, RU 31, SIA.
231 *As soon as he learned*: SPL prepares to write history of aerodrome experiments, SPL diary, 3/19/1904, SIA.
232 *"thought how changed he was"*: Eliza Orne White to Cyrus Adler, 4/7/1907, box 22b, RU 7003, SIA.
232 *"While you have chided me"*: Charles M. Manly to SPL, 3/28/1905, box 45, RU 31, SIA.
233 *The Army requested the return*: J. G. Bates to SPL, 7/31/1905, box 96, RU 31, SIA.
233 *He explained that at first*: Discovery of W. W. Karr's embezzlement, conversation between Rathbun and Karr, [Richard Rathbun], draft report, "Actions, when embezzlement was first made known," box 79, RU 31, SIA; see also "Smithsonian Funds Gone," *New York Times*, 6/8/1905.
233 *It could not have been*: Dates and details of W. W. Karr's tenure at the Smithsonian appear in

"Memorandum, Case of William W. Karr, Disbursing Agent, Smithsonian Institution," box 79, RU 31, SIA.

233 *Langley informed the regents:* SPL gives up salary, Charles G. Abbot, *Adventures in the World of Science* (Washington D.C.: Public Affairs Press, 1958), 27.

234 *The craft was more imposing:* Description of 1905 flyer, McFarland, ed., *Papers*, vol. 2, appendix 5, "Aeroplanes and Motors," 1190–92. Key changes, including the controls, are described by Crouch, *The Bishop's Boys*, 295.

234 *Orville's attempt of July 14:* OW's accident of 7/14/1905, entry of same date, WW's Diary F, 1905, McFarland, ed., *Papers*, vol. 1, 501.

234 *"The good humor of Wilbur":* William Werthner, "Personal Recollections of the Wrights." *Aero Club of America Bulletin*, July 1912.

234 *more heft to the horizontal rudder:* Crouch, *The Bishop's Boys*, 297.

234 *jump from hummock to hummock:* WW to OC, 8/16/1905, McFarland, *Papers*, vol. 1, 506.

234 *"a very comical performance":* Entry of 8/30/1905, WW's Diary F, 1905, McFarland, ed., *Papers*, vol. 1, 507.

234 *As they had hoped:* Improvement in rudder control, Fred C. Kelly, *The Wright Brothers* (New York: Harcourt, Brace and Co., 1943), 134.

235 *On September 7:* Flights of 9/7 and 9/8/1905, entries of same dates, WW's Diary F, 1905, McFarland, ed., *Papers*, vol. 1, 509.

235 *They altered the shape:* WW's diary F, 9/23/1905, McFarland, *Papers*, vol. 1, 510; Kelly, *The Wright Brothers*, 133–34.

235 *"Our experiments have been progressing quite satisfactorily":* WW to OC, 9/17/1905, McFarland, ed., *Papers*, vol. 1, 511.

235 *Will went up and began:* WW's eleven-mile flight of 9/26/1905, entry of same date, WW's diary F, McFarland, *Papers*, vol. 1, 512.

235 *the brothers began to invite:* Witnesses to 1905 flights, WW and OW to Augustus Post, 3/2/1906, McFarland, ed., *Papers*, vol 2, 702. See also WW to Georges Besançon, 11/17/1905, GC, WBP, LC.

235 *At one point Will asked:* Werthner, "Some Personal Recollections of the Wrights."

236 *"we were frightened":* "The Aeroplane Problem," *London Daily Mail*, 12/12/1906, scrapbooks, WBP, LC.

236 *In one of the last flights:* OW's glide back to shed, "The Aeroplane Problem," *London Daily Mail*, 12/12/1906, scrapbooks, WBP, LC. This apparently happened on either 9/30/1905 or 10/31/1905. See WW's Diary F, entries of 9/30/1905 and 10/3/1905, McFarland, ed., *Papers*, vol. 1, 513.

236 *"It looked like a monstrous bird":* "Secret of Aerial Flight Wrested from the Birds," *St. Louis Republic*, 3/11/1906, scrapbooks, WBP, LC.

236 *"I wouldn't believe it":* "North Pole Can Not Be Reached," *Dayton Daily News*, 10/1/1905, scrapbooks, WBP, LC.

237 *"They walked down the field":* "Real Flying Machine," *Weekly Enquirer*, 5/24/1906, scrapbooks, WBP, LC. Waddell's account is apparently the first description of an airplane taxiing—though obviously he was exaggerating at least slightly when he said the machine "lifted itself clear of the ground."

237 *"It was beyond my comprehension":* "Fly Over St. Louis at 50 Miles an Hour," *St. Louis Post-Dispatch*, 4/21/1907, scrapbooks, WBP, LC.

237 *He got a single word:* Torrence Huffman's question about practical use of airplane and WW's answer, "War," interview with William Sanders, 3/15/1967. Wright Brothers–Charles F. Kettering Oral History Project, University of Dayton. Sanders, an Ohio newsman, told an interviewer he had discussed the Huffman Prairie flights with William Huffman, son of Torrence Huffman. According to Sanders, William Huffman "said that he remembers his father asking Wilbur what the use of the plane would be, and Wilbur said just one word, war."

237 *"In forenoon, at home writing":* Entry of 10/5/1905, Bishop Milton Wright, *Diaries, 1857–1917* (Dayton: Wright State University Libraries, 1999), 633.

237 *"We put many months of study":* "Successful in Flying, the Wrights Guard their Secret Well," *New York Herald*, 11/25/1906, scrapbook WBP, LC.

238 *"I am afraid I cannot describe":* "The Aeroplane Problem," *London Daily Mail*, 12/12/1906, scrapbooks, WBP, LC.

238 *had left him "intoxicated":* "Secrecy," *Cincinnati Enquirer*, 6/16/1907, scrapbooks, WBP, LC.

238 *"There is no sport in the world"*: WW to Aldo Corazzo, 12/15/1905, reproduced in Mario Cobianchi, *Pioneri dell'aviazione in Italia* (Rome: Editoriale Aeronautica, 1943), table lxxxviii.

238 *"There is a sense of exhilaration"*: WW, "Flying as a Sport—Its Possibilities," *Scientific American*, 2/29/1908, reprinted in Jakab and Young, *Published Writings*, 194–95. Another description, attributed to the brothers, but without citing either of them specifically, appeared in the *New York Herald* late in 1906. I have not included it in the text because it appears to be the reporter's paraphrase, and it may be embellished. Nonetheless, it is interesting: "When we are tacking into the wind we realize it only by finding we have to point up just as if we were in a sailboat in order to reach a given point. If we aim at a certain tree and see ourselves being carried past it we know that the wind is blowing across our course, but otherwise we have no means of knowing that the air is not perfectly still. You see, we are racing along at forty miles an hour ourselves, and it is a forty mile an hour wind which is in our faces, no matter what the velocity of the air currents themselves may be at the time. Our most acute sensations are during the first minute of flight, while we are soaring into the air and gaining the level at which we wish to sail. Then for the next five minutes our concentration is fixed on the management of the levers to see that everything is working all right, but after that the management of the flyer becomes almost automatic, with no more thought required than a bicycle rider gives to the control of his machine. The whole thing is worked by reflex action, based upon experience, for there is no time while you are in the air to give conscious thought to what has to be done in an emergency. Action becomes and must be instinctive, otherwise it would be all over before you could consciously think out what to do to save yourself. But when you know, after the first few moments, that the whole mechanism is working perfectly, the sensation is so keenly delightful as to be almost beyond description. Nobody who has not experienced it for himself can realize it. It is a realization of a dream so many persons have had of floating in air. More than anything else the sensation is one of perfect peace, mingled with an excitement that strains every nerve to the utmost, if you can conceive of such a combination." "Successful in Flying, the Wrights Guard Their Secret Well," *New York Herald*, 11/25/1906, scrapbooks, WBP, LC.

239 *One day that fall*: Manly's visit to Huffman Prairie, Kelly, *The Wright Brothers*, 136–37.

240 *Ernest Archdeacon announced*: quoted in Gibbs-Smith, *The Rebirth of European Aviation*, 173.

240 *"The Wright brothers refuse"*: Quoted in Gibbs-Smith, *The Rebirth of European Aviation*, 179.

240 *Weaver had no idea*: Henry Weaver report to Frank Lahm, 12/6/1905, copy in GC, WBP, LC.

241 *One evening in Washington*: Bell's dinner with Langley, Chanute, Newcomb, and David Fairchild is described in Fairchild's memoir, *The World Was My Garden*, 333–34. Writing in the 1930s, Fairchild placed the dinner in the winter of 1906–07. But if Langley was indeed present, that timing cannot be correct, since Langley died in February 1906. It is possible the dinner occurred in the winter of 1904–05, since Chanute had seen Orville Wright fly at Huffman Prairie in October 1904. But that would mean Fairchild's memory was off by two years, not just one. Also, Fairchild refers to the appearance before the dinner of newspaper accounts of the Wright flights. Such reports were far more numerous in January 1906, when the press was covering French efforts to purchase the Wright airplane, than they had been the previous winter. Thus, though not certain, it appears likely that the dinner occurred in the first two or three weeks of 1905. Bell had just returned to Washington from Nova Scotia and would have been eager to see and entertain Langley, who had suffered his stroke while Bell was at Beinn Bhreagh, yet was still capable of seeing friends. On January 20 Bell gave a speech in New York that seems to reflect the news he heard from Chanute at the dinner. In late January, Langley departed Washington for South Carolina, where he died several weeks later. It is also possible that Fairchild was correct about the dinner occurring in the winter of 1906–07, but mistaken about Langley being present. But this is unlikely, since other documents make it clear that by then Bell harbored no doubts at all about the Wrights' claims, and therefore would not have asked Chanute about the evidence for them.

241 *"See that hawk up there?"*: Bailey Millard, "Death and the Airmen," *Technical World Magazine*, September 1910, copy in box 134, AGB papers, LC.

241 *On January 20, 1906*: Bell's address to Automobile Club of America, "Dr. Bell Predicts Airships," *New York Sun*, January 21, 1906, scrapbooks, WBP, LC.

242 *At about noon*: SPL's stroke, "Diary, 1905," box 13, RU 7003, SIA; Cyrus Adler, *I Have Considered the Days*, 260.

243 *"His usefulness as an executive chief"*: W. B. Pritchard to Cyrus Adler, 1/2/1906, box 22b, RU 7003, SIA.

243 *But Langley rallied*: J. J. Reilly to W. B. Pritchard, 1/31/1906.

243 *"his interest in all matters"*: Untitled typescript beginning "A meeting of the officials and employees of the Smithsonian," 3/1/1906, box 42, RU 31, SIA.
243 *"Mr. Langley will never be our old friend"*: Copy, W. B. Pritchard to Cyrus Adler, undated letter, box 22b, RU 7003, SIA.
243 *began a memoir of his childhood:* Typescript copy, "Uncompleted Memoirs of S.P. Langley . . . ," box 22b, RU 7003, SIA.
244 *"Publish it"*: "Prof. Langley's Death," *Washington Star*, 3/2/1906; McFarland, *Papers*, vol. 2, 1092–93, footnote 11.
244 *When Representative Edgar:* Crumpacker-Hull exchange regarding money wasted on flying machines, "Laughed as Langley Died," *Kansas City Times*, 2/28/1906.
244 *"No doubt disappointment shortened his life"*: WW to OC, 3/2/1906, McFarland, *Papers*, vol. 2, 697–98.
244 *"We are parting"*: AGB's eulogy for SPL, "Ridicule Wounded Langley," *Boston Post*, 3/4/1906, "Funeral of Prof. Langley," *Washington Post*, 3/4/1906.
245 *"the facts that prevented his last ship"*: Copy, Julia H. Goodrich to AGB, box 225, AGB papers, LC.

## *Chapter Ten:* "A Flying Machine at Anchor"

246 *"As to our being abnormal"*: "Fly Over St. Louis at 50 Miles an Hour," *St. Louis Post-Dispatch*, 4/21/1907, scrapbooks, WBP, LC.
247 *"I have discovered that my interest"*: AGB to Mrs. Gardiner Hubbard, 6/24/1875, quoted in Robert V. Bruce, *Bell: Alexander Graham Bell and the Conquest of Solitude* (Ithaca, N.Y.: Cornell University Press, 1973), 151.
248 *"I will help you if you will not hate me"*: AGB to Mabel Bell, [undated] AGB papers, LC.
248 *"Let us lay it down"*: Quoted in Bruce, *Bell*, 322.
248 *"You are the mainspring"*: Quoted in Bruce, *Bell*, 323.
249 *"retire into myself"*: Bruce, *Bell*, 319.
249 *"I have found by experience"*: Quoted in Bruce, *Bell*, 328.
249 *"I wonder"*: Quoted in Bruce, *Bell*, 323.
249 *"I can't bear to hear"*: Quoted in Bruce, *Bell*, 334.
249 *"indefinable sense of largeness"*: David Fairchild, *The World Was My Garden* (New York: Charles Scribner's Sons, 1938), 290.
249 *"clear, crisp articulation"*: Thomas Watson, *Exploring Life* (New York: D. Appleton and Co., 1926), 58.
249 *"He always made you feel"*: Fairchild, *The World Was My Garden*, 290.
249 *"the tendency of your mind"*: Gardiner Hubbard to AGB, quoted in Bruce, *Bell*, 160.
250 *filled with devices:* Frank Carpenter, "The Telephone," *Detroit Free Press*, 10/20/1895.
250 *"articles not too long"*: Quoted in Bruce, *Bell*, 333.
250 *"A new invention"*: George Wise, unpublished essay, "General Impressions Gained from the Lab Notebooks of Alexander Graham Bell," AGB Collection, Special Collections, Boston University.
250 *"I have got work to do"*: Quoted in Bruce, *Bell*, 355.
250 *"From my earliest association"*: *Exploring Life: The Autobiography of Thomas A. Watson* (New York: D. Appleton and Co., 1926), 153–54.
251 *"I shall have to make experiments"*: Quoted in Bruce, *Bell*, 358–59.
251 *"kindly, loveable, simple"*: And Mabel Bell's letter to Kelvin, Bruce, *Bell*, 363; copy of Kelvin to Mabel Bell, 4/20/1898, box 152, AGB papers, LC.
251 *Bell tested many forms:* Bell's aeronautical experiments in 1890s, Bruce, *Bell*, 359–61.
251 *"I am finding out in the laboratory"*: Quoted in Bruce, *Bell*, 361.
251 *whom Bell's daughters called:* SPL as Bell daughters' "grandfather," SPL to Marian Bell Fairchild, box 73, AGB papers, LC.
252 *"Poor dear man"*: Quoted in Bruce, *Bell*, 407.
252 *"He goes up there"*: Quoted in J. H. Parkin, *Bell and Baldwin: Their Development of Aerodromes and Hydrodromes at Baddeck, Nova Scotia* (Toronto: University of Toronto Press, 1964), 6.
252 *"There is a great fascination"*: Untitled typescript beginning "For a good many years past," box 152, AGB papers, LC.
252 *"surprised, indeed overwhelmed"*: AGB to SPL, 4/2/1901, box 71, RU 31, SIA.

253 *"The great difficulty"*: Quoted in Dorothy Harley Eber, *Genius at Work: Images of Alexander Graham Bell* (New York: Viking Press, 1982), 88.
253 *Often Bell imitated:* Hargrave's box kites, Charles Gibbs-Smith, *The Invention of the Aeroplane, 1799–1909* (New York: Taplinger Publishing Co., 1966), 19–21; Valerie Moolman, *The Road to Kitty Hawk* (Alexandria, VA.: Time-Life Books, 1980), 95.
254 *"Do not increase the size"*: And "let everything be built up," Eber, *Genius at Work*, 91.
256 *"the brick, as it were"*: Quoted in Bruce, *Bell*, 437.
256 *"Eagle flight is marked"*: Gilson Gardner, "When Men Wear Wings," *Technical World Magazine*, December 1905, scrapbooks, WBP, LC.
256 *"In multi-cellular kites"*: Copy, AGB to OC, 9/29/1903, box 150, AGB papers, LC.
256 *"Other kites, in a gust"*: Typescript copy, "Dr. A. Graham Bell's Address Before the Canadian Club of Ottawa, March 27, 1909," *Bulletin of the AEA*, XXXIX, 4/12/1909, NASM Library, SI. It would later be learned, according to Robert V. Bruce, that "the center of pressure on each small cell moved only a fraction of the cell's length as the angle of flight changed. In a single-celled kite of the same overall dimensions, the pressure center would move by a comparable fraction of the whole kite's length and therefore many times as far by absolute measure. So a small cell unit gave Bell's kites more stability." *Bell*, 437.
256 *Soon after that, a tetrahedral:* Successful flights of early tetrahedral kites, J. H. Parkin, *Baldwin and Bell: Their Development of Aerodynamics and Hydrodromes at Baddeck, Nova Scotia* (Toronto: University of Toronto Press, 1964), 10–11.
256 *Workers from Baddeck:* Baddeck workers building kites, Eber, *Genius at Work*, 94–101.
257 *"There were so many kites"*: Mayme Morrison Brown, quoted in Eber, *Genius at Work*, 105.
257 *"continually more wrought up"*: Quoted in Bruce, *Bell*, 367.
257 *"I do so appreciate"*: Quoted in Bruce, *Bell*, 367.
257 *"You simply must get that machine to fly"*: Copy, Mabel Bell to AGB, 9/1/1903, box 148, AGB papers, LC.
257 *"He looked gray"*: Quoted in Bruce, *Bell*, 438.
258 *"more and more desperate"*: Copy, Mabel Bell to Arthur McCurdy, 12/28/1905, box 60, AGB papers, LC.
258 *Every morning the family:* Trials of the *Frost King*, including Neil McDermid's "flight," copy, Mabel Hubbard Bell to Arthur McCurdy, 12/28/1905, box 60, AGB papers, LC, and "Copy from Mrs. Alexander Graham Bell's Journal that Contains Only This Kite Episode," box 149, AGB papers, LC.
260 *"At the present time"*: WW and OW to Georges Besançon, 1/17/1906, GC, WBP, LC.
261 *"The idea of selling to a single government"*: WW to OC, 12/27/1905, McFarland, ed., *Papers*, vol. 1, 539–40.
261 *"We do not wish to take this invention abroad"*: WW and OW to secretary of war, 10/9/1905, McFarland, ed., *Papers*, vol. 1, 514–15.
261 *"the experimental development of devices"*: Major General J. C. Bates to WW and OW, 10/16/1905, reprinted in Fred C. Kelly, ed., *Miracle at Kitty Hawk: The Letters of Wilbur and Orville Wright* (New York: Farrar, Straus & Young, 1951), 148–49.
261 *"We have no thought"*: WW and OW to Board of Ordnance and Fortification, 10/19/1905, McFarland, ed., *Papers*, vol. 1, 518.
262 *"not care to formulate any requirements"*: Minutes of Board of Ordnance and Fortification meeting of 10/24/1905, reprinted in Kelly, ed., *Miracle at Kitty Hawk*, 151–52.
262 *"so insulting in tone"*: WW to Godfrey L. Cabot, 5/19/1906, GC, WBP, LC.
263 *a British engineer's letter:* Letter, *The Times*, 1/24/1906.
263 *"If such sensational and tremendously important experiments"*: "The Wright Aeroplane and Its Fabled Performances," *Scientific American*, 1/13/1906.
263 *"It seems strange"*: Copy, AGB to Mabel Bell, 6/26/1906, box 48, AGB papers, LC.
263 *"If they will not take our word"*: WW and OW to Georges Besançon, 1/17/1905, GC, WBP, LC.
264 *In the spring of 1906:* Santos-Dumont experiments of 1906, Peter Wykeham, *Santos-Dumont: A Study in Obsession* (London: Putnam, 1962), 203–15.
265 *"It seems to be the opinion"*: "Have an Aeroplane to Sell," *New York Sun*, 1/25/1907, scrapbooks, WBP, LC.
266 *"take an unexpected part"*: OW to Commander Holden C. Richardson, 2/17/1926, McFarland, ed., *Papers*, vol. 2, 1137–39. On the Wrights' preparations for a Jamestown flyover, see also

"Newest Invention of Wright Brothers will Carry Their Aeroplane on Water," *Dayton Herald*, 3/21/1907, scrapbooks, WBP, LC.

266 *"I do hope"*: KW to WW, 6/27/1907, FC, WBP, LC.

267 *"It is desperate to be so in the dark"*: And "Sister is sick enough," KW to WW and OW, 7/28/1907, FC, WBP, LC.

268 *"though I sometimes restrained myself"*: WW hears of Chanute's role in Wrights' work, WW to OC, 1/29/1910, McFarland, ed., *Papers*, vol. 2, 984.

268 *Orville was in Paris:* Meeting of OW, Berg, and Archdeacon, OW to MW, 11/19/1907, FC, WBP, LC.

268 *"One could just force it"*: quoted in Curtis Prendergast, *The First Aviators* (Alexandria, Va.: Time-Life Books, 1980), 30.

269 *"nature makes no distinctions"*: Parkin, *Bell and Baldwin*, 25.

269 *"worked it out far enough to demonstrate"*: Mabel Bell, private memorandum, copied as "Notes from an Old Note Book of Mabel G. Bell Evidently Written in Fall of 1907 . . ." 7/25/1923, box 143, AGB papers, LC.

269 *"For years [Mabel told him]"*: Mabel Bell to AGB, 1/24/1907, box 143, AGB papers, LC.

270 *"You have tried struggling"*: Ibid.

270 *One day in 1885:* Bell fixes McCurdy's telephone, Grosvenor and Wesson, *Alexander Graham Bell*, 137.

270 *"casual, friendly and unworried"*: H. Gordon Green and J. A. D. McCurdy, "I Flew the Silver Dart," *Weekend Magazine*, vol. 9, no. 6, 1959.

272 *because Curtiss's sister was deaf*: Bruce, *Bell*, 446–47.

272 *She ticked off responses:* Mabel Bell's case for forming AEA, Mabel Bell to AGB, 1/24/1907, 1/24/1907, box 143, AGB papers, LC.

272 *The AEA's first project*: Bruce, *Bell*, 447–48.

273 *"It seems unfortunate"*: Quoted in Crouch, *The Bishop's Boys*, 341.

## *Chapter Eleven:* "A World of Trouble"

275 *"a sidelong pair of parentheses"*: C. R. Roseberry, *Glenn Curtiss: Pioneer of Flight* (New York, Doubleday, 1972), 91.

275 *"It is a beautiful machine"*: Copy, AGB to Mabel Bell, 3/9/1908, box 49, AGB papers, LC.

275 *On March 12, a good portion:* Design, construction, and first flight of *Red Wing*, Parkin, *Bell and Baldwin*, 50–54; Grosvenor and Wesson, *Alexander Graham Bell: The Life and Times of the Man Who Invented the Telephone*, 221–22; C. R. Roseberry, *Glenn Curtis*, 90–93.

276 *"This is declared"*: Bell's characterization of the *Red Wing*'s first flight as "the first successful public flight in America" appears in "Work of the Aerial Experiment Association as Recorded in Associated Press Dispatches Written by Dr. Alexander Graham Bell," 5/17/1908, [Internet] AGB papers, LC. "Bell's Aerodrome Tested," *New York Times*, 3/13/1908; "New Airship Flies," *Washington Post*, 3/13/1908.

276 *The men released:* Second flight and crash of *Red Wing*, Glenn Curtiss to AGB, 3/19/1908, box 143, AGB papers, LC; Charles R. Cox, "Lecture on Aviation at Baddeck," Bulletin of the Aerial Experiment Association, no. XXXVIII, 3/29/1909, National Air and Space Museum Library, SI.

277 *"moveable surfaces"*: quoted in Roseberry, *Glenn Curtiss*, 96–97.

277 *"We were familiar"*: Roseberry, *Glenn Curtiss*, 97.

278 *"if, as I have reason"*: Roseberry, *Glenn Curtis*, 97.

278 *Every month brought fresh news:* French flights in March 1908, "The First Two-Passenger Aeroplane," *Scientific American*, 4/11/1908.

279 *"There is a great deal of interest"*: Stanley Y. Beach to OW and WW, 4/23/1908, GC, WBP, LC.

279 *The Army wanted:* U.S. Army requirements for Wright flying machine trials, "Articles of Agreement," U.S. Signal Corps and Wilbur and Orville Wright, 2/10/1908: WW and OW, "Our Recent Experiments in North Carolina," *Aeronautics*, June 1908, reprinted in Jakab and Young, *Published Writings*, 19–21; WW and OW, "Our Aeroplane Tests at Kitty Hawk," *Scientific American*, 6/13/1908, reprinted in Jakab and Young, *Published Writings*, 21–23.

280 *"a little practice"*: WW to OC, 4/8/1908, McFarland, ed., *Papers*, vol. 2, 861.

280 *At the Kill Devil Hills:* Scenes and preparations in WW's first days at Kitty Hawk in April 1908, WW to OW, 4/9/1908, WW to KW, 4/14/1908, KW to MW, 4/16/1908, FC, WBP, LC; WW's Diary T, 4/10/1908, McFarland, ed., *Papers*, vol. 2 862.

280 *He introduced himself:* WW-Newton exchange regarding Delagrange flight of 5/13/1908, "Wilbur Wright on Delagrange Flight," *New York Herald*, 5/13/1908, scrapbooks, WBP, LC.
281 *And the eight or ten:* Reporters and their tactics at Kitty Hawk in 1908, OW to KW, 5/13/1908, FC, WBP, LC.
281 *"Collier's is deeply interested":* Arthur Ruhl to WW and OW, 1/28/1907, GC, WBP, LC.
282 *The men at the lifesaving station:* Lifesaving crew as sources for reporters, WW to OW, 5/20/1908. Of his accident, Wilbur told Orville: "If the life savers had not been crazy to spread it so soon it would have attracted no attention in the papers."
282 *"like faint smoke":* Arthur Ruhl, "History at Kill Devil Hill," *Collier's*, 5/30/1908.
283 *"a runaway street car":* Gilson Gardner, "Wright Brothers' Airship Steers Like Car; Flies Fast," *Chicago Journal*, 5/15/1908, scrapbooks, WBP, LC.
283 *"to the world waiting":* Ruhl, "History at Kill Devil Hill."
283 *"hardly write an ordinary sentence":* Arthur Ruhl to Eliza Drane, [5/14/1908], Robert Drane papers, #2987, folder 7A, Southern Historical Collection, Wilson Library, University of North Carolina at Chapel Hill.
284 *"hanging about in the woods":* WW to OW, 5/17/1908, FC, WBP, LC.
284 *"be as civil as you can":* KW to WW, 5/17/1908, FC, WBP, LC.
284 *"It is a good thing sometimes":* WW to OW, 5/20/1908, FC, WBP, LC.
284 *"beat any roller coaster":* OW to KW, 5/10/1908, FC, WBP, LC.
284 *After the midday meal:* WW's experience of flight that ended in wreck, WW's Diary T, 5/14/1908, McFarland, ed., *Papers*, vol. 2, 877–79.
285 *A mile away, Orville:* OW's view of WW's crash of 5/14/1908, OW to MW, 5/15/1908, FC, WBP, LC. A good summary of the Wrights' flights at Kitty Hawk in 1908 appears in their draft press statement attached to a letter, OW to Stanley Y. Beach, 6/2/1908, GC, WBP, LC.
285 *"I do not think the accident":* WW to KW, 5/17/1908, FC, WBP, LC.
285 *"all Hammondsport seems to be alive":* Copy, AGB to Mabel Bell, 5/11/1908, box 49, AGB papers, LC.
285 *"fine bright-looking boys":* Copy, AGB to Mabel Bell, 5/11/1908, box 49, AGB papers, LC.
285 *"for fear the wind would go down":* Presence of inventors, reporters, and Aero Club officials, copy, AGB to Mabel Bell, 5/11/1908, box 49, AGB papers, LC.
286 *He remained just as avid:* Bell's thoughts on tetrahedral experiments at Hammondsport, copy, AGB to Mabel Bell, 5/12/1908, box 49, AGB papers, LC.
286 *"I have not the heart":* Copy, AGB to Mabel Bell, 5/3/1908, box 49, AGB papers, LC.
286 *Rolling the machine out: White Wing*'s early false starts, "Work of the Aerial Experiment Association as Recorded in Associated Press Dispatches Written by Dr. Alexander Graham Bell," 5/17/1908, [Internet] AGB papers, LC.
287 *On May 18, just:* Baldwin's and Selfridge's first flights of *White Wing* and Bell's comment to reporters, Roseberry, *Glenn Curtiss*, 99–100.
287 *To bring the machine's:* Curtiss's alterations on *White Wing*, Roseberry, *Glenn Curtiss*, 100.
287 *On the second run:* Curtiss's successful flight in *White Wing*, G. H. Curtiss, "Ideas on Aviation," *Bulletin of the Aerial Experiment Association*, 8/10/1908, NASM Library, SI; Roseberry, *Glenn Curtiss*, 100–101.
287 *A day later, barely:* McCurdy's crash in the *White Wing*, J. A. D. McCurdy, "Experiences in the Air," *Bulletin of the Aerial Experiment Association*, 8/10/1908, NASM Library, SI. For a photograph of the ruined *White Wing*, see Eber, *Genius at Work*, 133.
288 *He was already pining:* That Curtiss "coveted" the *Scientific American* trophy is asserted in Roseberry, *Glenn Curtiss*, 102–3.
288 *"I will talk to you":* "Wright Brothers Silent," *Washington Post*, 5/18/1908.
289 *"They are so afraid":* "Wright Brothers Avoid Curiosity of Outsiders," *New York Herald* (Paris edition), 5/5/1908, scrapbooks, WBP, LC.
289 *"On account of the mystery":* "Secret of the Wright Air-Ship," *Literary Digest*, [June 1908], scrapbooks, WBP, LC.
290 *"Today's performance, while not equal":* "Wright Brothers in Great Flight," *New York Herald* article reprinted in *Philadelphia Inquirer*, 5/14/1908, scrapbooks, WBP, LC.
290 *"the brothers Wright have certainly made serious flights":* "Mr. Farman's Challenge Unanswered by Wrights," *New York Herald*, 5/18/1908, scrapbooks, WBP, LC.
290 *When the New York reporters:* WW's exchanges with reporters in New York, "Wright, Conqueror of the Air, Is Silent on Challenge," *New York Herald*, 5/20/1908; "Wrights Make a Denial," *New*

*York Journal*, 5/18/1908, scrapbooks, WBP, LC; WW's Diary T, 5/20/1908, McFarland, ed., *Papers*, vol. 2, 882; WW to OW, 5/20/1908, FC, WBP, LC.

291 *"We were worried"*: KW to WW, 5/17/1908, FC, WBP, LC.

292 *"To read what Anderson had to say"*: KW to WW, 5/26/1908, FC, WBP, LC.

292 *"a very nice letter"*: And "makes me sick," KW to WW, 5/31/1908, FC, WBP, LC.

292 *Columbus, Robert Fulton, "and the Wright Brothers"*: MW to OW, 5/18/1908, FC, WBP, LC.

292 *"the Wrights and their Flyer"*: MW, *Diaries*, 1857–1917, entry of 5/29/1908, 677.

292 *"It says that you are 'slim, sedate and placid' "*: MW to OW, 5/18/1908, FC, WBP, LC. Milton said the article was from the *Philadelphia Press*.

293 *"We need to have our true story"*: WW to OW, 5/20/1908. McFarland, ed., *Papers*, vol. 2, 882–83.

293 *Inside the Flint offices*: WW's meeting with Nelson Miles and canceled meeting with Thomas Edison, WW's Diary T, 5/20/1908, McFarland, *Papers*, vol. 2, 882.

293 *"I do sometimes wish"*: WW to KW, 5/19/1908, FC, WBP, LC.

294 *"in a fog so dense"*: WW to MW, 5/28/1908, FC, WBP, LC.

294 *"I have never been able to discover"*: OW to Arthur Ruhl, 6/8/1908, GC, WBP, LC.

294 *"it always takes more time"*: WW to OW, 5/20/1908, McFarland, ed., *Papers*, vol. 2, 882–83.

294 *"Do not attempt,"* WW to OW, 5/29/1908, FC, WBP, LC.

294 *"If at any time Orville is not well"*: WW to KW, 5/19/1908, McFarland, ed., *Papers*, vol. 2, 881.

295 *"The more I think of the circumstances"*: WW to OW, 5/29/1908, FC, WBP, LC.

295 *"The fact that the newspapers"*: WW to OW, 6/3/1908, McFarland, ed., *Papers*, vol. 2, 886–87.

295 *"Does he not intend"*: WW to KW, 6/09/1908, FC, WBP, LC.

295 *"He doesn't [write] as much"*: KW to WW, 6/11/1908, FC, WBP, LC.

296 *"I am sure that with a scoop shovel"*: WW to OW, 6/17/1908, McFarland, ed., *Papers*, vol. 2, 900–901.

296 *Will was forced*: WW's frustrations in repairing and preparing machine, WW's Diary T, 6/9/1908, 6/19/1908, 6/20/1908, 6/22/1908, 6/24/1908, 6/25/1908, McFarland, ed., *Papers*, vol. 2, 901–2; WW to OW, 6/20/1908, FC, WBP, LC.

296 *"I have had an awful job"*: WW to OW, 6/20/1908, FC, WBP, LC.

296 *a spout of boiling water*: McFarland, ed., *Papers*, vol. 2, 906 note.

297 *"If you had permitted me"*: WW to OW, 7/9/1908, FC, WBP, LC.

297 *"The trouble comes at the Customs house"*: OW to WW, 6/30/1908, FC, WBP, LC.

298 *"an anxiety lest something"*: David Fairchild, "The Coming of the Winged Cycle," *Bulletin of the Aerial Experiment Association*, 9/14/1908, NASM Library.

298 *"it haunted me"*: David Fairchild, *The World Was My Garden* (New York: Scribner's, 1938), 343.

298 *"seemed to be pleased"*: G.H. Curtiss, "Winning the Scientific American Trophy, July 4, 1908," *Bulletin of the Aerial Experiment Association*, 7/27/1908, NASM Library.

299 *"In spite of all"*: Mrs. David G. Fairchild, "Winning the Scientific American Trophy July 4, 1908," *Bulletin of the Aerial Experiment Association*, 7/27/1908, NASM Library.

299 *The thing is done*: Fairchild, "The Coming of the Winged Cycle," *Bulletin of the AEA*, 9/14/1908, NASM Library.

## *Chapter Twelve*: "The Light on Glory's Plume"

300 *He wasted little time*: WW's instructions to OW on warning Curtiss about patent infringement, WW to OW, 7/10/1908, FC, WBP, LC.

300 *"We believe it would be very difficult"*: OW to Glenn H. Curtiss, 7/20/1908, McFarland, *Papers*, vol. 2, 907.

301 *Curtiss ducked*: Curtiss plans no exhibition flights, OW reported Curtiss's response in letter to WW, 7/29/1908, McFarland, *Papers*, vol. 2, 909–11.

301 *In France, Wilbur slept*: WW's living circumstances at Le Mans, WW to KW, 8/2/1908, WW to MW, 8/5/1908, FC, WBP, LC.

301 *"I always tremble"*: Hart Berg to OW, 9/21/1908, GC, WBP, LC.

301 *Will promised her*: Ivonette Wright Miller, comp., *Wright Reminiscences* (Air Force Museum Foundation, 1978).

301 *"he is a pretty nice sort of fellow"*: WW to OW, 6/17/1908, FC, WBP, LC.

301 *"On my part,"*: Louis Blériot to WW, 8/1/1908, GC, WBP, LC.

302 *"caused them to be despised"*: "Aeronauts of France Are Confident," *Cincinnati Enquirer*, 5/17/1908, scrapbooks, WBP, LC.
302 "Le bluff continue": Fred C. Kelly, *The Wright Brothers*, 234.
302 *"I think they have had their patience"*: WW to KW, 8/2/1908, FC, WBP, LC.
302 *"When it became known"*: "Wright Aeroplane Nearly Ready for Trial Flights," *Paris Herald*, 7/5/1908, scrapbooks, WBP, LC.
302 *"Mr. Brewer, let's go and have some dinner"*: Griffith Brewer, *Fifty Years of Flying* (London: Air League of the British Empire, 1946), 93–94.
303 *"They have been hard at work"*: McMechan's speech and exchange between reporter and Farman, "Throws Brick at the Wrights," *New York Globe*, 7/29/1908, scrapbooks, WBP, LC.
304 *"It is now generally conceded"*: Chanute's article in *The Independent*, 6/4/1908, 1287–88.
304 *"I think I will write him"*: OW to WW, 7/19/1908, FC, WBP, LC.
304 *"The Bell outfit"*: OW to WW, 6/7/1908, FC, WBP, LC.
304 *"a fool or a knave"*: KW to WW, 6/1/1908, FC, WBP, LC.
304 *"If you don't hurry"*: OW to WW, 7/29/1908, McFarland, ed., *Papers*, vol. 2, 911.
304 *"I am so tired"*: KW to WW, 6/17/1908; FC, WBP, LC.
304 *"If the trials are successful"*: WW to OW, 8/2/1908, FC, WBP, LC.
305 *"While I was operating alone"*: WW to MW, 9/13/1908, FC, WBP, LC.
305 *"You should get everything ready"*: "Be exceedingly cautious," WW to OW, 8/30/1908, FC, WBP, LC.
306 *The air was calm*: Spectators and decision for WW's first flight at Les Hunaudières, WW to OW, 8/9/1908, FC, WBP, LC; Kelly, *Wright Brothers*, 236–37.
306 *For several hours he worked*: WW's preparations for 8/8/1908 flight, Fred C. Kelly, *The Wright Brothers*, (New York: Harcourt, Brace, 1943), 236–37; WW to OW, 8/9/1908, FC, WBP, LC.
306 *One of those watching*: François Peyrey's account of WW's 8/8/1908 flight, "Aerial Navigation," *The Times* (London), 8/14/1908. See also Fred Howard, *Wilbur and Orville* (New York: Knopf, 1987), 256–59. Howard's account of WW's flights at Le Mans is especially thorough and vivid.
308 *Paul Zens, who*: Newspaper coverage of WW's first flights at Les Hunaudières, with comments from French aviators, Charles Gibbs-Smith, *The Rebirth of European Aviation* (London: Her Majesty's Stationery Office, 1974), 286–87.
309 *"For a long time"*: Quoted in Crouch, *The Bishop's Boys*, 368.
309 "Nous sommes battus": Gibbs-Smith, *Rebirth of European Aviation*, 288.
309 *"violent gusts"*: Frantz Reichel in *Le Figaro*, quoted in Gibbs-Smith, *Rebirth of European Aviation*, 304.
309 *"Wright is a titanic genius!"*: Gibbs-Smith, *Rebirth of European Aviation*, 287.
309 *"could turn very short curves"*: WW to OW, 8/9/1908, FC, WBP, LC.
310 *"Blériot & Delagrange"*: WW to OW, 8/15/1908, McFarland, ed., *Papers*, vol. 2, 912.
310 *"I believe that our machines"*: Gibbs-Smith, *Rebirth of European Aviation*, 289.
310 *The great European newspapers*: European press reports of WW's flights at Le Mans, Gibbs-Smith, *Rebirth of European Aviation*, 279–85.
310 *"You never saw anything"*: WW to OW, 8/15/1908, McFarland, *Papers*, vol. 2, 911–13.
311 *"partly because I have not felt"*: WW to KW, 9/10/1908, FC, WBP, LC.
312 *"they were packed exactly"*: OW to WW, 8/23/1908, McFarland, *Papers*, vol. 2, 914–15.
312 *The only mishap*: Horses uneasy at sight of Baldwin's dirigible, "Aeroplane Hard Target for Guns," *New York Herald*, 8/29/1908, Claudy scrapbook, WBC, WSU.
312 *Brigadier General James*: General Allen's skepticism, "Wright, Sure of Victory, Ready to Fly Monday," *New York Herald*, Claudy scrapbook, WBC, WSU.
312 *"Would you really think"*: "Wright Almost Ready for Flight," *New York Herald*, 8/25/1908, C. H. Claudy scrapbook, 1908–09, WBC, WSU.
312 *"They talked only to each other"*: Benjamin D. Foulois and C. V. Glines, *From the Wright Brothers to the Astronauts: The Memoirs of Major General Benjamin D. Foulois* (New York: McGraw-Hill, 1968), 52.
312 *"The Signal Corps does not hesitate"*: OW to WW, 8/23/1908, McFarland, *Papers*, vol. 2, 914–15.
312 *"You have probably seen the photos"*: Glenn Curtiss to AGB, 8/29/1908, in *Bulletin of the Aerial Experiment Association*, 9/7/08, NASM Library, SI.
313 *he immediately conceived*: OW's private reports on Selfridge, OW to WW, 9/6/1908; OW to MW, 9/7/1908, FC, WBP, LC.

313 *"We haven't any secrets now"*: "Orville Wright Bares Secrets of Big Aeroplane," *New York Herald*, Claudy scrapbook, WBP, WSU.
313 *"The reporters seem to think"*: OW to KW, 8/29/1908, FC, WBP, LC.
313 *People from the neighborhood:* Onlookers' questions at Fort Myer, "Aeroplanes for Transportation," *New York Herald*, 8/30/1908, Claudy scrapbook, WBC, WSU.
313 *"stacks of prominent people"*: OW to KW, 8/29/1908, FC, WBP, LC.
313 *"I am meeting some very handsome young ladies!"*: OW to KW, 8/29/1908, FC, WBP, LC.
314 *"Work goes slowly"*: OW to WW, 8/24/1908, FC, WBP, LC.
314 *"Don't go out"*: WW to OW, 8/25/1908, FC, WBP, LC.
314 *"turning the entire time"*: OW to MW, 9/7/1908, FC, WBP, LC.
315 *"I didn't say I was going to land"*: "Balky Motor Delays Fort Myer Tests," 9/1/1908, *New York Herald*, Claudy scrapbook, Wright Brothers Collection, WSU.
315 *On the morning of September 3:* OW's 70-second flight at Fort Myer of 9/3/1908, "Fly? Why, Of Course," *Washington Star*, 9/4/1908, scrapbooks, WBP, LC; "Orville Wright Flies at Fort Myer," *New York Herald*, 9/4/1908, Claudy scrapbook, WBC, WSU. The account of the *Star* is the more detailed and vivid. The account of tears on the faces of newsmen is from Ivonette Wright Miller in her compilation, *Wright Reminiscences*.
318 *"Have I anything to say?"*: "Wright's Flight Pleases Squier," *Washington Times*, 9/9/1908, scrapbooks, WBP, LC.
318 *General Nelson Miles:* Comments on OW's record flights of 9/9/1908, "Mr. Orville Wright, in Fifty-Seven and Sixty-Two Minute Flights, Far Exceeds the World's Record for Aeroplane Navigation," *New York Herald*, 9/10/1908, Claudy scrapbook, WBC, WSU.
318 *"make them feel until their dying days"*: "Let Dayton Honor Orville and Wilbur Wright," *Dayton Herald*, 9/10/1908, scrapbooks, WBP, LC.
319 *"Well, it was fine news"*: WW to KW, 9/13/1908, FC, WBP, LC.
319 "Très bien": WW to OW, 9/10/1908, FC, WBP, LC.
319 *"ON BEHALF OF AERIAL EXPERIMENT ASSOCIATION"*: AGB to OW, 9/11/1908, GC, WBP, LC.
319 *"They treat you in France"*: MW to WW, 9/9/1908, FC, WBP, LC.
319 *"What would you think"*: KW to OW, 9/12/1908, FC, WBP, LC.
319 *"The world is all a fleeting"*: Quotation from Moore poem, MW to WW, 9/9/1908, FC, WBP, LC.
320 *"I do not think I will make many more practice flights"*: OW to KW, 9/15/1908, FC, WBP, LC.
320 *After three smooth rounds:* OW's key account of the accident of 9/17/1908 appears in OW to WW, 11/14/1908, McFarland, ed., *Papers*, vol. 2, 936–39.
320 *A photographer named W. S. Clime:* Key accounts of the 9/17/1908 crash at Fort Myer include "Airship Falls; Lieut. Selfridge Killed, Wright Hurt," *Washington Post*, 9/18/1908, scrapbooks, WBP, LC; "Mr. Orville Wright Hurt, Lieutenant Selfridge Killed as Aeroplane Falls One Hundred Feet," *New York Herald*, 9/18/1908, Claudy scrapbook, WBC, WSU; W. S. Clime, "The Wright Accident by an Eyewitness," *Bulletin of the Aerial Experiment Association*, 11/16/1908, NASM Library, SI.
322 *"I cannot help thinking"*: WW's reaction to Fort Myer crash, WW to KW, 9/20/1908, McFarland, ed., *Papers*, vol. 2, 925–27; WW to MW, 9/22/1908, FC, WBP, LC.
324 *"looking pretty badly"*: KW to Lorin Wright, 9/19/1908, FC, WBP, LC.
324 *"school can go"*: KW to WW, 9/24/1908, FC, WBP, LC.
324 *"I am awfully sorry"*: WW to KW, 9/20/1908, McFarland, ed., *Papers*, vol. 2, 925–27.
324 *When she was not:* KW's experiences at Fort Myer, KW to MW, 9/21/1908; KW to MW, 10/25/1908; KW to WW, 11/13/1908, FC, WBP, LC; KW to Agnes Beck, Wright Brothers Collection, Special Collections, Wright State University.
325 *"The natural inference"*: MW to KW, 9/26/1908, FC, WBP, LC.
325 *"I am sorry that I could not come home"*: WW to OW, 1/1/1909, FC, WBP, LC.
325 *"The airship which, within ten years"*: G. H. Curtiss, "Future Air Travel," *Bulletin of the AEA*, 10/26/1908, NASM Library, SI.

## *Chapter Thirteen:* "The Greatest Courage and Achievements"

327 *In the time remaining:* Bell's determination to continue tetrahedral experiments in last six months of AEA, C. R. Roseberry, *Glenn Curtiss: Pioneer of Flight* (New York, Doubleday, 1972), 140–41.
328 *"has only to look at the engine"*: Quoted in Roseberry, *Glenn Curtiss*, 156.
328 *"What we want to know"*: Quoted in Roseberry, *Glenn Curtiss*, 144–45.

328 *"We have read so much"*: Glenn Curtiss to AGB and Mabel Bell, 11/24/1908, *Bulletin of the AEA*, 12/16/1908, NASM Library, SI.
329 *Bell wired:* Bell-Curtiss exchange about January 1909 meeting of AEA, Roseberry, *Glenn Curtiss*, 147–48.
329 *"I am very much averse"*: Quoted in Roseberry, *Glenn Curtiss*, 150.
329 *he had been thinking about converting:* AGB's and Curtiss's views on future of AEA, Roseberry, *Glenn Curtiss*, 150–52; J. H. Parkin, *Bell and Baldwin: Their Development of Aerodromes and Hydrodromes at Baddeck, Nova Scotia* (Toronto: University of Toronto Press, 1964), 155–61.
330 *Bell said there was no point:* Discussions of company to succeed AEA, Roseberry, *Glenn Curtiss*, 147–52, Parkin, *Bell and Baldwin*, 155–61.
330 *"Alec is not discouraged"*: Quoted in Dorothy Harley Eber, *Genius at Work: Images of Alexander Graham Bell* (New York: Viking, 1982), 126.
331 *On the ice, Bell:* Accounts of *Cygnet II* trials, Roseberry, *Glenn Curtiss*, 153–54; Parkin, *Bell and Baldwin*, 96–97.
331 *"Although at times it seemed"*: Copy, Mabel Bell's draft article on the AEA, box 143, AGB papers, LC.
331 *"Papa feels down"*: Mabel Bell to "little girl," [Marion Bell Fairchild], 2/24/1909, box 154, AGB papers, LC.
331 *Bell immediately began:* Adjustments to *Cygnet II*, AGB to Mabel Bell, 3/6/1909, [online citation, in "AEA dissolves" folder].
332 *Curtiss weighed his options:* Negotiations for formation of Herring-Curtiss Company, Roseberry, *Glenn Curtiss*, 152–53, 156–60.
333 *"America has taken the lead"*: Announcement of Herring-Curtiss Company, "Aeroplane Factory for this Country," *New York Times*, 3/4/1909; Roseberry, *Glenn Curtiss*, 157.
334 *"would like to have Mr. Bell"*: Glenn Curtiss to Mabel Bell, 3/12/1909, box 59, AGB papers, LC.
334 *Casey Baldwin, who had:* Casey Baldwin's remarks on Herring-Curtiss venture, quoted in Roseberry, *Glenn Curtiss*, 158; Baldwin's souring on Curtiss as engineer, Frederick W. Baldwin to Mabel Bell, 3/11/1909, AGB papers, LC.
334 *"high appreciation"*: AGB to Charles J. Bell, 3/31/1909, in *Bulletin of the AEA*, 4/12/1909, NASM Library, SI.
334 *"The Aerial Experiment Association is now a thing of the past"*: "Dissolution of the A.E.A.," *Bulletin of the AEA*, 4/12/1909, NASM Library, SI.
335 *"Can't you come over"*: WW to OW, 11/14/1908, McFarland, *Papers*, vol. 2, 939–40.
335 *"I know that you love 'Old Steele' "*: WW to KW, 12/8/1908, FC, WBP, LC.
335 *She was more than ready:* Katharine's complaints about MW and OW at home, late 1908, KW to WW, 11/30/1908, KW to WW, 12/7/1908, 12/17/1908, FC, WBP, LC.
336 *"I am so tired"*: KW to WW, 12/17/1908, FC, WBP, LC.
336 *At the Paris station:* Reception of OW and KW in Paris, KW to MW, 1/13/1909, FC, WBP, LC.
336 *For three months Kate:* KW's activities and observations in France, KW to MW, 1/17/1909, 1/20/1909, 1/24/1909, 2/1/1909, 2/8/1909, 2/11/1909, 2/16/1909, 2/22/1909, FC, WBP, LC.
337 *Will assembled the demonstration:* Setting of assembly and flights at Rome, WW to MW, 4/4/1909; WW to OW, 4/4/1909, FC, WBP, LC.
337 *Kate now found the company:* KW's activities and observations in Italy, KW to MW, 4/11/1909, 4/14/1909, 4/20/1909, 4/25/1909, FC, WBP, LC.
338 *"Not long ago little old Suzanne"*: Quoted in Robert Wohl, *A Passion for Wings: Aviation and the Western Imagination, 1908–18* (New Haven, Conn.: Yale University Press, 1994), 104.
340 *"A horse race, an automobile contest"*: "The Rheims Aviation Week," *The Times*, 8/13/1909.
340 *Blériot raised a ruckus:* Blériot's protest of evening flights, "The Rheims Aviation Week," the *Times*, 8/16/1909.
341 *"Those few packages?"*: Owen S. Lieberg, *The First Air Race* (New York: Doubleday, 1984), 1.
341 *"a dangerous competitor"*: "Curtiss Ready to Fly," *New York Times*, 8/15/1909.
341 *"The event was billed"*: Lieburg, *The First Air Race*, 126–44; Roseberry, *Glenn Curtiss*, 189–97; Glenn H. Curtiss and Augustus Post, *The Curtiss Aviation Book* (New York: Frederick A. Stokes Co., 1912), 65–75.
344 *"It is evident the serious effect"*: Copy, Julien A. Ripley to AGB, 8/30/1909, box 143, AGB papers, LC.
344 *"great land parades"*: "Fourth Annual Report on the Hudson-Fulton Celebration Commission to the Legislature of the State of New York," 5/20/1910.
344 *"striking, even dramatic"*: Hudson-Fulton Celebration Commission, *The Hudson-Fulton Celebra-*

*tion* (Albany, N.Y.: J. B. Lyon, 1910), 486–89. This passage covers the Wright and Curtiss contracts.

345 *"special attention to this contest"*: "The Fulton Airship Flight Contest," *Scientific American*, 4/17/1909.

346 *"I have not come here to astonish the world"*: "Says He Could Fly Over Skyscrapers," *Dayton Herald*, 9/20/1909, Hudson-Fulton scrapbook, WBC, WSU.

346 *One of them observed closely*: Comparison of Wright and Curtiss and atmosphere at work sheds on Governors Island, undated newspaper article in Hudson-Fulton scrapbook, WBC, WSU.

347 *"How do you do, Mr. Curtiss"*: Wright-Curtiss exchange upon Curtiss's arrival at Governors Island, Roseberry, *Glenn Curtiss*, 212.

347 *Loening got through*: Loening meets WW, Grover Loening, *Our Wings Grow Faster* (Garden City, N.Y.: Doubleday, Doran & Co., 1935), 3–10.

348 *"The winds will probably be rather strong"*: WW to OW, 9/26/1909, FC, WBP, LC.

348 *"I wouldn't fly over the buildings"*: "Strong Wind Stops Flight," *Dayton Journal*, 9/29/09, Hudson-Fulton scrapbook, WBC, WSU.

348 *Years later, Loening said*: Grover Loening's claim that Curtiss did not fly on the morning of 9/29/1909, Loening, *Our Wings Grow Faster*, 10–11; WW's comment, "Wright Circles Liberty in Second of 3 Flights," *New York Press*, 9/30/1909, Hudson-Fulton scrapbook, WBC, WSU.

349 *"Within minutes Will and Charlie"*: accounts of WW's Statue of Liberty flight include "Liberty Statue Encircled by Wright in Aeroplane," *New York Evening Sun*, 9/29/1909; "Wright in Daring Flights Rounds Liberty Statue . . . ," *The Globe*, 9/29/1909; "After Twice Circling 'Liberty,' Wright Makes Third Flight," *New York Evening Telegram*, 9/29/1909; "Wright Flies Around Liberty Statue," *Evening Mail*, 9/29/1909; "Wright Circles Liberty in Second of 3 Flights," *New York American*, 9/30/1909; also coverage in *New York World*, Hudson-Fulton scrapbook, WBC, Special Collections, WSU.

353 *"these exciting times"*: And "You know I have the greatest interest," MW to WW, 10/2/1909, FC, WBP, LC.

354 *"like stopping the battle"*: Arthur Ruhl, "In the Crowd," *Collier's*, 10/16/09.

354 *Curtiss's contract was due*: Roseberry, *Glenn Curtiss*, 214–15.

354 *"I didn't like it up there"*: "Curtiss Flies for 45 seconds, Then, Fearful, Alights," *New York World*, 10/4/1909, scrapbooks, WBP, LC.

355 *"as if they were being driven on wheels"*: "Air Voyage Past Skyscrapers Most Remarkable Ever Made," *New York World*, 10/5/1909, scrapbooks, WBP, LC; "Wright Does a Fine Flight," *New York Sun*, 10/5/1909, scrapbooks, WBP, LC; "Mr. Wilbur Wright Makes Twenty-four Mile Flight," *New York Herald*, 10/5/1909, scrapbooks, WBP, LC.

355 *"It was an interesting trip"*: WW to MW, 10/7/1909, FC, WBP, LC.

356 *"It is all over, gentlemen"*: "A Talk with Wilbur Wright," *Scientific American*, 10/23/1909, 290, reprinted in Jakab and Young, eds., *Published Writings*, 206–08.

356 *They chatted for only a few minutes*: Cass Gilbert meeting with WW, "Cass Gilbert and Wilbur Wright," *Minnesota History*, September 1941.

## Epilogue

359 *"I have brought to a close"*: SPL, "The 'Flying-Machine,' " *McClure's Magazine*, June 1897, 660.

359 *he rode as Orville's passenger*: WW and OW fly together, entry, 5/25/1910, MW, *Diaries, 1857–1917*, 714.

359 *Once, finding himself*: OW to Paul Brockett, 11/13/1917, GC, WBP, LC.

359 *"He was entirely too daring"*: Hart Berg to WW, 9/10/1909, GC, WBP, LC.

359 *"we are too thirsty for air"*: quoted in Wohl, *Passion for Wings*, 110.

360 *"to ask if you do not think"*: copy, George Spratt to WW and OW, 9/29/1909, GC, WBP, LC.

360 *"a trifle hurt"*: WW to Spratt, 10/16/1909, McFarland, ed., *Papers*, vol. 2, 967–68.

360 *"I do not see how you"*: Spratt to OW, 11/27/1922, GC, WBP, LC.

360 *"not wishing to add"*: undated draft, OW to Charles S. Foltz, quoted in McFarland, ed., *Papers*, vol. 2, 1135, note 1.

360 *"I am afraid, my friend"*: OC to WW, 1/23/1910, McFarland, ed., *Papers*, vol. 2, 980–82.

360 *"As to inordinate desire"*: WW to OC, 1/29/1910, McFarland, ed., *Papers*, vol. 2, 982–86.

360 *"his labors had vast influence"*: WW, "The Life and Work of Octave Chanute," *Aeronautics*, January 1911, reprinted in Jakab and Young, *Published Writings*, 168–69.

360 *"specially meritorious investigations"*: Charles Walcott to WW and OW, 2/18/1909, GC, WBP, LC.

360 *He congratulated the Wrights:* "Historical Address of Dr. Alexander Graham Bell," attached to Charles Walcott to WW and OW, 2/17/1910, "Smithsonian" file, box 20, GC, WBP, LC.

361 *One carrier of the legend:* C. B. Veal, "Manly, the Engineer," *S.A.E. Journal*, April 1939.

361 *"still confident":* "Langley Airplane in New Dispute," *New York Times*, 10/25/1931, reprinted in Robert B. Meyer, Jr., ed., *Langley's Aero Engine of 1903* (Washington, D.C.: Smithsonian Institution Press, 1971), 166–69.

361 *"We had hoped in 1906":* WW to M. Hévésy, 1/25/1912, McFarland, ed., *Papers*, vol. 2, 1035.

362 *"A short life":* entry of 5/30/1912, Bishop Milton Wright, Diaries: 1857–1917 (Dayton, OH: Wright State University Libraries, 1999), 749.

362 *Curtiss responded with an extraordinary maneuver:* The struggle between OW and the Smithsonian over SPL's aerodrome is thoroughly reported in Tom D. Crouch, "Capable of Flight: The Feud Between the Wright Brothers and the Smithsonian," *American Heritage of Invention and Technology*, spring 1987, 34–46.

363 *"There were thousands of men":* OW to Lloyd D. Bower, 3/30/1928, GC, WBP, LC.

363 *"I had thought that truth":* OW, "Copy of statement for aeronautical magazine," GC, WBP, LC.

363 *she accepted his approach:* KW to Henry Haskell, 6/16/1925, Katharine Wright Haskell papers, Western Historical Manuscript Collection, University of Missouri at Kansas City [hereafter WHMC].

364 *"It hasn't been the usual":* KW to Henry Haskell, 6/18/1925, WHMC.

364 *"Altogether, Harry, my world":* *ibid.*

364 *"I can't go to Dayton yet":* KW to Agnes Osborne Beck, 3/11/1927, Agnes Osborne Beck/Katharine Wright letters, Special Collections and Archives, Wright State University.

365 *He watched the proliferation:* OW to George Burba, 10/12/1917, GC, WBP, LC.

365 *"I once thought the aeroplane":* OW to Major Lester D. Gardner, 8/28/1945, McFarland, ed., *Papers*, vol. 2, 1176.

365 *others did too, including Charles Lindbergh:* *The Wartime Journals of Charles A. Lindbergh* (New York: Harcourt, Brace, Jovanovich, 1970), 276–77, 383.

366 *"The aeroplane means many things":* "Remarks by Milton Wright . . . ," Ivonette Wright Miller, comp., *Wright Reminiscences* (Air Force Museum Foundation, 1978), 68–71.

# A NOTE ON SOURCES

## *The Wrights' Published Writings*

The central source for any study of the Wright brothers is their own mass of personal and business correspondence and technical diaries. Most of these are housed in the Manuscript Division of the Library of Congress. Hundreds of the brothers' letters and key diary entries are collected in Marvin W. McFarland's richly annotated *The Papers of Wilbur and Orville Wright* (vol. 1, 1899–1905; vol. 2, 1906–1948) (New York: McGraw-Hill, 1953), a great treasure of public scholarship. McFarland's main purpose was to document the invention of the airplane; thus the collection is weighted toward the technical and away from the personal. The technical appendices on the Wrights' nomenclature, wind tunnel, airplanes, motors, and propellers remain the best explanations of their topics some fifty years after Fred Howard (then a member of McFarland's staff at the Library of Congress, later a distinguished Wright biographer) composed them. The *Papers* must be supplemented by Fred C. Kelly, ed., *Miracle at Kitty Hawk: The Letters of Wilbur and Orville Wright* (New York: Farrar, Straus & Young, 1951), which includes excerpts of numerous important letters that McFarland omitted.

A much more recent and now indispensable addition to the published corpus is Peter L. Jakab and Rick Young, eds., *The Published Writings of Wilbur and Orville Wright* (Washington, D.C.: Smithsonian Institution Press, 2000). This is a comprehensive collection of sixty-nine articles, press statements, and interviews, including many heretofore available only to the most diligent miners of periodical libraries. Five important statements by associates of the Wrights are included as an appendix.

Another essential source is Milton Wright's published *Diaries, 1857–1917* (Dayton: Wright State University Libraries, 1999), compiled by Dawne Dewey, head of Special Collections and Archives at Wright State University. The bishop's nightly discipline of sixty years did not leave us a very readable memoir, but it did provide an important source of insight on the life of his family and a crucial record of his sons' daily doings.

## *Manuscript Sources*

Most of the brothers' personal papers—letters, diaries, legal documents, clippings, photographic negatives—are housed in the Manuscript Division of the Library of Congress in Washington, D.C. This narrative relies heavily on the personal letters of Wilbur, Orville, Katharine, and Milton Wright in the "Family Correspondence" series, only a small number of which are published in the McFarland volumes mentioned above. The much larger "General Correspondence" series contains much crucial material as well, including the brothers' exchanges with such figures as Octave Chanute, Samuel Langley, Alexander Graham Bell, Glenn Curtiss, George Spratt, Ferdinand Ferber, Griffith Brewer, and hundreds of others, some of whom Orville Wright favored with small but important nuggets of information in the years after his brother's death. The family's scrapbooks of newspaper and magazine articles, available on microfilm, are a priceless source of details, large and small.

An equally fascinating source is the Wright Brothers Collection at Wright State University in Dayton, Ohio, which includes the brothers' aeronautical library; the leading collection of Wright photographs (many of which are available on the Internet); the diaries and papers of Milton Wright; copies of the Wrights' correspondence with Octave Chanute and George Spratt; and hundreds of other documents reflecting the family's life and their community of Dayton, Ohio.

An absorbing collection of Katharine Wright Haskell's letters to her friend, fiancé, and husband, Henry Haskell, is held by the Western Historical Manuscript Collection at the University of Missouri at Kansas City. This is available on microfilm.

The Library of Congress holds the papers of Octave Chanute—including microfilm copies of Chanute's extensive correspondence with aeronautical enthusiasts around the world—and the enormously rich collection left by the family of Alexander Graham Bell. Many Bell letters and documents are available in electronic form through the library's website. The privately published *Bulletin of the Aerial Experiment Association and Beinn Bhreagh Recorder* are important sources for Bell's aeronautical work and the short but prolific career of the AEA.

Many of Samuel Langley's papers were lost to fire long ago, but many others remain in two repositories at the Smithsonian Institution: the Smithsonian Institution Archives and the Ramsey Room of the National Air and Space Museum. The former tends more to reflect Langley's term as the administrative leader of the Institution, while the latter includes more material from his aeronautical work, including the voluminous "waste books" kept by Langley and his aides. The Air and Space archives and the museum's library contain much additional material that sheds light on the Wrights; their contemporaries, and the birth of powered flight. At the Smithsonian Archives, the papers of Charles Walcott and Richard Rathbun, Langley's associates, were helpful on several points.

A collection of Wright materials at the United Brethren Historical Center at Huntington College, Huntington, Indiana, mostly pertaining to Milton Wright's career, has been little mined by scholars. Especially interesting is a lengthy series of letters from Bishop Wright to his friend, the Reverend Jacob Howe.

## *Newspapers and Magazines*

Several series of scrapbooks ease the task of finding contemporary published accounts—especially in periodicals that are defunct or without indexes—of the Wrights and their rivals. These include the Wrights' scrapbooks at the Library of Congress; Langley's scrapbooks at the Ramsey Room of the National Air and Space Museum; and several additional Wright scrapbooks at Special Collections, Wright State University, including an album devoted to the Fort Myer flights that was compiled and presented to Orville Wright by C. H. Claudy, an admiring newsman.

Among mainstream periodicals, *Scientific American* covered the dawn of aviation more thoroughly than most, though its editors were skeptical about the Wrights early on. A good collection of the harder-to-find aeronautical press is housed in Special Collections at Wright State.

## *Biographies, Memoirs, and Monographs*

The comprehensive biographies of the Wrights are Fred Howard, *Wilbur and Orville: A Biography of the Wright Brothers* (New York: Knopf, 1987; revised edition, Mineola, N.Y.: Dover Publications, 1998); and Tom D. Crouch, *The Bishop's Boys: A Life of Wilbur and Orville Wright* (New York: Norton, 1989). Crouch delves more deeply into the family's history. An essential supplement to these fine books is Peter L. Jakab, *Visions of a Flying Machine: The Wright Brothers and the Process of Invention* (Washington, D.C.: Smithsonian Institution Press, 1990), a fascinating contribution to the history of science and a deeply insightful study of the brothers as thinkers. Also helpful on technical matters is Harry Combs (with Martin Caidin), *Kill Devil Hill: Discovering the Secret of the Wright Brothers* (Boston: Houghton Mifflin, 1979). John Evangelist Walsh, *One Day at Kitty Hawk* (New York: Crowell, 1975), though tendentious in his attack on Orville Wright, offers interesting views and information on the brothers' relative contributions and Orville's attempts to encourage history to judge them in his favor. The only biography that Orville authorized, Fred C. Kelly's *The Wright Brothers* (New York: Harcourt, Brace & Co., 1943), now seems trite, outdated, and not entirely trustworthy. But it remains an important source of details that Orville revealed to Kelly and no one else. The first biography, John R. McMahon's *The Wright Brothers: Fathers of Flight* (Boston: Little, Brown, 1930), which Orville suppressed in an earlier form, is interesting chiefly for that reason.

A fine reference work that helped me on countless factual points is Arthur G. Renstrom's *Wilbur & Orville Wright: A Chronology* (Washington, D.C.: Library of Congress, 1975). Similar help on Dayton came from Mary Ann Johnson's carefully researched *A Field Guide to Flight: On the Aviation Trail in Dayton, Ohio* (Dayton: Landfall Press, rev. ed., 1996). A good source on Dayton as a center of invention is Mark Bernstein's *Grand Eccentrics: Turning the Century: Dayton and the Inventing of America* (Wilmington, Ohio: Orange Frazer Press, 1996).

The only full study of Milton Wright's life and career—and a good one—is Daryl Melvin Elliott, "Bishop Milton Wright and the Quest for a Christian America" (Ph.D. dissertation, Drew University, 1992). For the life of the Wrights' largely invisible but important assistant, see H. R. DuFour, *Charles E. Taylor: The Wright Brothers' Mechanician* (Dayton, Ohio: Prime Printing, 1997). A rich source of family anecdotes is Ivonette Wright Miller's compilation of *Wright Reminiscences* (Wright-Patterson Air Force Base: Air Force Museum Foundation, 1978).

The Wrights' extraordinary photographic legacy is captured in several attractive books, including Spencer Dunmore and Fred E. C. Culick, *On Great White Wings: The Wright Brothers and the Race for Flight* (New York: Hyperion, 2001); Russell Freedman, *The Wright Brothers: How They Invented the Airplane* (New York: Holiday House, 1991); and Lynanne Wescott and Paula Degen, *Wind and Sand: The Story of the Wright Brothers at Kitty Hawk* (New York: Harry N. Abrams, 1983).

Insightful essays by important historians of aviation (and a full version of Amos Root's 1905 eyewitness account) appear in Richard P. Hallion, ed., *The Wright Brothers: Heirs of Prometheus* (Washington, D.C.: Smithsonian Institution Press, 1978). Technical reviews of the Wrights as engineers appear in Howard S. Wolko, *The Wright Flyer: An Engineering Perspective* (Washington, D.C.: Smithsonian Institution Press, 1987).

The authoritative work on Bell, by the great historian of science Robert V. Bruce, is *Bell: Alexander Graham Bell and the Conquest of Solitude* (Ithaca, N.Y.: Cornell University Press, 1973), which shaped my view of the inventor and Mrs. Bell and supplied a number of anecdotes. Excellent and well-illustrated chapters on Bell's aeronautical aspirations appear in Edwin S. Grosvenor and Morgan Wesson, *Alexander Graham Bell: The Life and Times of the Man Who Invented the Telephone* (New York: Harry N. Abrams, 1997). J. H. Parkin, *Bell and Baldwin* (Toronto: University of Toronto Press, 1964), is useful on technical points. Ann J. Bishundayal's *Mabel Hubbard Bell: A Biography* (Protea Publishing, 2002) is the only book-length treatment of Mrs. Bell's life. For the life of Glenn Curtiss, I relied on C. R. Roseberry's thorough *Glenn Curtiss: Pioneer of Flight* (Garden City, N.Y.: Doubleday, 1972).

There is no full study of Samuel Langley's life and work. A summary is provided in J. Gordon Vaeth, *Langley: Man of Science and Flight* (New York: Ronald Press, 1966). The most complete source on the aerodrome work—though occasionally impenetrable—is the two-volume *Langley Memoir on Mechanical Flight* (Washington, D.C.: Smithsonian Institution, 1911). Authorship was divided between Langley and Charles Manly. Key letters and diary entries are collected in Robert B. Meyer, Jr., ed., *Langley's Aero Engine of 1903* (Washington, D.C.: Smithsonian Institution Press, 1971). For Langley's career in astronomy, including his travails with colleagues at the Western University of Pennsylvania, by far the best source is Donald Obendorf, "Samuel Pierpont Langley: Solar Scientist, 1867–1891" (Ph.D. dissertation, University of California at Berkeley, 1969). Langley's drive for support from the U.S. War Department is well covered in Russell Jay Parkinson, "Politics, Patents and Planes: Military Aeronautics in the United States, 1863–1907" (Ph.D. dissertation, Duke University, 1963).

Several studies in the early history of aviation have been my valued guides. Tom D. Crouch, *A Dream of Wings: Americans and the Airplane, 1875–1905* (Washington, D.C.: Smithsonian Institution Press, 1981), paints a detailed picture of the context in which Langley and the Wrights did their work. Among several foundational works by Charles Gibbs-Smith, the most important for this book was *The Rebirth of European Aviation, 1902–1908: A Study of the Wright Brothers' Influence* (London: Her Majesty's Stationery Office, 1974), which tells its story with nearly day-by-day documentation and detail. A second superb contribution to the European side of the story is Alfred Gollin, *No Longer an Island: Britain and the Wright Brothers, 1902–1909* (London: Heineman, 1984).

The invention of the airplane has been treated, of course, largely in the realms of biography and the history of science and engineering. But technology is only part of the story. I benefited from these considerations of flight as a cultural phenomenon: Lawrence Goldstein, *The Flying Machine and Modern Literature* (Bloomington, Ind.: Indiana University Press, 1986); Robert Wohl, *A Passion for Wings: Aviation and the Western Imagination, 1908–1918* (New Haven, Conn.: Yale University Press, 1994); and Joseph J. Corn, *The Winged Gospel: America's Romance with Aviation, 1900–1950* (New York: Oxford University Press, 1983).

## *Websites*

Among numerous Wright-oriented websites, the standout is Nick Engler's Wright Brothers Aeroplane Company and Museum, a vast storehouse of Wrightiana at www.first-to-fly.com. Also highly interesting is the First Flight website (http://firstflight.open.ac.uk/), a project in virtual science education at the United Kingdom's Open University. The site includes fascinating interactive simulations of the Wrights' wind tunnel balances and 1903 powered flyer.

# INDEX

Note: Illustrations are indicated in italics.

# ILLUSTRATION CREDITS

*Note:* Credits below are for in-text photos only; photos in insert are credited alongside each image.

The A. I. Root Company, page 212
David Messing, page 55
David Messing, David MacArthur, page 110
*Harper's Weekly*, pages 339, 351
*Langley Memoir on Mechanical Flight*, pages 7, 22, 50, 57, 167, 169, 182
Langley, *The New Astronomy*, page 13
Library of Congress, pages 1, 81, 88, 132, 193, 221, 246, 255, 259, 271, 274, 298, 343
Santos-Dumont, *My Airships*, pages 133, 134
*Scientific American*, page 26
Smithsonian Institution, pages 188, 358
*The Washington Post*, page 199
Wright State University, pages 36, 115, 204, 282, 300, 305, 307, 317, 327